Environmental Science

Environmental Science

V. K. Ahluwalia
Sunita Malhotra

CRC Taylor & Francis
Taylor & Francis Group
Boca Raton London New York

CRC is an imprint of the Taylor & Francis Group,
an informa business

Ane Books India

Environmental Science

First Published in 2008 by

Ane Books India
4821 Parwana Bhawan, 1st Floor
24 Ansari Road, Darya Ganj, New Delhi -110 002, India
Tel: +91 (011) 2327 6843-44, 2324 6385
Fax: +91 (011) 2327 6863
e-mail: anebooks@vsnl.com
Website: www.anebooks.com

For

CRC Press
Taylor & Francis Group
6000 Broken Sound Parkway, NW, Suite 300
Boca Raton, FL 33487 U.S.A.
Tel : 561 998 2541
Fax : 561 997 7249 or 561 998 2559
Web : www.taylorandfrancis.com

For distribution in rest of the world other than the Indian sub-continent

ISBN-10 : 1 42007 069 x
ISBN-13 : 978 1 42007 069 9

British Library Cataloguing in Publication Data
A catalogue record for this book is available from the British Library

Printed at Brijbasi Art Press, Noida

Preface

Environment is becoming a matter of concern for one and all today. There is a tremendous increase in the developmental activities all around the world and their impact on the environment, in the last few decades, has been very alarming. A need for conservation of the environment is being raised at all national and international fora.

In this scenario, the awareness about various aspects of environment becomes very important. Every citizen needs to be educated about the environmental issues before a meaningful contribution is envisaged. Thus, environmental education has become the prime concern today. Every curriculum, be it the school education or higher education, gives due emphasis on environmental aspects, even a lot is being published in newspapers, journals and magazines. Many governmental and non-governmental organizations are working in the area of environmental education. In India, a separate Ministry, i.e. Ministry of Forests & Environment, is looking after various issues related to the environment.

Every contributor or worker in the area of environment has his/her own perception and looks at environment from his/her own point of view. Also, the environmental issues are many in number and complex in nature, so a comprehensive text is needed to give a complete overview of all the aspects of environment. With this objective in mind, this book has been developed to sensitize the users to realize the importance of different aspects of environment and the related current issues.

V K Ahluwalia
Sunita Malhotra

Acknowledgements

We thankfully acknowledge the following sources of the information included in this book:

msn.Encarta
MNES Annual Report 2004
www.marinesturbines.com
www.wilkipedia.com
www.news.bbc.co.uk
www.greenfactors.com
www.wii.gov.in
www.teriin.org
www.rainharvesting.org
www.cameco.net
www.metroplanet.com
ISO

V K Ahluwalia
Sunita Malhotra

About the Book

'Environmental Science' gives an exhaustive coverage to a variety of issues related to the environment. The theme on 'Energy' has been chosen as Chapter 1 as it is the prime requirement of developed as well as developing countries. The energy demands have already gone high and would go on increasing in future also. Hence, harnessing of various sources of energy is a focal point for our relationship with the environment.

Chapter 2 on 'Environment' covers fundamentals of environment, its segments and components as well as the general terminology which is used throughout the book.

Chapters 3 to 9 include a detailed discussion on various components of environment including various pollutants, their sources and effects. A due coverage has been given to the effects of these pollutants on human health. The treatment of these pollutants and their control have been dealt at appropriate places.

In addition to the pollution of air, land, soil and water, another serious threat to the environment is the 'hazardous waste' generated by various human activities. The developmental activities are responsible for the generation of hazardous waste, noise pollution, climate change, and are threatening the biodiversity. In fact, the survival of human beings and various other species is under threat. All these issues have been elaborated in the later half of the book in Chapters 10 to 13.

A better control and management is the need of today. Various legal provisions, agreements, conventions and role of agencies involved in the protection of the environment are described in Chapter 14 on 'Environmental Management'.

Chapter 15 includes some case studies which have a direct relevance to the environment and the people. It emphasizes how we need to change our life styles in this age of development and modern technology while maintaining the healthy surroundings.

Unit 16 is a window to the cleaner future and discusses the role of newer, cleaner, greener techniques of chemistry in making the environment pollution free. It also throws light on the paths and initiatives taken by various segments of the society to protect the environment.

The main thrust of this book is on the chemistry of environmental processes and pollutants. In addition to this, a wide coverage on various others aspects of environment is purposely included to make it relevant for a large number of students studying chemistry, environmental sciences, life sciences, engineering and other disciplines at undergraduate and post graduate levels. The interdisciplinary approach makes the book quite informative and interesting for all readers.

The contents cover the topics suggested in the syllabii of several universities at undergraduate and postgraduate levels.

A special feature of this book is that the environmental issues have been dealt from the global perspective, in general and with the Indian focus, in particular. The data included has been obtained from the original and reliable sources which we gratefully acknowledge. The latest and current information for various environmental parameters have been included. Thus, this book aims to serve the students of all Indian and other universities.

Finally, the style is reader friendly and the language used is quite simple. A lot of illustrations have been included to make the coverage more interesting and informative. To facilitate the readers, an index of contents and the glossary have also been incorporated.

V. K. Ahluwalia
Sunita Malhotra

CONTENTS

Preface v
About the Book vii

Chapter - 1. Energy and Environment **1**

1.1 Introduction 1
1.2 Sources of Energy 2
1.3 The Energy Scenario 3
1.4 Future Projections of Energy Demand 4
1.5 Utilisation of various Energy Sources 5
1.5.1 Conventional Sources 6
Coal 6
Oil and Natural Gas 8
Oil Shale 9
1.5.2 Non-Conventional Sources 10
Solar Energy 10
Wind Energy 13
Geothermal Energy 16
Hydropower 16
Biomass 17
Biogas 18
Tidal Energy 20
Energy from Waste 21
Hydrogen – The Future Fuel 23
Nuclear Energy 25

Chapter - 2. Environment **29**

2.1 Introduction 29
2.2 Management of Environment 30
2.3 Components of Environment 31
2.4 Types of Environment 31
2.5 Segments of Environment 32
2.5.1 Atmosphere 32
2.5.2 Hydrosphere 32
2.5.3 Lithosphere 33

2.5.4 Biosphere 33
2.5.4.1 Biomes 35
2.6 Environmental Pollution 37
2.6.1 Types of Pollution 37
2.7 Ecosystems 37
2.7.1 Types of Ecosystems 38
2.7.2 Components of Ecosystems 38
2.7.3 Energy in Ecosystems 39
2.8 Various Cycles of Environment 40
2.8.1 Gaseous Cycles 41
2.8.1.1 Water Cycle 41
2.8.1.2 Carbon Cycle 43
2.8.1.3 Nitrogen Cycle 45
2.8.1.4 Oxygen Cycle 48
2.8.2 Sedimentary Cycles 50
2.8.2.1 Sulphur Cycle 50
2.8.2.2 Phosphorus Cycle 51
2.8.2.3 Sodium Cycle 51
2.8.2.4 Potassium Cycle 52
2.8.2.5 Magnesium Cycle 52
2.8.2.6 Iron Cycle 52

Chapter - 3. Atmosphere **53**

3.1 Introduction 53
3.2 Major Regions of Atmosphere 54
3.3 Composition of Atmosphere 57
3.4 Radiation Balance on Earth 58
3.5 Chemical and Photochemical Reactions in Atmosphere 59
Oxygen 59
Nitrogen 60
Nitrous Oxide 61
Carbon Dioxide 61
Ozone 62

Chapter - 4. Air Pollution **63**

4.1 Introduction 63
4.2 Classification of Pollutants 63
4.2.1 Natural Pollutants 63
4.2.2 Primary Pollutants 64
4.2.3 Secondary Pollutants 64
4.3 Major Air Pollutants 65
4.3.1 Oxides of Carbon 65
4.3.1.1 Carbon Monoxide 66
4.3.1.2 Carbon Dioxide 68
4.3.2 Oxides of Nitrogen 69

4.3.3 Oxides of Sulphur 72
4.3.3.1 Sulphur Dioxide (SO_2) 72
4.3.4 Hydrogen Sulphide 74
4.3.5 Chlorine 75
4.3.6 Ozone 75
4.3.7 Hydrocarbons 76
4.3.8 Particulates 77
4.4 Sources of Air Pollution 80
4.4.1 Pollutants from Industry 81
4.4.2 Pollution by Petroleum 88
4.4.3 Mining Industries 88
4.4.4 Power Plants 89
4.4.5 Automobiles 90
4.4.6 Agrochemical Based Industries 90
4.4.6.1 Fertilizers 90
4.4.6.2 Pesticides 92
4.4.6.2.1 Insecticides 92
4.4.6.2.2 Fungicides 95
4.4.6.2.3 Herbicides 98
4.5 Effects of Air Pollution 100
4.5.1 Acid Rain 100
4.5.2 Green House Effect 103
4.5.3 Global Warming 104
4.5.4 Depletion of Ozone (Ozone Hole) 105
4.5.5 Smog 108
4.6 Control of Air Pollution 111

Chapter - 5. Water **121**

5.1 Introduction 121
5.2 Availability and Sources 121
5.2.1 Ground Water 122
5.2.2 Desalination of Sea Water 123
5.2.3 Conservation of Water 123
5.3 Characteristics of Water 126
5.4 Water Quality Parameters 127
5.4.1 Dissolved Oxygen (DO) 129
5.4.2 Biochemical Oxygen Demand (BOD) 130
5.4.3 Chemical Oxygen Demand (COD) 131
5.4.4 Alkalinity 132
5.4.5 Most Probable Number (MPN) 132
5.4.6 Total Solids 133
5.4.7 Oxidation State 133
5.4.8 Transparency 133
5.4.9 Silica Content 134

5.4.10 Hardness 134
5.4.10.1 Units of Hardness of Water 135
5.4.10.2 Water Softening 137
5.4.11 Dissolved Inorganic Impurities 139
5.4.12 Toxic Metals in Water 139
5.4.13 Microbial Contaminats in Sewage 148

Chapter - 6. Water Pollution **151**
6.1 Introduction 151
6.2 Types of Water Pollutants 151
6.2.1 Biological Agents 151
6.2.2 Chemical Agents 152
6.2.3 Physical Agents 152
6.3 Types of Water Pollution 153
6.3.1 Ground Water Pollution 153
6.3.2 Surface Water Pollution 154
6.3.3 Sea Water Pollution 156
6.4 Sources of Water Pollution 156
6.4.1 Sewage and Domestic Wastes 157
6.4.2 Industrial Effluents 158
6.4.2.1 Textile Industry 158
6.4.2.2 Paper and Pulp Industry 158
6.4.2.3 Leather Tanning Industry 160
6.4.2.4 Agrochemical Industry 161
6.4.2.5 Petrochemical Industry 162
6.4.2.6 Food Processing Industries 166
6.4.2.7 Dairy Industry 170
6.4.2.8 Soaps and Detergents 171
6.4.2.9 Hazardous Wastes 175
6.5 Impact of Water Pollution on Human Health 177
6.5.1 Water as a Carrier for the Transmission of Diseases 177
6.5.2 Chemical Toxicants in Water Affecting Health 178
6.5.3 Water as a Vector Habitat in Disease Transmission in Humans 181
6.5.4 Treatment of Waste Water 181
6.5.5 Eutrophication 182

Chapter - 7. Marine Pollution **183**
7.1 Introduction 183
7.2 Sources of Marine Pollution 183
7.2.1 Hazardous Materials 183
7.2.2 Oil Pollution 183
7.2.2.1 Composition of Crude Petroleum 184
7.2.2.2 Some Major Oil Tanker Accidents 185
7.2.2.3 Effects of Oil Pollution 186
7.2.2.4 Control of Oil Pollution in Seas 187

7.2.3 Pollution by Radioactive Materials 187
7.2.3.1 Types of Radionuclides 187
7.2.3.2 Effects of Radionuclides 188
7.3 Thermal Pollution 188

Chapter - 8. Soil **191**
8.1 Introduction 191
8.2 Formation of Soil 191
8.3 Characteristics of Soil 193
8.4 Soil Profile 193
8.5 Composition of Soil 194
8.5.1 Inorganic Components of Soil 195
8.5.2 Organic Components of Soil 195
8.5.3 Water and Air in Soil 196
8.6 Macronutrients and Micronutrients in Soil 196
8.7 Sources of Essential Elements in Soil 197

Chapter - 9. Soil Pollution **199**
9.1 Introduction 199
9.2 Sources of Soil Pollution 199
9.2.1 Industrial Wastes 200
9.2.2 Urban and Domestic Wastes 200
9.2.3 Radioactive Pollutants 201
9.2.4 Agrochemicals 201
9.2.5 Chemical and Metallic Pollutants 202
9.2.6 Biological Agents 202
9.2.7 Acid Rain 202
9.3 Harmful Effects of Soil Pollutants 203
9.3.1 Effects of Industrial Pollutants 203
9.3.2 Effects of Urban and Domestic Waste Products 203
9.3.3 Effects of Radioactive Pollutants 204
9.3.4 Effects of Agrochemicals 204
9.4 Soil Erosion 205
9.4.1 Causes of Soil Erosion 205
9.4.2 Effects of Soil Erosion 206
9.4.3 Control of Soil Erosion 206
9.5 Reclamation of Soil 206
9.6 Treatment of Soil Pollutants 207
9.7 Control of Soil Pollution 208
9.8 Indicators of Soil Pollution 209
9.8.1 Plants as Indicators of Soil Pollution 209
9.8.2 Organisms as Bio -indicators 210
9.8.3 Micro-organisms as Bio-indicators 210

Chapter - 10. Hazardous Waste and its Management 211
10.1 Introduction 211
10.2 Radioactive Waste 212
10.2.1 Types of Radiation and Their Characteristics 212
10.2.2 Sources of Radioactive Waste and Radiation 213
10.2.2.1 Accidents at Nuclear Power Plants 216
10.2.3 Effects of Radiation 218
10.2.4 Management of Radioactive Waste 219
10.3 Hazardous Waste 221
10.3.1 Incidents of Improper Disposal and Accidents 221
Love Canal 221
Minamata 222
Bhopal Gas Tragedy 223
10.3.2 Biomedical Waste 223
10.4 Management of Hazardous Waste 224
10.5 Trans-boundary Issues 226
10.6 International Conventions 228
10.7 Hazardous Waste in India 229

Chapter - 11. Noise Pollution 231
11.1 Introduction 231
11.2 Sound and Noise 231
11.3 What is Sound? 232
11.4 The Decibel Scale 232
11.5 The Structure and Functioning of Ear 234
11.6 Sources of Noise 236
11.7 Effects of Noise Pollution 237
11.8 Control of Noise Pollution 238
11.9 Legal Provisions and their Implementation 239

Chapter - 12. Climate Change 243
12.1 Introduction 243
12.2 Role and Importance of Climate 243
12.3 Factors Affecting Climate 244
12.4 The Global Climate 245
12.5 Climate Change 246
12.6 Effects of Climate Change 247
12.6.1 Climate Change in Asia 248
12.6.2 Impacts of Climate Change on India 251
12.7 Monitoring Climate Change 253
12.8 UNFCCC and India 254
12.9 Other Aspects of Climate Change 256

Chapter - 13. Biodiversity **259**

13.1 Introduction 259
13.2 The Richness of Biodiversity 259
13.3 The Importance of Biodiversity 260
13.4 Biodiversity of India 262
13.5 The Threatened Biodiversity 267
13.5.1 The Indian Scenario 268
13.6. Causes of Extinction 269
13.7. Conservation of Biodiversity 270
13.7.1 Conventions, Protocols and Acts 272
13.7.2 Other National Initiatives 275

Chapter - 14. Environmental Management **277**

14.1 Introduction 277
14.2 Important Environmental Laws and Acts 277
14.3 Environmental Impact Assessment 284
14.4 Environmental Auditing 285
14.5 Environmental Management Systems: ISO-14000 Standards 285
14.6 Agencies and Institutions involved in the Environmental Management 291
14.6.1 The Central Pollution Control Board 291
14.6.2 State Governments 292
14.6.3 Role of Governmental Organisations (NGOs) 293
14.6.4 Other Bodies 293

Chapter - 15. Some Case Studies **295**

15.1 Introduction 295
15.2 Bhopal Gas Tragedy 295
15.3 Vehicular Pollution in Delhi 301
15.4 Industrial Pollution in Delhi 311

Chapter - 16. Towards Cleaner Future **317**

16.1 Introduction to Green Chemistry 317
16.2 Designing a Green Synthesis 318
16.2.1 Choice of Starting Materials 318
16.2.2 Choice of Reagents 318
16.2.3 Choice of Catalysis 318
16.2.4 Choice of Solvents 318
16.3 Basic Principles of Green Chemistry 318
16.3.1 Prevention of Waste /by Products 320
16.3.2 Maximum Incorporation of the Reactiants (Starting Materials and Reagents) in the Final Product 320
16.3.2.1 Rearrangement Reactions 321

16.3.2.2 Addition Reactions 321
16.3.2.3 Substitution Reactions 322
16.3.2.4 Elimination Reactions 323
16.3.3 Prevention or Minimization of Hazardous Products 324
16.3.4 Designing Safer Chemicals 324
16.3.5 Energy Requirement for Synthesis 324
16.3.6 Selection of Appropriate Solvent 324
16.3.7 Selection of Starting Materials 325
16.3.8 Use of Protecting Groups 325
16.3.9 Use of Catalyst 326
16.3.10 Products Designed Should be Biodegradable 327
16.3.11 Designing Manufacturing Plants 327
16.3.12 Re-use of Waste or By-products 328
16.4 Rain Water Harvesting 328
16.5 Sustainable Development 334
16.6 Some Success Stories 336

Glossary *341*

Inxex *347*

Energy and Environment

CHAPTER 1

1.1 Introduction

As we all know, "energy is the capacity to do work". Energy consumption, economic growth and population increase are interlinked. The energy demands of developing and developed countries are increasing day by day. India, being a developing country is not an exception. There is a rapid increase in energy consumption in India. In 1980, the consumption was 4.16 quadrillion Btu (quads) which increased to 12.8 quads in 2001. The increase of 208% in energy consumption is due to the increase in population, industrialization, urbanization and other developmental activities. The population of India crossed one billion mark in the year 2000. The average growth rate of population being 2.1% per year over the last three decades, though in 2001 the growth rate declined a little to 1.9%.

With about 16% of the world's population, India consumes only 3.4% of the total energy produced in the world. This energy consumption is still lower than that of Germany (14.35 quads), Japan (21.92 quads), China (39.67 quads) and Unites States (97.05 quads). The United States consumed 24.5% of world's commercial energy with its population share of only 4.6% in the world. In India, the per capita energy consumption in 2001 was 12.6 million Btu which is one of the lowest in comparison to the rest of other Asian countries. It is $1/5^{th}$ of world average and $1/26^{th}$ of that of USA.

Energy consumption is also one of the indicators of the development of a country. India has to go a long way to be at par with the developed nations but at the same time, there is also a need to keep a watch on the environment. The development, of course, should not be at the cost of environmental degradation. We should keep in mind the resources available to us and use them optimally and very cautiously for the sustainable development.

1.2 Sources of Energy

The energy resources can be classified as follows:

Energy Resources

- Conventional (Non-renewable)
  - Coal
  - Oil
  - Natural gas
- Non-Conventional (Renewable)
  - Solar
  - Wind
  - Geothermal
  - Tidal
  - Fuel Cells
  - Biomass
- Nuclear

Coal accounts for a little more than 50% of India's energy consumption. Majority of the energy generated by coal is used by power sector which is followed by heavy industry. **Petroleum** meets 34.4% of the energy demand while **natural gas** accounts for 6.5% followed by **hydroelectricity** which is 6.3%. **Nuclear energy** has a very small share of 1.7% whereas **solar**, **geothermal** and **wind energy** – all combined together yield only 0.2% of the total amount. Fig 1.1 Below depicts the contribution of various energy resources.

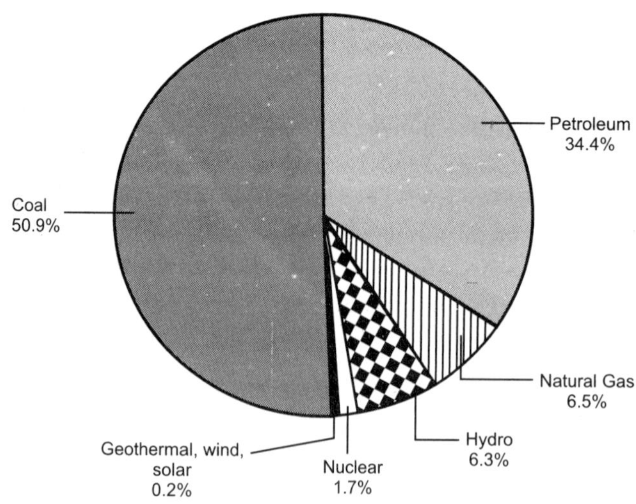

Fig 1.1: Contribution of various energy resources as in 2001

1.3 The Energy Scenario

The total energy use has witnessed an expansion over last five decades and there is a shift from non-commercial to commercial energy sources. Table 1.1 shows the trends in production of commercial energy.

Table 1.1 Production of Commercial Energy

Sources	1970-71	1980-81	1990-91	2001-02
Coal (MT*)	72.95	114.01	211.73	325.65
Lignite (MT*)	3.39	4.80	14.07	24.3
Crude Oil (MT*)	6.82	10.51	33.02	32.03
Natural Gas (BCM**)	1.44	2.35	1.79	26.69
Hydropower (BKWh***)	22.25	46.54	71.66	82.8
Nuclear Power (BKWh***)	2.42	3.00	6.14	16.82
Wind Power (BKWh***)	—	—	0.03	1.70

*MT - Million Tonnes,

**BCM - Billion Cubic metre,

***BKWh - Billion kilo watt hour

Table 1.1 indicates that ***coal*** is the most abundant commercial energy source and during 1970-2001, its production has shown an average annual growth rate of 5.11%. However, its share in energy supply has decreased which used to be 60.47% in 1970-71 and is about 44.8% in 2001-2002.

The coal will remain in the centre stage of energy scenario of India because power generation favours this fuel as its availability is quite high. About 70% of coal produced is consumed in the power sector. Other industries like steel, cement, fertilizer, chemicals, paper and other small and medium industries also depend upon coal for their energy demands. As far as the transport is concerned, the railways is phasing out steam locomotives but the energy requirement for electric traction is still dependent on coal (*Ministry of Coal, 2003*).

About 75% of the total coal reserves in India are located in Bihar, Madhya Pradesh and West Bengal. The total coal reserves, as per Geological Survey of India (January 2003), are estimated to be 240.78 billion tonnes. Out of this, 37.42% is in the proven category, 46.78% in the indicated category and 15.8% is in the inferred category. The coking coal resources amount to 13.28% whereas non-coking coal resources form 86.72% of the total resources.

The coal resources of India account for 8% of the world's total resources. India is the third leading coal producer in the world after China and the United States.

Although, the domestic production of ***petroleum*** and ***natural gas*** has increased over the years, India depends heavily upon crude oil imports. During 1970-2000, the oil production has increased at the rate of 5.33% per annum. Parallel to this, there is an increase from 12.43 MT to 75 MT in the import of crude oil and petroleum products (*Ministry of Petroleum and Natural Gas, 2002*). This is mainly because the consumption of petroleum products has increased manifold in all sectors, particularly in industry and transport.

During 2000-2001, the transport sector consumed nearly 23.24 MT of petroleum products while the industrial sector consumed 12.32 MT. Petroleum products such as high speed diesel, diesel, light diesel oil, fuel oil etc. for power generation amounted to 5 MT whereas the agricultural sector consumed nearly 7.7 MT of petroleum products (*Ministry of Petroleum and Natural Gas, 2002*).

The recoverable resources of crude oil and natural gas were 732 MT and 763 BCM (billion cubic metre), respectively as on March 2002. In 2001-02, the production of crude oil and natural gas was 32 MT and 29.7 BCM, respectively. However, the country still imported 78.7 MT of crude oil which amounted to 73.3%. The discovery of 7 TCF (trillion cubic feet) of natural gas by Reliance in Krishana-Godavari basin in 2002 has improved the energy situation.

As far as the power sector is concerned, the consumption was 1.3 GW (gigawatt) at the time of independence which has grown to more than 100 GW at present. The installed generating capacity is nearly 107 GW as on January 2003 as per the Central Electricity Authority. Out of the total installed capacity, 90% is owned by the public sector (60% under State Governments and 30% under the Central Government) and rest 10% is with the private sector.

1.4 Future Projections of Energy Demand

From 1981 to 2001, the annual rate of growth of primary commercial energy demand was 6%. It is expected that the demand would be 6.6% by the end of Present Plan period (2002-2007) while it would be 6.1% by the end of next plan period (2011-2012). Table 1.2 shows the energy demand for various fuels in the coming years.

Table 1.2 An Estimate of the Energy Demand

Primary Fuel	Demand in Original		Demand (MTOE)	
	2006-2007	**2011-2012**	**2006-2007**	**2011-2012**
Coal (MT)	460.50	620.00	19.00	254.93
Lignite (MT)	57.79	81.54	15.51	22.05
Oil (MT)	134.50	172.47	144.58	185.40
Natural Gas (BCM)	47.45	64.00	42.70	57.60
Hydropower (BkWh)	148.08	215.66	12.73	18.54
Nuclear Power (BkWh)	23.15	54.74	6.04	14.64
Wind Power (BkWh)	4.00	11.62	0.35	1.00
Total Commercial Energy	—	—	411.91	553.68
Non Commercial Energy	—	—	151.30	170.25
Total Energy Demand			**563.21**	**723.93**

*MTOE= Million Tonnes of Oil Equivalent
Source : Ministry of Non-Conventional Energy Sources

The data in Table 1.2 shows that coal dominates the energy demand scenario in India. Its share being 46.13% and 46.04% in the final years of two Plans, respectively. The demand for oil is expected to grow at an average rate of 3.6% over next five years. The average global energy demand growth rate is, however, 2.1% (*Planning Commission, 2002*).

Natural gas is visualised as the fuel of the future as per the *India Hydrocarbon Vision 2025*.

1.5 Utilisation of Various Energy Sources

Various sources of energy can be classified as **primary** or **secondary sources**. **Primary sources** are those which are obtained directly from the environment. These include conventional sources, nuclear sources, geothermal sources and various forms of energy such as solar, wind and tidal. On the other hand, **secondary sources** include sources which are obtained from primary sources. The secondary sources themselves are not present in nature but are derived from other sources. Their examples being petrol, electricity, hydrogen etc.

1.5.1 Conventional Sources

The conventional sources of energy are coal, oil and natural gas. They are also known as **fossil fuels**. Fossil fuels were formed from the incomplete biological decomposition of dead organic matter (mostly plants and marine organisms). They are also called *carbon* or *hydrocarbon fuels* because they are based on organic compounds which contain the elements carbon and hydrogen.

a. Coal

When partially decomposed vegetation got deeply buried in a sedimentary environment, it slowly transformed into solid, brittle, carbonaceous rock called **coal**. The formation of coal is shown below in Fig. 1.2.

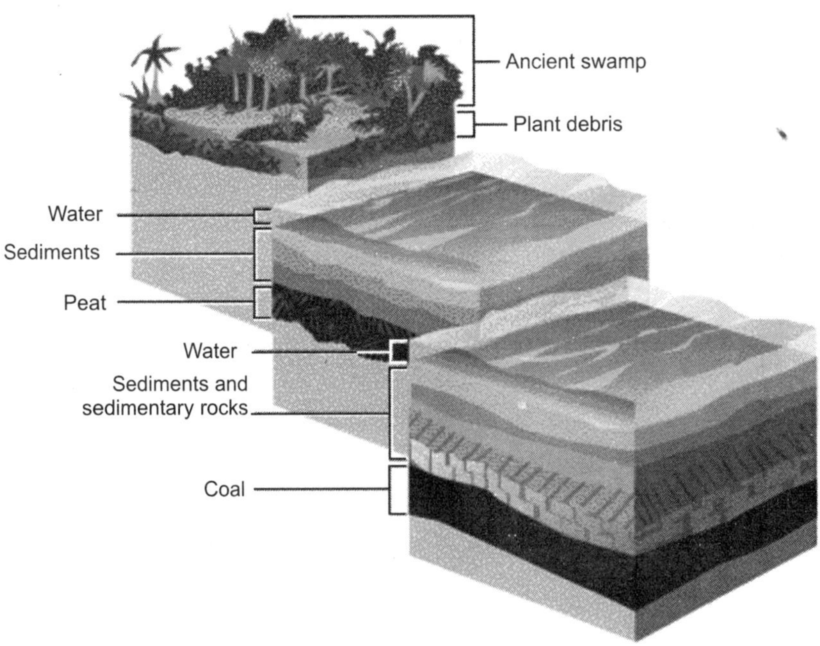

Fig 1.2: Formation of Coal

Coal can be classified into various types according to its carbon and sulphur content. The characteristics of different types of coal are given in Table 1.3.

Table 1.3 Types of Coal

Type	% Composition				Heat value or energy content *million of joules* kg^{-1}
	C	O	Volatile matter	Moisture	
1. Lignite	60-70	18	45	35	13-16
2. Sub-bituminous	75-83	15	40	17	16-23
3. Bituminous	78-90	10	20	6	23-34
4. Anthracite	92-98	3	5	1	30-34

In the due course of time, peat (a type of coal) gets compacted to lignite (a sedimentary rock with woody texture) because more partially decayed organic material accumulated on it. After further compaction and cementing (lithification), lignite transformed to bituminous coal which is soft black coal that burns with a smoky yellow flame.

Anthracite is a hard black coal. It burns slowly but gives off intense heat. It is formed by folding and hardening of sedimentary strata containing bituminous coal. Anthracite is the highest ranking coal whereas lignite is the lowest ranking. Anthracite has relatively low volatile constituents (oxygen, hydrogen and nitrogen) and its moisture content is also less.

The quality of coal is also rated according to its sulphur content. The low sulphur is advantageous because it causes less environmental pollution. Thus, it is a more desirable fuel for power plants.

Coal is considered by many people to be a dirty fuel as there are many environmental impacts associated with mining, processing, transporting and use of coal.

Coal deposits are exploited mainly by **strip mining**. It involves surface mining in which overlying layer of soil and rock is stripped off to get the coal. *Strip mining* is easier than underground mining and hence, it has become a common practice for mining the coal. In *underground mining*, about 50% of the coal must stay as such to support the mine roof.

There are several environmental problems associated with strip mining. It involves removal of vegetation and topsoil using bulldozers, shovels and scrapers. Then, the exposed coal beds are drilled and blasted to fracture coal. The coal is then removed using trucks. Finally, the land is filled, topsoil is replaced and the vegetation is restored. This land reclamation minimizes the damage caused by strip mining. The damage would otherwise include total destruction of the ecosystem and turning of such a land into a barren desert.

Also, rainwater or surface water can infiltrate into rock debris left after removal of the coal. This water on reaction with sulphide minerals such as *pyrite* (FeS_2) produces sulphuric acid which pollutes streams and ground water resources. Such acid drainage is a serious problem around the coal mine areas. Their strip mining can destroy scenic, water, biological and other land resources.

Underground mining, on the other hand, has hazards of collapse, explosion and fire. Respiratory illnesses are common amongst miners. Land subsidence over mines can also occur.

Transport of coal from mines to areas of actual usage over long distances is presently handled by railways. It also has its own associated problems. Although coal can be converted to more easily transportable synthetic oil, synthetic gas or electricity, but these alternatives require large amount of water, good technology and are very expensive also.

Burning of coal also produces sulphur dioxide, nitrogen oxides and carbon dioxide. Sulphur dioxide leads to *acid rain* whereas carbon dioxide is a significant contributor to *global warming*.

The harmful effects of these gaseous pollutants are discussed in detail in Chapter 4 on air pollution. The combustion of coal yields ash which ranges from 5% to 20% of the amount of coal burnt. Also, the *scrubbing* of coal, i.e. treatment of gases obtained after combustion with calcium carbonate to remove sulphur dioxide, yields calcium sulphate as *sludge*.

The disposal or use of ash, sludge or boiler slag (cinder produced in the furnace) is also of environmental concern.

b. Oil and Natural Gas

Crude oil or *petroleum* is found along with natural gas primarily along tectonic belts. Oil and natural gas are derived from organic materials which got buried with marine or lake sediments. The high temperature and pressure are responsible for the conversion of organic materials into oil and gas.

India draws most of its oil from Bombay High, Upper Assam, Cambay, Krishna-Godavari and Cauvery basins. The oil reserves are estimated to be 4.7 billion barrels.

The consumption of petroleum products rose from 57 million tonnes in 1991-92 to 107 million tonnes in 2000. It is estimated to be 163.8 million tonnes in 2005. The *India Hydrocarbon Vision* puts future refinery demand at 368 million tonnes by 2005.

Natural mineral oil is a thick greenish-brown flammable liquid. It can be refined to produce a number of valuable products including oil and petrol. The primary production method involves pumping the oil from wells under the natural pressure whereas the secondary recovery method involves the injection of steam, water or gases such as carbon dioxide or nitrogen into the reservoir to push the oil. About two-thirds (65%) of the world's total reserves are located in the Middle East. These reserves are going to be depleted in about 80 years.

Natural gas is a mixture of flammable hydrocarbon gases in which methane is the main component. It is mostly found in association with oil reserves. It is a convenient and clean fuel as far as the environmental aspects are concerned. It produces very low amounts of oxides of nitrogen on combustion. The only problem in its utilization involves the construction of pipelines to transport it to the places of consumption. About 7% of India's energy needs are met by natural gas. Natural gas is mainly used in areas of power production, petrochemical production and fertilizer production.

The use of oil, however, poses environmental problems at the stage of extraction, refining, transportation and combustion. Some of these problems are as follows:

- Disruption of land to construct wells, laying pipelines or storage tanks and other production facilities.
- Pollution of surface waters and ground water from runoff or leakage.
- Land sinking after removal of oil and gas.
- Damage to ecosystem.
- Release of drilling muds.
- Seepage and spilling of oil.

Oil spills due to accidents while transport have serious environmental impacts. They kill thousands of seabirds, spoil beaches and affect fish and other marine organisms. Amongst these disasters, oil spills from Torrey Canyon (March, 1967), Exxon Valdez (March, 1989), Braer (1992) and Nakhodka (January, 1997) are a few to mention.

In addition to above, the combustion of various fractions obtained from oil such as petrol, diesel etc, produce many primary and secondary pollutants. This can also lead to the formation of smog which has ill effects on both human health and vegetation.

c. Oil Shale

It is fine sedimentary rock containing organic matter in the form of kerogen. The destructive distillation of oil shale at 500^0C yields hydrocarbons. One ton of shale can give upto 60 litres of oil.

The mining and processing of oil shale produces waste which is 20-30% in excess of the original volume of the mined shale. Thus, its disposal causes a problem.

Shale oil, although a fuel source, has not been developed to its full utilization because of the other available options. It, however, remains as one of the alternatives in the times of crisis.

The fossil fuels will not exist forever. They are depleting at a very fast rate. Thus, we need to consume them and use them judiciously. Also, their formation in nature took over millions of years and they will not be replaced so soon. They are, thus, called **non-renewable resources**. The need of the hour is, therefore, to develop alternative sources of energy which are discussed below under *non-conventional sources*.

1.5.2 Non-Conventional Sources

India is the only country that has a separate government ministry i.e. Ministry of Non-conventional Energy Sources (MNES) meant exclusively to look after the non-conventional energy sources. The Ministry is involved in the implementation of programmes for development, demonstration and utilisation of technologies for *renewable energy sources* such as *solar*, *wind*, *biomass*, *biogas*, *small hydropower* and other emerging technologies in the areas of *geothermal* and *tidal energy*.

Although, the technologies of biogas plants and improved cooking stoves were available in India since late 1940's, the Commission on Additional Sources of Energy (CASE) started in 1980 and the Department of Non-conventional Energy Sources (DNES) was established in 1982. This Department was converted to full-fledged Ministry in July 1992. The MNES has sectoral groups of rural energy, urban/industrial energy and power generation. Each sector has integrated programmes to serve different energy needs.

a. Solar Energy

Sun is the earliest source of energy known to the mankind. The other forms of energy are derived from solar energy directly or indirectly.

Fossil fuels represent solar energy which has been captured by the process of photosynthesis and has been stored for million of years. The photosynthesis, in present times, is responsible for the biomass available as an energy source. Similarly, wind power and hydropower generation is possible by circulation of air and water, respectively which in turn is governed by solar energy. The Sun is the ultimate source of energy and the solar energy has made possible the existence of life on earth.

The positive features of solar energy are – its *wide-spread distribution*, a virtually *inexhaustible supply* and the *pollution free nature*. Tremendous amount of solar energy reaches earth's surface. India receives solar energy equivalent to over 5000 trillion KWhr per year which is more than the total energy consumption of the country. Depending upon the location, the daily average solar energy incident over India varies from 4-7 KWhr/m^2. India has one of the world's largest programmes to promote the use of solar energy. It includes *solar thermal programme* and *solar photovoltaic programme*.

Solar energy is being used in many parts of India for cooking, heating, lighting and cooling purposes both in homes and industries. It is also used for electricity generation. Fig. 1.3 below shows the use of solar energy for different purposes.

(a) Solar Lantern

(b) Solar Panels for Heating Water

(c) Solar Power Plant in a Village in Rajasthan

(d) Solar Photovoltaic Water Pumping System in Uttar Pradesh

Fig 1.3: Various uses of Solar Energy

Solar energy may be used directly through *passive solar systems* or *active solar systems*. The **passive solar systems** are designed to enhance the absorption of solar energy without the use of mechanical power. The **active solar systems**, however, use mechanical power in the form of pumps etc. for circulating air, water or other fluids from solar collectors to a heat sink meant for heat storage.

Passive solar heating uses solar energy to heat buildings directly by trapping the heat directly within the structure of the building followed by its slow release. This is similar to greenhouse effect. In active solar heating, building are heated indirectly by circulating heated water using pumps and pipes.

Solar Collectors

Since solar radiation reaching earth has fairly low temperature, its energy requires concentration which is done using solar collectors. They are usually flat panels consisting of a glass cover plate over a black background where water is circulated through tubes. They can heat water from 38° to 93°C by acting as greenhouses.

Such collectors require a lot of sunshine for efficient working. The energy produced is relatively expensive due to high capital costs involved in the installation of heating system.

Photovoltaics

It is the technology for converting sunlight directly into electricity using solid semiconductor material such as silicon. The solar cells are made of silicon or other materials such as gallium arsenide. When sunlight strikes these cells, the electrons from the cell flow through the electrical wires. The electricity so produced can be used for lighting, running electrical appliances, in calculators and other such devices. The sector wise application of such energy includes household, agriculture, telecommunication, defence, railway etc. The range of products running on solar energy includes solar lanterns, home lighting systems, stand alone power plants, solar water pumping systems. Desalination of water, powering of remote telecommunication stations and railway signals are also done by using solar energy.

The efficiency of conversion of solar energy to electricity varies from 10% to 25% which is not very high. Its use is being encouraged and over 61,00,00 systems, aggregating to over 20 MW, have been installed according to MNES (2002).

Solar energy offers a great potential but suitable technologies are yet on the way to fully exploit this non-depleting source.

Solar Energy and Environment

As far as the environment is concerned, there is hardly any direct adverse impact of the use of solar energy. But, there are certain indirect ways by which the technology used for harnessing solar energy can have harmful effects on the environment. It involves the use of a variety of materials such as metals, plastics, fluids etc. The manufacture of these materials produces toxic wastes, and can also accidentally release these materials in the environment.

The general disadvantage of solar energy, however, is that the solar energy being relatively dispersed, requires large area. This can be taken care of by using solar collectors on the roof tops of buildings.

b. Wind Energy

Wind energy has been in use since early Chinese and Persian civilizations. It has been used for sailing, grinding grain or pumping water for irrigation. Lately, it has been also used for the generation of electricity.

It is a renewable source of energy like solar energy. There is a tremendous potential for its use but wind is highly variable in time, place and intensity which offer limitations to its use.

Wind energy can be obtained by using wind mill, wind turbines or wind farms. A wind mill, as shown in Fig 1.4, involves rotation of blades or vanes with the help of wind energy. This, then, generates mechanical energy for lifting of water from well or rivers, for grinding or for generation of electricity. The use of wind energy for the generation of electricity was first done in Denmark in 1890s.

Fig 1.4: Wind Mills and Wind Turbine at Davengere, Karnataka

The use of wind power has increased all over the world because of rising prices of fossil fuels and their fast depletion. The energy crisis in 1970s and

1980s also encouraged the use of wind energy on a large scale. This has been possible by the use of wind turbines.

Although, turbines operate on the principle of windmills, they are much sophisticated in design. They use propeller type rotors mounted on high towers. The amount of wind energy captured by a wind turbine, depends upon the size of the blades as well as the speed of the wind. The largest wind turbine is in Hawaii in the Pacific Ocean. It has two blades of 50 m length and the 20 storeyed tower.

Wind turbines can be used individually to meet the energy demand for smaller towns or villages in isolated or remote areas. They can also be grouped together or clubbed as *wind farms,*see Fig. 1.5.

Fig 1.5: A Wind Farm

Wind farms require a large area of land. Many wind farms (about 17000) were installed in California in 1980s and they have a capacity of about 1400 MW. The Altamont wind farm of California is one of the largest and best-known wind farms. Other than USA, other countries exploiting wind energy are France, Netherlands, Denmark, Germany and parts of Britain.

India is the fifth largest producer of wind power in the world. The wind energy potential of India is 45,000 MW. The states of Tamil Nadu, Gujarat, Andhra Pradesh, Karnataka, Kerala, Madhya Pradesh and Maharashtra have high wind potential. The state wise gross and technical potentials have been given in Table 1.4.

Table 1.4 Wind Power Potential

State	Gross Potential (MW)	Technical (MW)
Andhra Pradesh	8275	1750
Gujarat	9675	780
Karnataka	6620	1120
Kerala	875	605
Madhya Pradesh	5500	825
Maharashtra	3650	3020
Orissa	1700	680
Rajasthan	5400	895
Tamil Nadu	3050	1750
West Bengal	450	450
Total	**45195**	**11875**

Source: Annual Report 2003-2004, MNES

For 2002-2007 period, India is exceeding its target of installing 1500 MW. There are 945 water pumping wind mills, aero-generators and hybrid systems of about 370 KW capacity and wind farm projects installed with a capacity of 2483 MW. There is a good manufacturing base of 12 manufacturers of wind turbines and allied equipment.

The Centre of Wind Energy Technology (C-WET), at Chennai, is a specialised institution for developing state of the art indigenous technology for wind power utilization. It also undertakes research and development, standardization, testing and resource assessment.

It is expected that the cost of wind energy production would decrease in coming years by improving grid connections and inviting greater participation of the private sector. A new concept of mega wind farms is being initiated by the private sector.

Wind Energy and Environment

It is a very clean source, still wind energy has (relatively small) adverse impacts on the environment. The effects include the following:

(i) Vibrations in the windmill cause objectionable noise.
(ii) Windmills can kill birds.
(iii) There can be interference with radio and television broadcasts by windmills.
(iv) Windmills can degrade the scenic environment.
(v) Windmills and wind farms require a large area of land which could have been otherwise required or used for roads, housing or other public buildings.

c. Geothermal Energy

Geothermal energy is the natural heat available in the interior of the earth. It was developed in Italy about 100 years back. It is used to generate electricity in Russia, Japan, New Zealand, Mexico, Hawaii and California. The Geysers (geothermal field) located north of San Francisco, California is the largest geothermal power operation in the world. It is shown below in Fig 1.6 and it produces energy directly from steam.

Fig 1.6: The Geysers

In India, the geothermal areas are the North Western Himalayas and Western Coast. The geothermal areas are located with the help of satellite. More than 350 hot spring sites have been identified by Geological Survey of India. They include Tattapani Geothermal fields in Chattisgarh where 300 KW power plant is planned by MNES and Puga Geothermal fields in Ladakh (J&K). In Puga valley, an experimental 1 KW generation project is being used for poultry farming, pashmina wool processing and mushroom cultivation.

Geothermal Energy and Environment

The exploitation of geothermal energy generates on-site noise, causes emission of gases and disturbance by drilling operations for laying pipes. It is also accompanied by thermal pollution resulting from hot wastewaters. These wastewaters may also be saline and hence can cause corrosion as well as disposal problems.

d. Hydropower

It is cheapest and cleanest source of energy. But, there are controversies about construction of mega dams. The small hydropower plants are, thus, emerging as alternatives for meeting the energy needs of remote and rural areas. The hydroelectric potential in India is estimated to be 600 KWh annually. The resources are mainly located in the northern and north-eastern region. In Tenth and Eleventh Plans, it is expected that 16 GWe and 19.3 GWe new capacity would be added, respectively.

The total for small hydro plants (generating upto 25 MW) is about 15000 MW. There are more than 420 such projects (*MNES, 2002*) spread all over the hilly regions of the country.

e. Biomass

Biomass is organic matter derived from plant materials, animal waste and the waste derived from various human activities. It is also generated from timber industry, agricultural crops, and raw materials from forests, household waste and wood.

Biomass accounts for about one-third of the fuel used in India. The usage is more than 90% by rural people and about 15% by urban people. They use wood, cow dung cakes, crop residues, sawdust etc. Firewood is the most widely used fuel and more than 1 billion people in the world still use wood as primary energy source.

Biomass can be directly burnt to produce heat or electricity. It can also be converted to gaseous fuel via **gasification** or can be converted to *biofuels* such as ethanol, methanol, methane etc. by distillation.

Biomass gasifiers convert biomass into producer gas through thermochemical gasification process. Biomass can be converted to blocks of different shapes, called *briquettes*, which are convenient to use. The briquettes can be used in traditional chulhas in place of coal or in the gasifiers for generation of gaseous fuel.

The efficiency of biomass energy generation is improving. Half kilo of a dry plant can yield upto 1890 kcal of heat which is nearly the same amount of heat available from 250 g of coal. For improving the efficiency, gas turbine engines are being used in biomass plants. These have an efficiency of about 35%.

The biomass potential in India is enormous because of large quantities of agricultural, forestry and agro products. The power generation by biomass accounts for 14% of the total energy supply worldwide. 38% of this energy is consumed in developing countries in rural and traditional sectors. In addition to heating, pumping water and power generation (standalone or grid connected), biomass powers is also used for village electrification and industrial uses. There are 34 cogeneration plants with capacity 210 MW in operation while 26 more such units having capacity 237 MW are being installed. The total biomass based power generation capacity as per MNES, 2000 is 358 MW.

Biomass gasifier systems are being used by various industries, such as rice and textile mills, steel rolling mills, ceramic industry, brick kiln industry, tyre and tube manufacturing companies, plywood industry etc., for reducing the energy cost. India also exports gasifiers to Africa, Europe, Bangladesh, Indonesia, Sri

Lanka, Maynmar and the USA. In countries like Finland, USA and Sweden, the per capita biomass energy used is higher than that in India, China or Asia.

There is a national level programme of MNES on biomass-based power generation, biomass/bagasse-based co-generation, research and development and biomass resource assessment. The potential for bagasse-based cogeneration in major sugar producing states is shown in Table 1.5.

Table. 1.5 State-by-state potential for bagasse cogeneration in India

State	Potential (MW)	Commissioned exportable capacity, as of 31 December 2003
Maharashtra	1,250	21.0
Uttar Pradesh	1,250	75.0
Tamil Nadu	500	105.0
Karnataka	500	125.0
Andhra Pradesh	300	49.3
Bihar	300	Nil
Gujarat	250	Nil
Punjab	150	Nil
Other	500	Nil
Total	**5,000**	**375.9 (7.5%)**

Source : MNES, New Delhi.

Biomass Energy and Environment

Biomass fuels although plentiful can cause several environmental problems. Burning of biomass fuels causes air pollution by generating smoke. Also, if we keep on using wood from forests and do not replace them, it causes the problem of deforestation. Deforestation can, in turn, accelerate the process of soil erosion which can further lead to water pollution. Burning of organic urban waste releases toxins and causes air pollution. Conversion of biomass to alcohol also adds to air pollution and is also not a very viable option economically. The inefficient burning of such fuels in traditional chulhas is associated with the problems of indoor pollution and health hazards. Thus, there is a need for proper management of biomass resources.

There are many efforts for improving technology to make biomass a clean and affordable resource. One such technology is the use of biogas plants.

f. Biogas

Biogas is derived from cattle dung and human waste. It contains 55-70% methane which is inflammable. It burns with a blue flame. It can be used for lighting when burnt in silk mantle lamps. It can be used in dual-fuel engines and substitutes upto 75% diesel oil.

Biogas plants involve the mixing of organic waste with water. Their decomposition in the absence of air produces biogas which can be used as fuel with the help of pipelines. The left over digested slurry is used as manure in agriculture and pisciculture.

Biogas can also be generated by using wood, straw or waterweeds like water hyacinth, hydrilla, duckweeds etc.

There are three main models used for the generation of biogas. They are as follows:

(i) Floating gasholder type – also known as Indian or Khadi and Village Industries Commission (KVIC) model.
(ii) Fixed dome model – known as Deenbandhu model
(iii) Bag type portable digester made up of rubberised nylon fabric.

There is an estimated potential of 120 lakh plants out of which 36.51 lakh plants have already been installed. During 2003-04, the target was to install 1.50 lakh plants.

The MNES is having the National Biogas and Manure Management Programme (NBMMP) which is a modified version of Ninth Plan scheme known as the National Project on Biogas Development (NPBD). KVIC and State nodal agencies are involved in the programme implementation. Several NGOs, Sustainable Development Agency (SDA) and BIOTECH are working in this area in collaboration with grass-root level voluntary agencies and self-employed trained workers. Fig 1.7 shows a biogas plant.

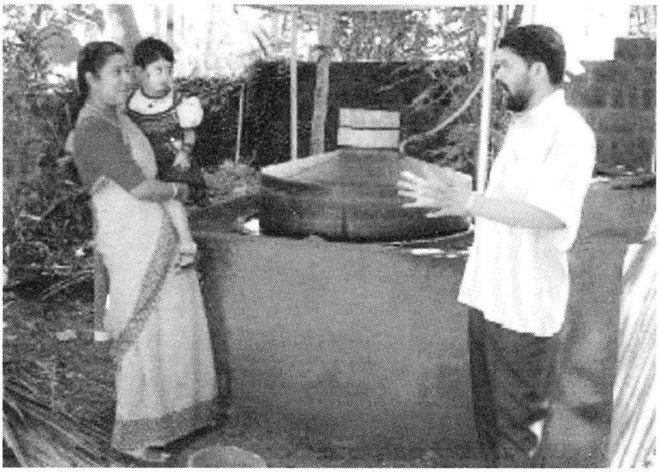

Fig 1.7 : (a) A Biogas Plant in Kerala

Fig 1.7: (b) A Deenbandhu Biogas Plant in Kerala

g. Tidal Energy

There are different forms of energy available from oceans. They are thermal energy, tidal energy and energy from ocean waves and currents. However, with the present technology, we are able to harness only tidal energy. The potential sites of tidal energy in India are Gulf of Kutch, Gulf of Khambat (Cambay) in Gujarat, Gurgaduani creek and delta of Ganga in sunderban area in West Bengal. Tidal energy can be harnessed for generating electricity in an area where sea water can move into a narrow cut. Such an area is naturally occurring at a point where rivers flow into the sea. Fig. 1.8 shows tidal power and its harnessing.

Fig 1.8: (a) Tidal Power **(b)** Harnessing of Tidal Energy

Both incoming and outgoing tides are held back by the dam. The difference in water levels generates electricity through reversible turbogenerators which can work both ways for power generation i.e. movement of water in opposite direction.

h. Energy from Waste

Worldwide, there is an increasing awareness about waste generated by human activities. Safe, efficient and scientific methods are being designed for treatment and disposal of wastes. The generation of energy from these wastes reduces the quantity of waste and also improves its quality to match with pollution control standards.

In India, the potential for generating power from urban and municipal waste is about 1700 MW whereas that from industrial waste is about 1000 MW at present.

The energy recovery from wastes is one of the thrust areas identified by MNES. The two programmes for implementing this objective are as follows:

- National Programme of Energy Recovery from Urban and Industrial Wastes.
- UNDP/GEF project on 'Development of High Rate Bio-methanation Processes'.

India's urban waste amounts to about 30 million tonnes of solid waste and 4400 million cubic meters of liquid waste per year. In addition to this, a large amount of industrial waste is also generated. This waste pollutes environment in a number of ways if not disposed off properly.

During the year 2003-04, five projects - two based on municipal solid waste (MSW) and one each on sewage, vegetable waste and starch industry solid waste, were commissioned with a total installed capacity of 15.65 MW.

The MSW plant is located at Lucknow. It is designed to process 500-600 tonnes of waste per day from Lucknow city. It is meant for power generation by producing biogas.

Table 1.6 shows the achievements of renewable energy programme in India (*Source: MNES*).

Table 1.6 Major programme of Renewable Energy sources and their cumulative achievements

S. No.	Source/Technology	Units	Achievement (Cumulative)	Position in the world
1.	Power Generation			
	(i) Wind Power	MW	2483.20	V
	(ii) Small Hydropower	MW	1603.32	X
	(iii) Biomass based power	MW	681.53	IV
	(iv) Biomass Gasifiers	MW	60.480	I
	(v) Solar Voltaic power	MW	2.54	V
	(vi) Urban & Industrial waste	MW	41.43	I
2.	Thermal Applications			
	(i) Biogas plants	lakhs	36.51	II
	(ii) Improved Biomass Chulha (Cook stove)	crores	3.52	II
	(iii) Solar water heating systems	0.80	Million sq. m collector area	
	(iv) Box Solar cookers (No.)	No.	555000	I
	(v) Schefler solar cookers (No.)	No.	2000	
	(vi) Solar steam cooking systems	No.	10	
3.	Water Pumping			
	(i) Wind pumps	No.	1087	
	(ii) Solar PV pumps	No.	6452	
4.	Solar Lighting Systems			
	(i) Solar street lighting systems	No.	52102	
	(ii) Home lighting systems	No.	307763	
	(iii) Solar lanterns	No.	538718	
5.	Other applications			
	(i) Integrated Rural Energy Programme	Distt	253	
	(ii) Energy parks	No.	376	
	(iii) Hybrid systems	KW	370	
	(iv) Solar photovoltaic power (stand alone)	kWp	851	
	(v) Aditya solar shops	No.	59	
	(vi) Battery operated vehicles	No.	300	
	(vii) Village electrification	No. of villages	1563	

i. Hydrogen– The Future Fuel

Hydrogen is the simplest and most common element found on the earth. It has the high energy content per unit of weight (i.e. 120.7 kJ g^{-1}). It is very light in weight. It is used as a fuel for rocket and space crafts due to its above said qualities.

Hydrogen is receiving world-wide attention because it is a clean and efficient fuel. It can be used in any of the ways in which fossils fuels are normally used. On combustion, it produces water as the by-product and hence, it is an environmentally safe fuel.

Hydrogen can be produced in a variety of ways as follows:

(i) Biological conversion of organic effluents like distillery starch, sugar processing etc.
(ii) Electrolysis of water
(iii) Thermal decomposition of water through solar or nuclear energy
(iv) Gasification of coal
(v) Steam reformation of natural gas, naphtha
(vi) Pyrolysis of biomass
(vii) Partial oxidation of heavy hydrocarbons or coal

Hydrogen can be used for power generation and transport applications. It can be used either directly in internal combustion engines or through fuel cells.

Fuel cells produce electricity by combining fuel and oxygen by an electrochemical reaction. Hydrogen and phosphoric acid are most common fuels but fuel cells are also based on methanol, ethanol and natural gas. Fuel cells directly convert the chemical energy to electricity without involving the combustion to generate heat and then mechanical energy which is further converted to electricity as is the case with conventional fuels. Thus, it is an efficient way of generating electricity. A fuel cell is shown in Fig.1.9.

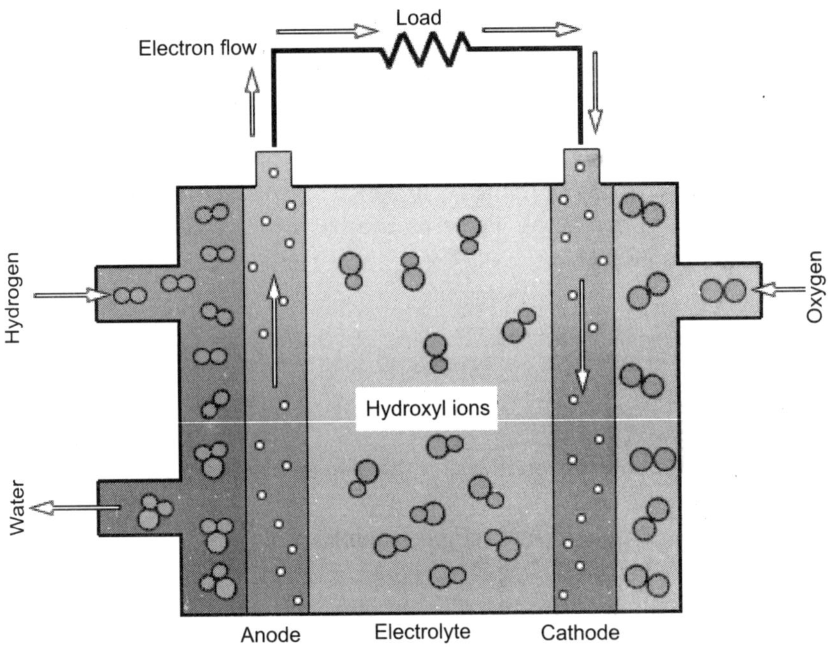

Fig.1.9: A Fuel Cell

Hydrogen (fuel) and oxygen are present in the electrolyte in the fuel cell. They remain separated from one-another. Upon ionisation, they migrate from one electrode to other through the electrolyte solution. The flow of electrons from negative to positive electrode is diverted through the electrical motor for supplying current.

Fuel cells are modular in nature construction wise and their efficiency is independent of size. A variety of fuel cells are available. They can be classified as *low temperature* and *Medium & High temperature* fuel cells. They are used for mobile, stationary and portable applications. Low temperature (upto 100°C) cells are useful for transport and small power generation applications while medium and high temperature (upto 1000°C) cells are preferred for power generation and combined heat and power applications. Fuel cells can also be used as UPS. Their applications cover domestic, industrial, transport and agricultural sectors.

j. Nuclear Energy

About 16% of the world's electrical energy is derived from nuclear energy. Nuclear energy is the energy stored in the atomic nucleus. Nuclear energy or atomic energy can be used for constructive purposes such as power or electricity generation. The two main modes of reactions used are fission and fusion reactions.

Nuclear fission involves splitting of an atomic nucleus into smaller fragments whereas in nuclear fusion, atomic nuclei combine to form heavier nuclei. Both these processes, release an enormous amount of energy. The fission reaction is shown in Fig 1.10.

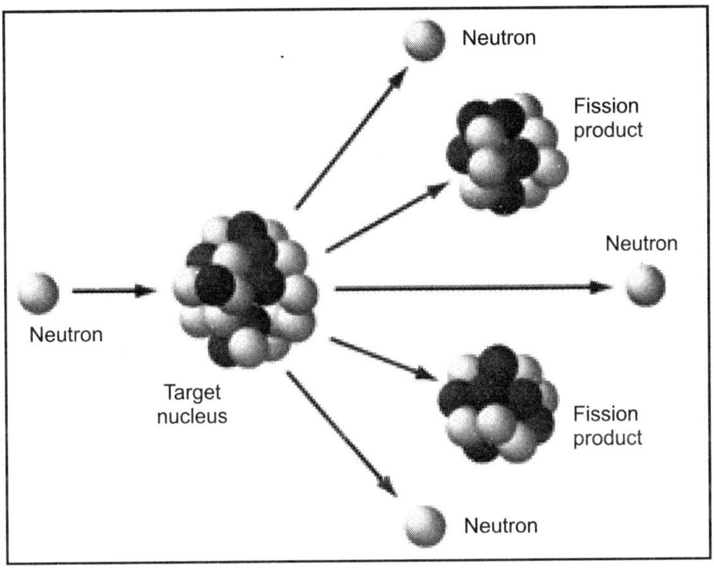

Fig 1.10: A Fission Reaction

Most nuclear power plants use **light-water nuclear reactors** which use ordinary water for transforming heat from reactor core to boilers. These reactors use U_{92}^{235} as fuel. The natural abundance of this isotope of uranium is only 0.7%. The process of enrichment concentrates this isotope to 3%. The fission of U_{92}^{235} using neutrons releases smaller fragments as ^{140}Ba and ^{93}Kr, a lot of energy and neutrons. These neutrons carry forward the chain reaction by further splitting the Uranium atoms present.

Another type of reactors called **fast breeder reactors**, can increase the nuclear fuel potential by 100 times. They do not require weapon-grade material and can use upto 60% of uranium in the ore.

The uranium reserves in India amount to about 95,000 tonnes of the metal. After accounting for losses due to mining (15%), milling (20%) and fabrication (5%), the amount available for power generation is 61,000 tonnes. But the thorium reserves are larger in quantity. The *monazite* reserves are about 8 million tonnes in which about 0.63 million tonnes of thorium is present. So far, only 17 deposits containing about 4 million tonnes of monazite have been identified as exploitable. The amount of thorium available for nuclear power is about 2,25,000 tonnes.

There are 14 operating atomic reactors in India. The power generation during 2001-2002 was 19193 Million Units. Table 1.7 lists the number and type of reactors.

Table 1.7 Nuclear Reactors in India

S.No	Place	Number	Capacity (Mwe)	Type
1.	Tarapur	2	160	BWR
2.	Rajasthan	2, 2	100 and 220	PHWR
3.	Kalapakkam	2	220	PHWR
4.	Narora	2	220	PHWR
5.	Kakrapara	2	220	PHWR
6.	Kaiga	2	220	PHWR

* BWR -Boiling Water Reactor
* PHWR-Pressurized Heavy Water Reactor

There are eight more reactors under construction at Tarapar (Maharashtra), Kaiga (Karnataka), Rawatbhata (Rajasthan) and Kudankulam (Tamilnadu). The nuclear power plants in India are owned, constructed and operated by Nuclear Power Corporation Ltd. (NPCL).

The PHWR type reactors use heavy water as moderator and coolant. They work one once-through-cycle of natural uranium. Nearly 330 GWe–yr of electricity can be produced using them. Using the same amount of uranium, the multiple recycling of the fuels by Fast Breeder Reactors (FBR) can provide about 42,200 GWe–yr assuming 60% utilization of heavy metal. However, the actual potential will be 150,000 GWe–yr. This much energy is good enough to meet our energy demands for many years to come.

The Department of Atomic Energy has worked out a three-stage nuclear power programme. The first stage involves the installation of a nuclear power plant of capacity 20GWe by 2020. The second stage involves building a chain of fast breeder reactors to enhance the capacity of fissile material and to produce power. Construction of first 500 Mwe Prototype Fast Breeder Reactor (PFBR) has been approved in September 2003, which would be completed in 2011. Four more such units would be constructed by 2020 to generate 20GWe power.

The third stage involves the exploitation of thorium reserves through fast or thermal critical reactors or accelerator driven sub-critical reactors (ADS). A 300 MWe Advanced Heavy Water Reactor (AHWR) is under development. It is designed to draw about two-third power from thorium fuel and will provide experience about the technology aspects related to thorium fuel cycle.

Energy production using nuclear fuels is relatively a clean and efficient option. Although 1 kg of uranium generates an equal amount of energy generated by 35,000 kg of coal, we have to weigh the two options in terms of their advantages or disadvantages given in Table 1.8.

Table 1.8 Energy from Coal and Nuclear Sources

	Coal	Nuclear sources
Advantages	1. Inexpensive	1. Small amount of solid waste is there
	2. Large availability	2. Large energy available
	3. Small capital investment	
Disadvantages	1. Leads to air pollution	1. Large operational costs
	2. Releases large amount of CO_2	2. Large capital investment
	3. Surface mining disturbs landscape	3. Problem of radioactive waste disposal
	4. Transportation-infrastructure is costly	4. Technology- availability and updating
	5. Ash-disposal	5. Risk factors -Radiation hazards -Accidents

In view of the disadvantages associated with nuclear power generation, it has been a controversial issue for several years.

The problems are site selection for nuclear power plants, strict safety controls, disposal of radioactive waste and effects on human health by radiation exposure, various stages of nuclear fuel cycle (discussed later) generate nuclear waste which poses a great problem as far as its disposal is concerned. This has been described under Hazardous Waste and its Management (Chapter 10).

In addition to this, weapon grade nuclear fuel is also smuggled for terrorist activities which is another kind of problem to be solved. The regulatory and safety functions of atomic energy are carried out by Atomic Energy Regulatory board. The President of India constituted it on 15th November 1983 under Atomic Energy Act, 1962.

Thus, nuclear power, if used sensibility and carefully,can solve the future energy problems.

Environment

CHAPTER 2

2.1 Introduction

The term *environment* (which means surroundings) is viewed differentially by different environmentalists. It is defined as the sum total of all living *(abiotic)* and non-living *(biotic)* components. In fact, environment creates favourable conditions for the existence and growth of living organisms. The environment has never been constant or static. It has been changing slowly, rapidly or drastically. All living organisms including human beings are affected by changes in the environment – the changes may be harmful or beneficial. It is well known that many species on the earth (which could not adapt with changing environment) vanished or are on the brink of extinction (vanishing) due to environmental changes or factors. A significant example is the extinction of dinosaurs due to hostile conditions of their environment. It is believed that the main cause of this was attributed to the newly appeared tiny primitive mammals, which ate the eggs of dinosaurs. Thus, for primitive mammals, the environment was favourable while for dinosaurs it was unfavourable.

For past more than two decades, human beings have primarily been responsible for the changes in the environment. Some of these activities are as follows:

- Devastating forests by cutting trees for wood and bringing forest land under cultivation. This caused erosion of the soil. This activity has been going on from the past thousands of years.
- Killing the gentle animals for food and the fierce animals due to safety reasons.
- The industrial and scientific revolution in the recent past has a tremendous effect on the environment. This aspect is mainly responsible for polluting the water bodies with chemicals from industries, polluting the atmosphere

by various poisonous and harmful gases discharged from industrial establishments including power plants using coal.

- A large number of episodes have affected the environment. The most important is the *London Smog* that killed about 4000 people in 1952, *nuclear catastrophes* including dropping of nuclear bombs on Hiroshima and Nagasaki and *Bhopal gas tragedy*. For a detailed discussion, see sec. 10.2.

To sum up, various changes in the environment have adversely affected the ecosystem as a whole. This aspect will be discussed subsequently (see sec. 2.7).

2.2 Management of Environment

Management of environment is of top priority for all nations. The main objectives of environment management include:

- Maintenance of environmental quality
- Balancing the ecosystem
- Regulating and restricting the exploitation of natural resources
- Preventing degradation of the environment
- Renewing natural resources as far as possible
- Preventing or at least regulating environment pollution
- Preventing or at least reducing various episodes or disasters
- Formulation of laws and regulations to achieve the above objectives

It is essential that the management of environment should be on long-term basis. An Environmental Education Conference (EEC) was held at Belgrade in 1975 and later by UNEP (United Nations Education Programme) at Tbilisi, USSR in 1977. Their main objective was creating awareness about degradation of the environment.

People have begun to recognise that the most important concern which is faced by the society is the problem of environmental pollution. There is absolutely no doubt that education plays a crucial role in solving such problems to achieve the so-called sustainable development. The Earth Summit held in Rio de Janeiro in 1992, declared that education is most critical in promoting sustainable development and improving the capability of the people for addressing environmental and developmental issues. Environmental education involves following three stages:

(i) ***Awareness***-It helps the people to acquire sensitivity to the environment and its problems. It helps people to acquire a basic understanding of how environment functions and how people interact with the environment. It is most important to understand how the environmental issues can be addressed or the environmental problems could be solved.

(ii) ***Attitude***-It helps people to develop values and feeling of concern for the environment and commitment to participate in environmental protection and improvement.

(iii) ***Action***-It helps the people to acquire skills for identifying and investigating the environmental problems and participating in their resolution.

The goal of the environmental education is to develop awareness, competence, attitude and values in the people, so that they can be effectively involved in the sustainable development at local, national and international levels. This is the only way to achieve a more equitable and sustainable future.

2.3 Components of Environment

The environment is comprised of the following three components:

(i) **Abiotic or Non-living components:** These include

Light	Atmospheric gases
Precipitation	Altitude
Humidity and water	Latitude
Temperature	Seasonal changes
Substrate	Topography

(ii) **Biotic or living components:** These include
Plants (Flora)
Animals including humans, parasites and microorganisms (Fauna)
Decomposers

(iii) **Energy Component:** It includes solar energy, geochemical energy, thermoelectric energy, nuclear energy etc. Energy from other sources including radiation also plays an important role to sustain the life processes.

2.4 Types of Environment

Broadly speaking, the environment is of the following two types:

(i) **Natural Environment:** The components of natural environment are air, water, soil, radiation, land, forest, wild life, flora and fauna etc.

(ii) **Man-made Environment:** These include transportation, housing, agricultural implements and industries, dams, energy such as hydro, thermal and nuclear energy, industries etc.

2.5 Segments of Environment

The environment comprises of the following four segments:

(i) Atmosphere
(ii) Hydrosphere
(iii) Lithosphere
(iv) Biosphere

2.5.1 Atmosphere

The cover of air that envelops the earth is known as the **atmosphere**. It is responsible for sustaining life on earth. It protects the earth from the harmful ultraviolet rays coming from the sun. It contains life-supporting gases like oxygen for human beings and animals and carbon dioxide for plants to help in carrying out the process of photosynthesis. The presence of nitrogen, oxygen and carbon dioxide is of great importance, as these gases keep on cycling in nature between organisms and the atmosphere through various cycles such as the carbon cycle, nitrogen cycle, water cycle etc. During the past 300 years or so, due to agricultural and industrial revolutions, human beings are responsible for dumping waste materials in the atmosphere which are creating problems for the survival of life on the earth. For a detailed discussion on atmosphere, see Chapter 3.

2.5.2 Hydrosphere

It covers more than two-thirds of the earth's surface either as oceans (salt water) or as fresh water and includes seas, rivers, oceans, lakes, ponds etc. The oceans account for about 95% of earth's water supply. This water is unfit for consumption due to its high salt content. The polar ice caps and glaciers account for about 2% of the remaining water. Fresh potable water is available only to the extent of about 1% in rivers, lakes and as ground water.

Water has a high value of latent heat and it is this property that moderates the temperature of the biosphere. It is a universal solvent. The solubility of oxygen in water is responsible for the survival of aquatic life. Though there is loss of water by evaporation but its amount remains constant. This is because there is continuous circulation of water involving water bodies and natural processes which use water (see water cycle sec. 2.8.1.1). Due to rapid industrialization, human beings are discharging waste materials into water bodies like seas, oceans, rivers etc. This has created serious problems for the survival of life not only on the land but also in water. For a detailed discussion on hydrosphere, see chapters 5 and 6.

2.5.3 Lithosphere

The upper most layer up to 400 km of the earth's surface is called **lithosphere**. It is the region which interacts with hydrosphere, atmosphere and biosphere (see biosphere in sec. 2.5.4). About three-fourth of the surface of lithosphere is covered by water. The earth comprises of three layers *viz. crust, mantle* and *outer* and *inner core* (*see* Fig.2.1). The core is the central fluid sphere having a diameter of about 2500 km. It is composed of mainly nickel and iron. The mantle extends about 2900-3000 km above the core and is also in the molten state. The crust is the outermost solid portion of the earth. It is about 8-40 km above the mantle. The surface of the earth is covered with the soil and it is the most important part of lithosphere from the point of view of environment.

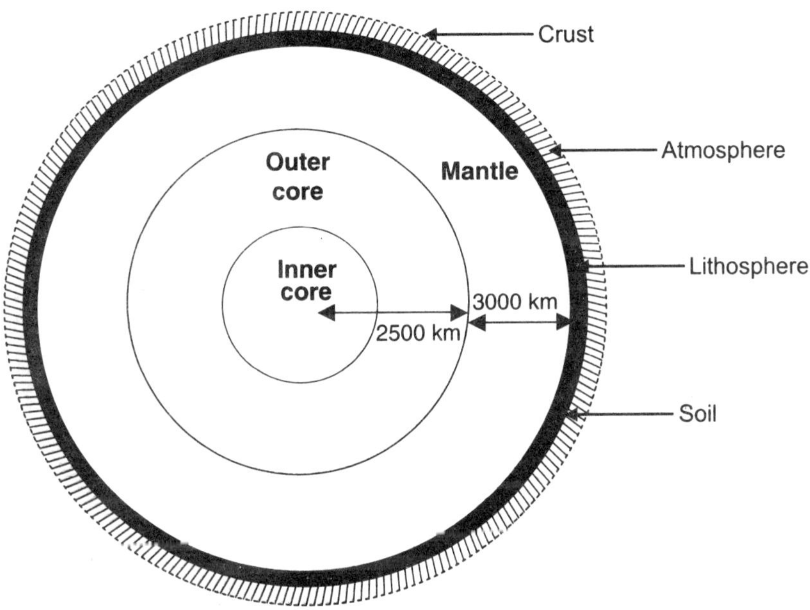

Fig. 2.1: Layers of Earth

The lithosphere is helpful in the metabolic processes of organisms by (i) providing most of the necessary minerals and (ii) forming soil needed by plants. For detailed discussion, see sec. 2.8.

2.5.4 Biosphere

Biosphere is that region where lithosphere, hydrosphere and atmosphere- all are present and where life (or living organisms) can exist. It is defined as the thin layer of soil, rock, water and air that surround the earth, along with the living organisms for which it provides support. The biosphere extends from about 11,000 m above the sea level to about 9,000 m below the sea level. A biosphere can be represented as shown below in Fig. 2.2.

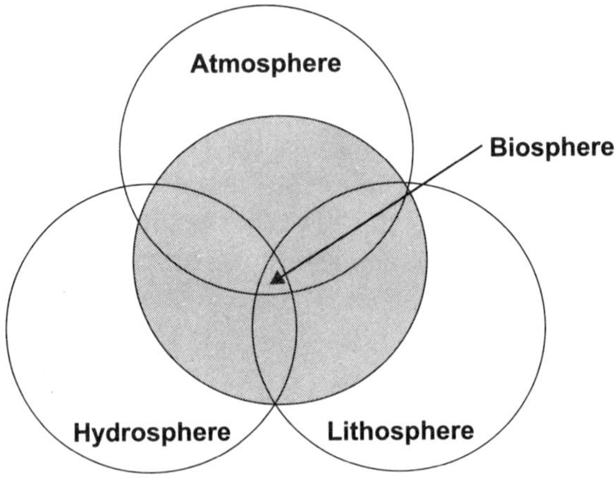

Fig. 2.2: Representation of the biosphere (shaded portion)

As seen in the Fig. 2.2, biosphere is the area of contact between the three components *viz, atmosphere, hydrosphere* and *lithosphere*. It is extremely important for life, as it is here that the entire life is contained and basic processes of life, the photosynthesis and respiration occur. The living organisms are confined to biosphere that receives solar radiation during the day. As already stated, biosphere extends from the floor of the ocean (some 11,000 m below the surface of the earth) to the top of the highest mountain (or about 9,000 m above the sea level). The most densely populated region is just above and below the sea level. The life in biosphere exists mostly between 600 m below the surface of the ocean and about 6000 m above the sea level. The energy required by the living organisms in the biosphere comes from the sun and the nutrients necessary for living organisms come from air, water and soil.

Biosphere is endowed with the necessary conditions that are essential to support life. Some of these conditions are:

- Availability of air
- Availability of water
- Availability of area or interface between the earth's crust, hydrosphere and atmosphere
- Availability of suitable temperature
- Supply of energy or radiation from the sun
- Availability of nutrients

The availability of nutrients in the biosphere in terms of macronutrients and micronutrients are given in Table 2.1.

Table 2.1 Availabilities of nutrients in biosphere

Type of nutrient	Element	Source
Macronutrients	Carbon	Atmosphere
	Hydrogen	Hydrosphere
	Oxygen	Atmosphere
	Nitrogen	Atmosphere and soil
	Phosphorous	Lithosphere
	Calcium	Lithosphere
	Chlorine	Lithosphere
	Copper	Lithosphere
	Iron	Lithosphere
	Magnesium	Lithosphere
	Sulphur	Lithosphere
	Sodium	Lithosphere
	Potassium	Lithosphere
Micronutrients or trace elements	Aluminium	Lithosphere
	Boron	Lithosphere
	Bromine	Lithosphere
	Zinc	Lithosphere
	Cobalt	Lithosphere
	Iodinc	Lithosphere
	Chromium	Lithosphere

2.5.4.1 Biomes

Biosphere is divided into various regions called **biomes**. Each biome is characterized by its climate, vegetation, type of soil and animal life. No two biomes are alike. The boundary of a biome is determined by the abundance of plants and animals found in each of them; these in turn are dependent on factors like temperature and precipitation or moisture. Table 2.2 gives various biomes in lithosphere along with their main features.

Table 2.2 Different biomes in lithosphere

Name of Biome	Region	Flora and Fauna
Tundra	Northern region adjoining the ice bound poles	Absence of trees; only shrubs are present. Flora includes lichen and mosses.
Taiga (Boreal forest)	North Europe, Asia and North America (Moderate temperature than that in tundra)	Vegetation includes coniferous evergreens-mostly spruce, pine and firs. The fauna includes small birds, hawks, fur bearing carnivores, Siberian tiger, wolves etc.
Temperate Deciduous Forest	Parts of Central and Southern Europe, North America, China, Japan, New Zealand etc. Temperature is moderate and abundant rainfall. These are the agricultural areas of the earth.	Flora – trees like peech, oak, maple and cherry etc. Fauna - the familiar vertebrates and invertebrates
Tropical Rain Forest	Tropical areas of higher rainfall in the equatorial region. Temperature is high.	About 40% of the world's plants and animal species exist. Broad-leaved evergreen plants
Savannah	Extensive in Africa (Tropical region)	Grasses, scattered trees and shrubs. Fauna includes antelopes, buffaloes, zebras, elephants and rhinoceros, lions, cheetah etc.
Grass Lands	North America, Midwest and Ukraine Temperate conditions, low rainfall.	Vegetation- grasses, Fauna include large herbivores like bison, antelope, cattle, rodents, wolves and birds
Desert	Continental interiors, low rainfall and low humidity	Flora like cacti, euphorbia etc. Fauna- reptiles, mammals and birds.

The hydrosphere (or the aquatic systems) is also divided into distinct life zones, which are not called biomes but are very similar, in that these are regions of distinct animal life and plants. The differences in these aquatic zones are due to salinity, levels of dissolved nutrients, temperature and depth of sun light penetration.

2.6 Environmental Pollution

The term environment as far as pollution is concerned, includes the air, the water and the land (soil). Pollution, broadly speaking is defined as the addition of harmful constituents or substances to water, air or land, which adversely alter the natural quality of the environment. The environmental pollution includes air, water and land (or soil) pollution. It is believed that there are various factors, such as population explosion, industrialization, deforestation combined with modern technology, which are responsible for serious environmental crisis. The industries are responsible for adding pollutants in the environment (land, soil, water and air) and thus, cause pollution.

Environmental pollution may be *natural* (which originates from natural processes such as forest fires, natural organic or inorganic decays, volcanic eruptions, earthquakes etc.) or it may originate due to human activities.

2.6.1 Types of Pollution

Most of the environmental pollution arises due to manufacturing processes, various industries, thermal power plants, automobiles, agrochemicals etc. Besides these, we come across noise pollution, thermal pollution and radioactive pollution. As far as environmental pollution is concerned, the worst pollutants are those which are not biodegradable; for example, polymers, plastics, asbestos, mica and some agrochemicals like DDT. Air pollution, water pollution and soil pollution are discussed in subsequent Chapters (see Chapters 4, 6, 7 and 8). The noise pollution has been explained in Chapter 11.

Environmental pollution is responsible for a number of hazards. For details see secs. 4.5, 6.5 and 9.3.

2.7 Ecosystems

The term ecosystem was coined by Prof. Arthur Tansley in 1935. The prefix ‘eco’ means environment, so an *ecosystem* is an environmental system. Ecosystem refers to a biotic community and its abiotic components put together. In other words, it includes both the living organisms, i.e. the biotic community and the non-living, i.e. abiotic environment.

Each of the biomes can be subdivided into smaller units; these are referred to as ecological systems or *ecosystems*. It can also be said that the ecosystems are the subdivisions of the biosphere. An ecosystem may be natural (e.g. a pond, a lake, a river, an ocean or a forest) or it may be artificial or man made (e.g. an aquarium, a dam, a garden, an orchard etc.).

2.7.1 Types of Ecosystems

Broadly speaking, the ecosystems are of following two types:

(i) **Terrestrial Ecosystems** –They include grassland, forest, desert, and cropland ecosystems.

(ii) **Aquatic Ecosystems** -They include pond, sea etc.

Most of ecosystems are self sufficient for the survival of life, i.e. they contain all the components necessary for life processes.

2.7.2 Components of Ecosystems

An ecosystem has two main components. These are - abiotic components and biotic components.

(i) Abiotic Components

The abiotic components or the non-living or the physical components include the following:

- Soil, water, air and light.
- Inorganic components like carbon, sulphur, calcium, phosphorus etc.
- Atmospheric gases like oxygen, carbon dioxide, nitrogen etc.
- Organic components like amino acids, proteins; these are links between abiotic and biotic components and are synthesised by plants and living organisms by various biochemical procedures.
- Climate and weather conditions like temperature, wind, humidity, heat etc.

(ii) Biotic Components

The biotic components of an ecosystem interact with abiotic components; this interaction is the basis of supporting life. These include the plants, animals and microorganisms. The organisms are of following types:

a. **Producers:** These are the autotrophic members of the ecosystem and are capable of synthesising their own food from simple substances like carbon dioxide and water in the presence of sunlight. Some examples include certain bacteria and algae etc. Of course, green plants can synthesise their own food by the process of photosynthesis.

b. **Consumers:** These organisms cannot make their own food but depend on other organisms for food to meet their energy requirements to survive. These are referred to as heterotrophic members of the ecosystem. Examples of consumers are animals and humans which depend on other organisms for the food.

Animals such as goat, cow, deer and insects etc. which eat green plants are called **primary consumers** or **herbivores**. Organisms that eat an herbivores like a bird that eats grasshoppers are called **secondary consumers**. The organisms that eat secondary consumers are called **tertiary consumers**. The secondary and tertiary consumers are **carnivores** while the primary consumers are **herbivores**.

c. **Decomposers:** Microorganisms which attack the dead bodies of producers and consumers and decompose them into smaller compounds are called **decomposers**. This process of decomposition also releases inorganic nutrients. Examples of decomposers are certain bacteria and fungi.

The producers, consumers and decomposers interact with each other in many ways. These interactions constitute the food chain.

2.7.3 Energy in Ecosystems

It is well known that energy used for all life processes is derived from sun in the form of solar energy. The flow of solar energy is unidirectional. In the absence of solar energy, the ecosystem is bound to collapse. We know that plants produce their food from carbon dioxide and water with the help of solar energy by the process known as photosynthesis. By this process, the solar energy finds its way as chemical energy into the plants. During photosynthesis, plants also incorporate in their protoplasm a number of inorganic elements and compounds. So the chemical energy is in the form of carbohydrates, fats and protein. A number of nutrients are also incorporated. When the animals (herbivores or carnivores) consume plant materials, the solar energy find its way into the animals. Finally, when the plants and animals (after death) decay, the nutrients etc. are returned to the soil by the decomposers and are again taken up by plants. In this way, the energy flows. This is exhibited in Fig. 2.3 below.

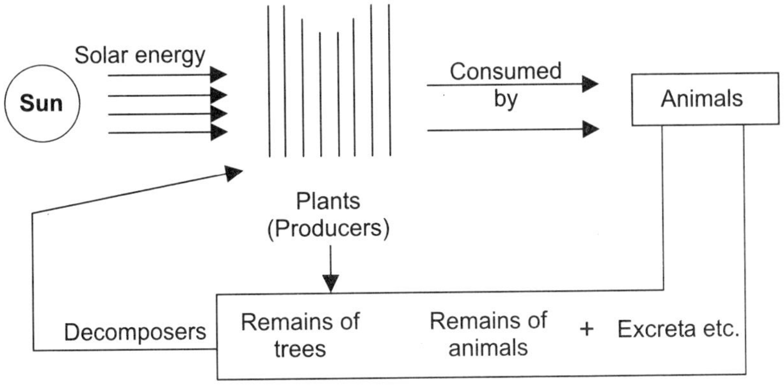

Fig. 2.3: Balancing the eco-system

When the trapped energy (in plants) is transferred to another system (e.g. animals etc.), it undergoes loss at each transfer along the chain. This is governed by the two laws of thermodynamics as given below:

According to the first law of thermodynamics, *the energy can neither be created nor destroyed but can only change from one form to another.* This is amply clear from the discussions in the above paragraph, wherein it is stated that solar energy is transformed into chemical energy in the plants in the form of carbohydrates etc. Thus, the energy is transformed but not lost. This chemical energy is transformed and used by the cells of the organisms via metabolism for numerous activities.

According to the second law of thermodynamics, *part of the useful energy is degraded into unusable waste as heat energy during every heat transformation; the waste energy escapes into the surroundings.*

On the basis of the above, it is concluded that the ultimate source of all energy, which is required for the ecosystem to function, is the solar energy. It has been found that about 30% of the total solar radiation which enters the atmosphere is reflected back by the earth-atmospheric system. The remaining 70% of the radiation is absorbed by the earth's atmosphere and the surface of the earth. The living world, thus, depends on the flow of energy and the circulation of nutrients through ecosystem.

2.8 Various Cycles of Environment

All living things need a number of chemical elements called *nutrients.* Of all the elements known, only 24 are required by human beings. These are hydrogen, sodium, potassium, magnesium, calcium, vanadium, molybdenum, manganese, iron, cobalt, copper, zinc, boron, aluminium, carbon, silicon, nitrogen, phosphorus, oxygen, sulphur, selenium, fluorine, chlorine and bromine. These are divided into macronutrients (elements required in large amount by all living organisms) and micronutrients (elements required in small amounts by all living organisms). The macronutrients include the 'Big Six'. These elements form the fundamental building blocks of life. These are carbon, nitrogen, oxygen, phosphorous, sulphur and calcium. Each element plays a special role in organisms. The elements sodium, vanadium, cobalt, boron, aluminium, silicon, fluorine, chlorine and iodine are required for some life forms.

For sustaining life in the biosphere, a number of nutrients are needed. These nutrients must be recycled so that once used they are made available again to the living organisms. During cycling, interchange of nutrients occurs among the biotic community and the abiotic environment. The passing of the nutrients between environment and organisms is referred to as **biogeochemical cycle** (*bio* for living and *geo* for atmosphere) or **nutrients cycle**. Each nutrient is normally concentrated in a source, e.g. nitrogen is basically concentrated in the atmosphere. Such a source is called **reservoir** or a pool. The biogeochemical cycles are of following two types:

(i) **Gaseous cycle :** In such cases, the reservoir is the atmosphere or hydrosphere. Examples of this type include water cycle, carbon cycle, nitrogen cycle, oxygen cycle etc.

(ii) **Sedimentary cycle :** In these cases, the reservoir is the earth's crust or lithosphere. Examples of this type include phosphorous, sodium, potassium, magnesium and iron cycles.

The following discussion is devoted to the various cycles of the environment.

2.8.1 Gaseous Cycles

2.8.1.1 Water Cycle

Water cycle, also known as **hydrological cycle,** is the most important of various cycles of the environment. This is because water carries along with it the materials moving through other nutrient cycles. Water, as we know, is one of the most important substances for the life processes. Living organisms contain about 75 percent water in them, though some plants contain less water. But algae and jellyfish may contain up to 95 percent water. Water covers about 75 percent of the earth's surface occurring in oceans, seas, rivers, lakes etc. Oceans alone contain 97 percent of all water present on the earth. Out of the remaining water, a sizable amount is locked in the frozen form in the Polar regions and glaciers. Only about 1 percent is available as fresh water. The global distribution of water is shown in Table 2.3.

Table 2.3 Global distribution of water

Source	Amount in percentage
Oceans	97.6
Ice in polar regions and glaciers	1.8699
Ground water	0.5
River, lakes, inland seas	0.02
Soil moisture	0.01
Atmosphere	0.0001

Water circulates between the living and non-living components of the biosphere in the form of an unending cycle. The hydrological cycle is depicted in Fig. 2.4 below. In hydrological cycle, there is movement of water. From the oceans to the atmosphere by evaporation and from atmosphere to oceans and land by precipitation, from land to oceans by run off and from streams and rivers and from land to atmosphere by evaporation. The cycle is driven by solar energy. In the above cycle, there is no involvement of living organisms and the water in oceans, seas and lakes etc. circulates between them and the atmosphere.

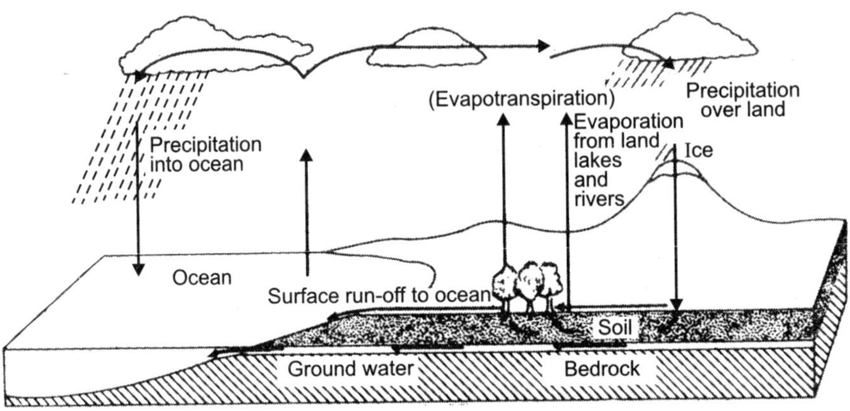

Fig. 2.4: Water (or hydrological) cycle

An alternative pathway for hydrological cycle involves soil water or the underground water which is circulated by plants, animals and the atmosphere. The soil water is taken by the plants through their root system. A part of the water is utilised by the plants for photosynthesis and the excess of water is given off into the atmosphere during transpiration (It is the loss of water from the leaves of the plants via evaporation). In fact, transpiration is the major source of water into the atmosphere. The decay of dead plants by microorganisms releases water back into the soil. Animals consume water by drinking from various sources and also through the plants they eat. Water vapour is released into the atmosphere by breathing and evaporation from the surface of animals (sweating) and excretion. Whatever the route is followed, the water vapour being lighter rises in the air, where it gets cool and condenses into tiny droplets to form a cloud. The water from the clouds ultimately falls on the earth in the form of rain, part of which gets absorbed by the soil and the remaining finds its way into water bodies.

Water is also used in considerable amounts by various industrial units engaged in manufacturing processes and power generation units. This water after use is finally discharged into the water bodies, though the discharged water is harmful and is responsible for pollution of the water bodies.

2.8.1.2 Carbon Cycle

Carbon is one of the most important constituents of all organic compounds. It is, in fact, the building block of life. It is present in carbohydrates, fats, proteins and nucleic acids. The main reservoirs of carbon are as follows:

(i) Atmosphere in which carbon is present as carbon dioxide.

(ii) Oceans in which carbon is present as carbon dioxide dissolved in water and also as calcium carbonate in various seashells and in rocks at the bottom of sea.

(iii) Land on which it is present in some ores like dolomite and other carbonates.

The carbon cycle involves circulation of carbon dioxide from the three sources as shown below in Fig. 2.5.

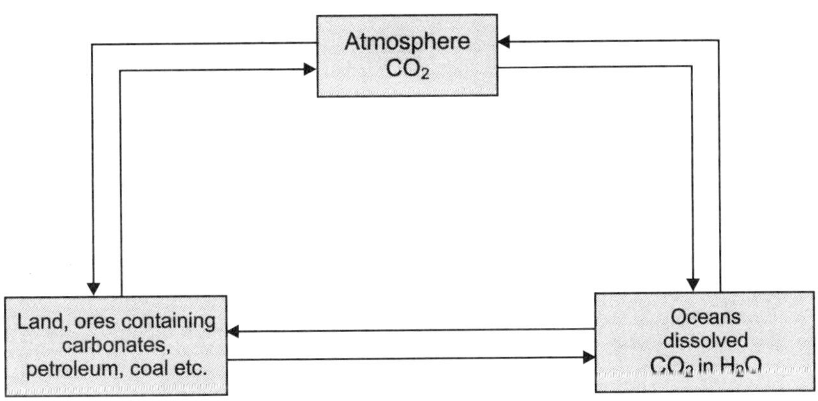

Fig. 2.5: Reservoirs of carbon

The various processes involved in carbon (or the so called carbon dioxide) cycle are given below:

(i) Carbon dioxide from the atmosphere enters the plants through photosynthesis where carbohydrates are produced.

(ii) From green plants, the carbon in the form of carbohydrates etc. enters into the bodies of animals.

(iii) The remains of dead animals and plants on putrefaction and decomposition by bacteria release carbon dioxide back to the atmosphere.

(iv) The atmospheric carbon dioxide gets dissolved in oceans etc. by simple diffusion process. Marine algae and photosynthetic bacteria obtain carbon dioxide from water. Some of the dissolved carbon dioxide is also trapped to form lime stone (calcium carbonate) and other carbonate containing rocks.

(v) Carbon dioxide returns to the atmosphere by respiration, combustion of fuels like coal, wood, diesel, natural gas, gasoline etc.

(vi) Weathering of carbonate containing rocks (particularly by water containing dissolved carbon dioxide as carbonic acid and by acid rain) releases carbon dioxide into the atmosphere. Volcanic eruptions are also responsible for the discharge of carbon dioxide to the atmosphere.

A word of caution - Release of excessive amounts of carbon dioxide into the atmosphere is responsible for green house effect and global warming (see Chapter 4).

The basic features of carbon cycle are depicted below in Fig. 2.6 (a) and 2.6 (b).

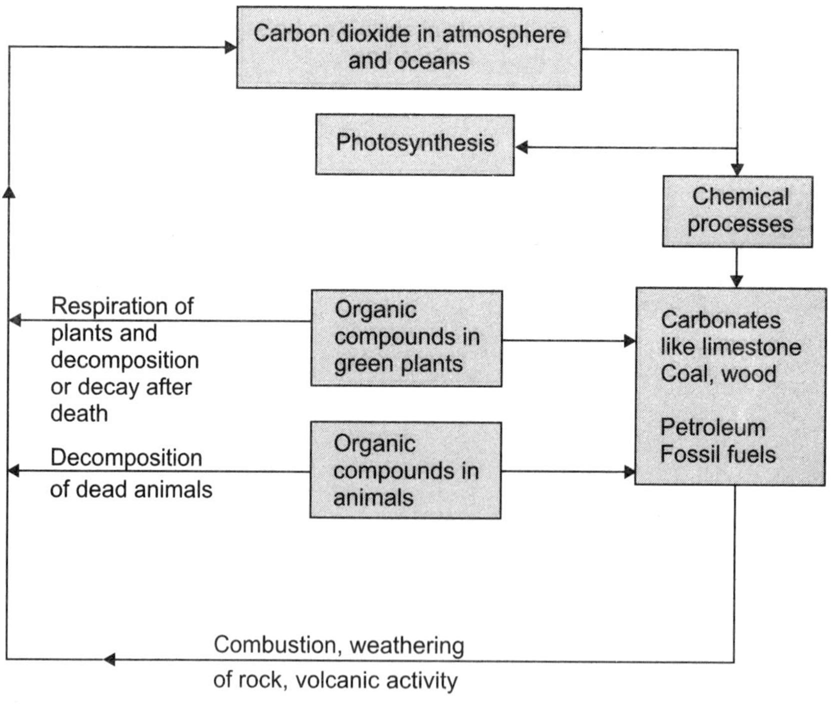

Fig. 2.6 (a): Carbon Cycle (Flow sheet)

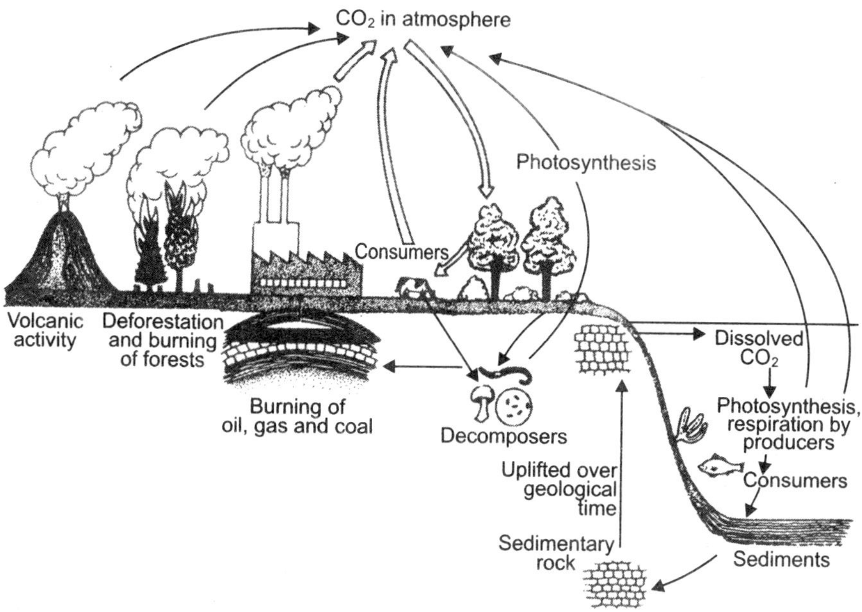

Fig. 2.6 (b): Carbon Cycle – various Features.

2.8.1.3 Nitrogen Cycle

Nitrogen is essential for the synthesis of amino acids and proteins in the animals and plants. Proteins are the building blocks of all living tissues. Nitrogen is present to the extent of 78-80% in the atmosphere. Atmosphere is the main source of nitrogen. However, the elemental nitrogen cannot be used by most of the living organisms. Plants can take up nitrogen either as nitrates or ammonia in the form of salts. This is achieved by the *fixation* of atmospheric nitrogen which in turn, is achieved either by a natural process or a synthetic process.

In the natural process of the fixation of atmospheric nitrogen, during periodic thunderstorms and lightening, the gaseous nitrogen is converted into nitric oxide, which gets oxidised into nitrogen dioxide. Both the oxides of nitrogen are washed down in rain and reach the soil, where the nitric acid formed combines with salts in the soil (e.g. sodium, calcium etc) to form the corresponding nitrates. During thunderstorms, atmospheric nitrogen may also combine with hydrogen in the atmosphere to produce ammonia, which is also washed down to the earth with rain and may combine with nitrates and sulphates in the soil to form ammonium salts. An alternate natural route for fixation of atmospheric nitrogen is with the help of certain microorganisms which can convert atmospheric nitrogen into ammonium ions. These microorganisms include nitrifying bacteria (e.g. aerobic *Azotobacter* and anaerobic *Clostridium*) and symbiotic nitrifying bacteria living in association with leguminous plants and symbiotic bacteria living in non-leguminous root nodule plants (e.g. Rhizobium) as well as blue-green algae (e.g. Anabaena, Spirulina). Some plants can take up directly ammonium ions as a source of nitrogen. Alternatively, ammonium ions can be oxidized to nitrites or nitrates by the specialized bacteria. For example, Nitrosomonas bacteria can promote oxidation of ammonia into nitrite, which is further oxidized into nitrate by the bacteria Nitrobacter. The nitrates, thus obtained in the soil are taken up by plants and converted into amino acids-the building blocks of proteins. The remains of plants decay and during the process, nitrogen is returned to the soil in the form of ammonia. Some of the nitrates (being soluble in water) find their way into the water bodies like rivers, oceans etc. The denitrifying bacteria, e.g. Pseudomonas is present in soil as well as oceans; these convert the nitrates into elemental nitrogen, which escapes into the atmosphere, thus completing the cycle.

In the synthestic process, fixation of atmospheric nitrogen is also achieved by industrial process. For example, in the Haber's process, nitrogen and hydrogen are made to react together in the ratio 1:3 at high temperature and pressure in presence of a catalyst. The ammonia thus obtained is subsequently converted into fertilizers like ammonium nitrate and urea; these fertilizers in the soil are converted by bacteria into nitrites and nitrates. Nitrogen and oxygen of the atmosphere can also be made to combine at high temperature in presence of a catalyst to give nitric oxide which further combines with oxygen to give nitrogen dioxide. The dissolution of nitrogen dioxide into water gives nitric acid. This is the basis of Oswald's process for the manufacture of nitric acid. The nitric acid, thus, obtained is converted into nitrates, which are used as fertilizers.

The flow sheet for fixation of atmospheric nitrogen is given below in Fig. 2.7 (a).

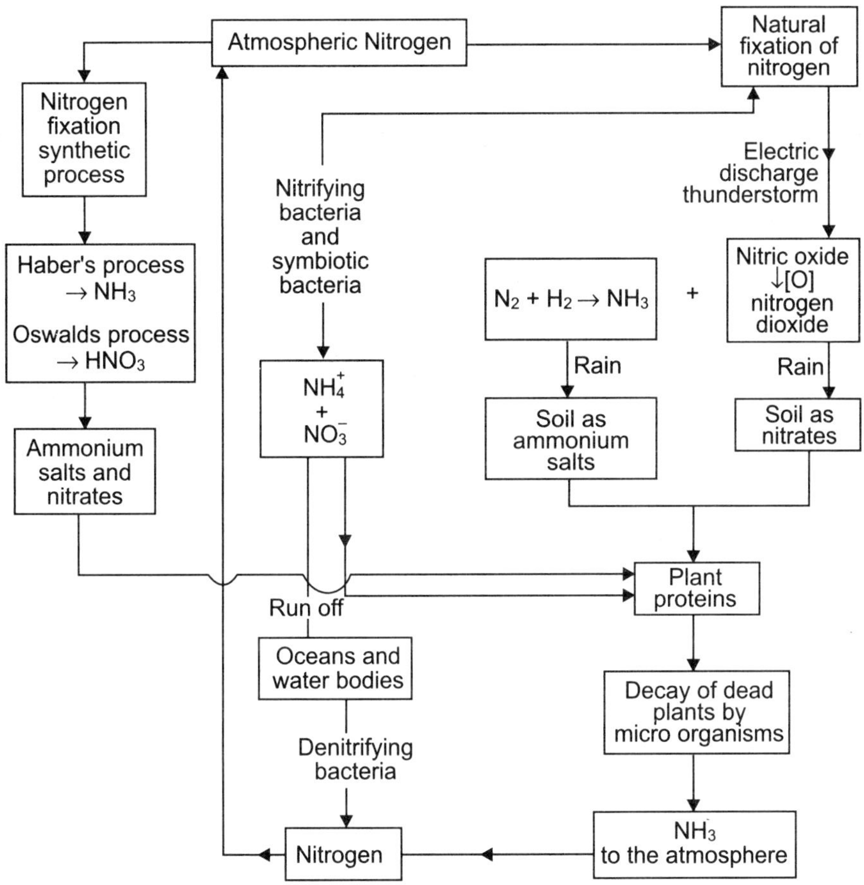

Fig. 2.7 (a): Flow Sheet for Nitrogen Fixation

Various features of nitrogen cycle are shown in Fig. 2.7 (b).

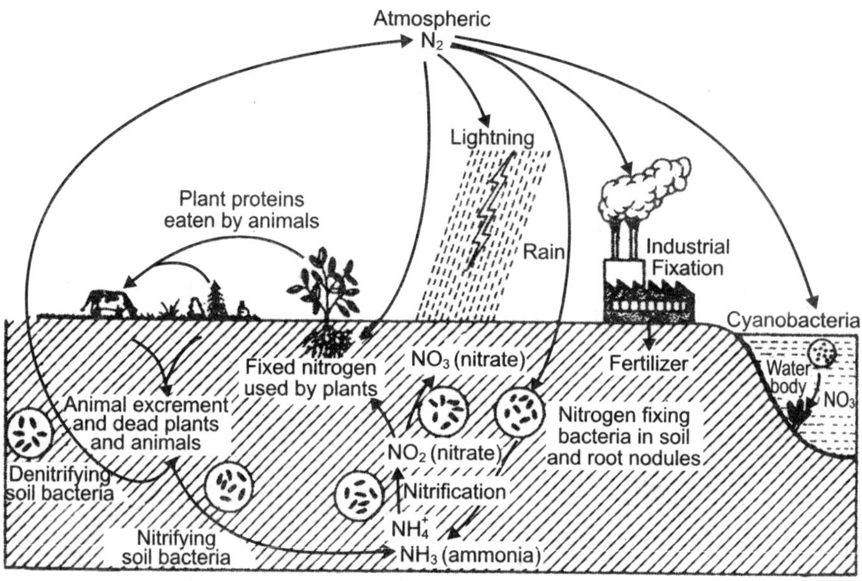

Fig. 2.7(b): Various features of Nitrogen cycle

There may be imbalance in nitrogen cycle due to following reasons:

(i) Soil erosion results in loss of nitrifying bacteria (present in top layer of the soil), which are essential component of nitrogen cycle.

(ii) The nitrogen compounds in the form of fertilizers may be washed away from the soil and may not be available for plants.

2.8.1.4 Oxygen Cycle

Oxygen is the second most abundant element in the atmosphere. It is present to the extent of about 21 percent and is essential for life processes. The main pools for oxygen are the atmosphere and oceans. The dissolved oxygen in water bodies is responsible for sustaining aquatic plants and animals. It is also a constituent of carbonate and oxide ores. The different steps in oxygen cycle are as follows:

(i) During respiration, organisms including animals and humans consume oxygen from the air and release carbon dioxide into the atmosphere.

(ii) Carbon dioxide is utilised by green plants for photosynthesis, in which carbon dioxide reacts with water in the presence of solar energy and a catalyst (chlorophyll) to produce carbohydrates and the oxygen formed is released into the atmosphere.

$$6\,CO_2 + 6\,H_2O + \text{Solar energy} \xrightarrow{\text{Chlorophyll}} C_6H_{12}O_6 + 6\,O_2$$

(iii) Oxygen is needed by bacteria and fungi during decomposition; the products of decomposition (CO_2 and H_2O) are released into the atmosphere. Oxygen is also needed for the burning of fossil fuels like wood, coal, petroleum etc; the products of combustion (CO_2 and H_2O) are released into atmosphere. The released CO_2 and H_2O are utilized during photosynthesis by the green plants.

(iv) In the upper atmosphere, solar energy splits H_2O to give H_2 and O_2, which are released in the atmosphere.

$$2\,H_2O + h\nu \longrightarrow 2\,H_2 + O_2$$

These processes complete the oxygen cycle in nature. The oxygen content in the atmosphere remains more or less constant. The flow chart for oxygen cycle is represented as shown below in Fig. 2.8.

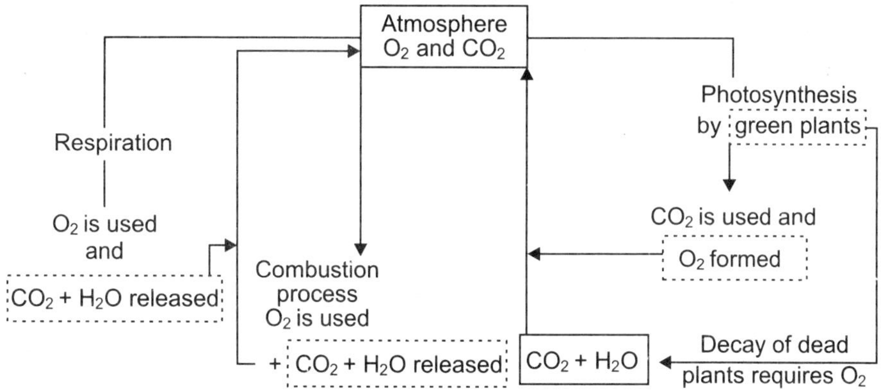

Fig. 2.8: Oxygen Cycle Flow Sheet

There may be imbalance in the oxygen cycle by deforestation which reduces the extent of photosynthesis (which is a major source of oxygen in the atmosphere) and hence the amount of oxygen in the atmosphere is reduced. However, this loss is compensated by sulphate reducing bacteria occurring in anaerobic environment; these bacteria use the sulphate ions as an oxygen source for oxidising organic matter.

$$SO_4^{2-} + 2\,C \longrightarrow 2\,CO_2 + S^{2-}$$

The above reaction gives CO_2 which is released into the atmosphere and utilized in photosynthesis leading to the release of oxygen.

2.8.2 Sedimentary Cycles

2.8.2.1 Sulphur Cycle

The sulphur cycle is basically sedimentary except for a comparatively smaller gaseous phase. The main source of sulphur is the lithosphere, where it is present as free sulphur and as sulphide ores of different elements like iron, copper, lead, tin, mercury, zinc etc. Sulphur is also present in coal and oil in the sedimentary rocks. Sulphur is a constituent of two of the twenty essential amino acids and is incorporated in several proteins and living organisms. It is released from the sedimentary rocks by weathering and decomposition by bacteria and fungi of organic matter. In the gaseous phase, sulphur is present as hydrogen sulphide and sulphur dioxide. Sulphur (as SO_2) enters the atmosphere from several routes like volcanic eruptions, combustion of fossil fuels and from various metallurgical operations. Hydrogen sulphide enters the atmosphere by the putrefaction of organic matter by bacteria and fungi.

In the atmosphere, SO_2 is oxidised to SO_3 which comes to the land in the form of dilute solution of H_2SO_4 (acid rain). Hydrogen sulphide in the atmosphere is oxidised to sulphur dioxide by ozone; the reaction is catalysed by particulates present in the atmosphere.

$$H_2S + O_3 \longrightarrow H_2O + SO_2$$

In anaerobic soils and sediments, hydrogen sulphide is formed by sulphate reducing bacteria like *Desulfavibrio*. Certain species of *Beggiatoa* oxidise hydrogen sulphide to elemental sulphur. Even green and purple sulphur photosynthetic bacteria oxidise hydrogen sulphide to elemental sulphur. The flow sheet of sulphur cycle is represented below (Fig 2.9).

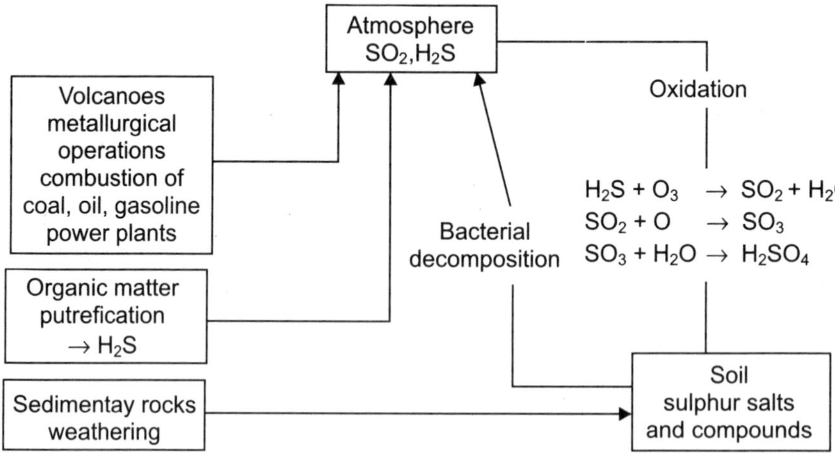

Fig. 2.9: Sulphur Cycle (Flow sheet)

2.8.2.2 Phosphorus Cycle

The content of phosphorous in living matter is very small, but it plays a vital and indispensable role. It is one of the most important element required by plants for their growth and development. It is a constituent of nucleic acids, phospholipids, ATP and ADP. In addition, significant quantities of phosphorus are present in the bones and teeth. The main reservoirs of phosphorous are rocks and other deposits. From these, phosphorous becomes available to the living organisms via a slow process of weathering of the rocks or by mining. Phosphorous is absorbed by the plants (through their root system) in its oxidized form, as phosphate, where phosphorous gets incorporated into different phosphorus containing compounds. From plants, the phosphorus finds its way to other living organisms. The decay of the dead organisms and plants releases the phosphorus content to the soil, which is again recycled. Some phosphorus finds its way to the oceans through the rivers. In oceans, phosphorous is consumed by fish and sea birds; any excess phosphorous gets deposited into the sediments.

Presence of excess of phosphorus causes environmental problems. This happens when too much of phosphate fertilizer is sprayed on the land. Excess of phosphorous is drained by rain into the water bodies.

A major problem associated with the presence of excess of phosphate fertilizers (and also other fertilizers) is *eutrophication* of lakes. The increase in the chemical elements or nutrients (by flowing of the excess fertilizers) in the water bodies is called **eutrophication**. These nutrients (particularly phosphorus) cause population explosion of photosynthetic bacteria and blue-green algae. With the passage of time, the whole of the lake or the pond is covered by a thick layer of algae resulting in the reduced availability of oxygen. This results in the death of algae, bacteria and fish.

2.8.2.3 Sodium Cycle

Sodium is an important micronutrient, though it is the sixth most abundant element in the lithosphere. It occurs as *rock salt* (sodium chloride), *Chile saltpetre* (sodium nitrate), *sodium sesquicarbonate* ($Na_2CO_3.NaHCO_3.2H_2O$) and *cryolite* (Na_3AlF_6). Oceans are a great source of sodium chloride, where it comes as run off from the mountainous areas. Sodium ions are absorbed by the roots of plants and utilised for various activities. Salts of sodium and potassium maintain the ionic and water balance of the body. Sodium chloride in moderate amounts is necessary for the healthy heart conditions. However, excess sodium is a curse and it leads to a variety of ailments. Sodium cycle includes isolation of salt from the oceans, its use by the plants and animals and its return back to the land or oceans by decay of dead plants and animals.

2.8.2.4 Potassium Cycle

Potassium (like sodium) is also a micronutrient and is vital to all living organisms. It is found in nature in the form of its salts like *carnallite* ($KCl.MgCl_2.6H_2O$), *polyhalite* ($K_2SO_4.MgSO_4.CaSO_4.2H_2O$), *kainite* ($KCl.MgSO_4.3H_2O$), *schonite* ($K_2SO_4.MgSO_4.6H_2O$), *Salt petre* ($KNO_3$) and *langbenite* ($K_2SO_4.2MgSO_4$). Potassium is commonly used as a fertilizer and is absorbed by the plants through root system. In plants, potassium is needed for various functions and is released after their death and decay.

2.8.2.5 Magnesium Cycle

Magnesium is also a micronutrient and is essential for plant and animal life. It is an essential constituent of chlorophyll, the green colouring matter of plants. Magnesium regulates the functioning of muscles and nerves in human beings who obtain it from leafy green vegetables and cereals. It is found in lithosphere in the form of its salts like *magnesite* ($MgCO_3$), *dolomite* ($MgCO_3.CaCO_3$), *carnallite* ($KCl.MgCl_2.6H_2O$), *asbestos* [$CaMg(SiO_3)_4$]. Halides of magnesium are found in seawater. In sea, magnesium is obtained by run off processes from rocks and sand. It is absorbed by the plants via their root system and performs various functions. After the decay of the plants, it is returned back to the soil.

2.8.2.6 Iron Cycle

Iron is essential for life since it is a constituent of haemoglobin in human blood. It is also needed for a number of cellular processes. Iron is one of the most common metals in the earth's crust and occurs in soil and granite rocks. It is also found in nature in the form of its salts eg, *magnetite* (Fe_3O_4), *haematite* (Fe_2CO_3), *limonite* ($Fe_2O_3.3H_2O$), *siderite* ($FeCO_3$) and *iron pyrites* (FeS_2). Plants absorb dissolved iron through their root system. Animals get iron through plants. After the death of animals and plants, it is released back to the soil and water.

Atmosphere

CHAPTER 3

3.1 Introduction

The region of air that envelops the earth is called *atmosphere*. It is capable of supporting life on earth and saves it from the ill effects of the ultraviolet rays from the sun. The air present in the atmosphere consists of a number of gases like N_2, O_2, CO_2, inert gases etc. The atmosphere extends up to a height of about 500 km from the earth's surface. The weight of the atmosphere is approximately 5×10^{15} metric tonnes, which is about one millionth of the total weight of the earth. The temperature, pressure and density vary considerably with altitude. The temperature of the atmosphere varies from about –100° to +1200° C depending on the altitude. Various parameters of the atmosphere are given below in Table 3.1.

Table 3.1 Various Parameters of the Atmosphere

Parameter	Amount
Total mass	~4-5x10^{15} metric tonnes
Temperature	–100° to 1200° C depending on the altitude
Atmospheric pressure	One atmosphere at sea level, 3x10^{-7} atmosphere at an altitude of 100 km above sea level
Atmospheric density	~0.0013 g/m^3 at the surface of earth. It decreases sharply with increasing altitude.

3.2 Major Regions of Atmosphere

The atmosphere extends up to about 500 km above the earth's surface. It is divided into four major regions depending on the height as shown below in Table 3.2.

Table 3.2 Major regions of atmosphere

Region	Height range (km)	Temperature range (°C)	Main constituents
Troposphere	0-11	15 to –56	O_2, N_2, CO_2, H_2O
Stratosphere	11-50	–56 to –2	O_3
Mesosphere	50-90	–2 to –92	O_2^+ NO^+
Thermosphere	90-500	–92 to 1200	O_2^+, O^+, NO^+

Besides the above major regions, the region of atmosphere lying between the altitudes of 50 km to100 km is called **ionosphere**. This region has appreciable levels of positive ions (e.g. O_2^+, O^+, NO_2^+ etc.) and electrons. These ions are formed due to the UV radiation from the sun. During night when there is no solar radiation, the ions combine with free electrons and give neutral species from which they originally originated. The division of atmosphere into various regions is shown in Fig 3.1.

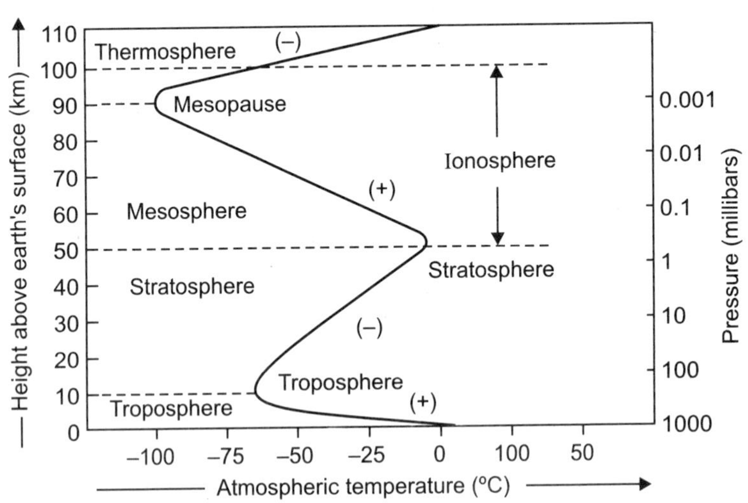

Fig. 3.1: Various Regions of the Atmosphere

(i) Troposphere

The region nearest to the earth, extending up to a height of about 11 km is called **troposphere**. About 70% of the total mass of air is contained in the troposphere. The temperature in this region steadily decreases from ground temperature of about 15° to about –56° (see Fig. 3.1). The fall in temperature is due to fall in pressure and density of air. The change in temperature is known as **lapse rate**. The decrease of temperature with increasing height in the troposphere is called **positive lapse rate**. This region is the most important zone of the atmosphere for sustaining life on earth since it contains mainly nitrogen and oxygen. It also contains argon, carbon dioxide and traces of He, Ne, CH_4, H_2, N_2O, CO, O_3, NO_2, SO_2, etc. Troposphere is a dusty zone containing water vapours.It is the region of strong air currents and cloud formation and is also referred to as a *turbulent zone*. The composition of air in the troposphere remains more or less constant provided there is no significant air pollution.

(ii) Stratosphere

The region above the troposphere is the stratosphere and extends up to an altitude of about 50 km. In stratosphere, the temperature rises from –56° to –2° C. The increase in temperature is due to absorption of solar radiation. The stratosphere is rich in ozone, which absorbs the harmful ultraviolet radiations and prevents it from reaching the earth, and raises the temperature causing negative lapse rate. Besides ozone, stratosphere also contains nitrogen and oxygen along with small amount of water vapour. In stratosphere, ozone is formed from oxygen by a photochemical reaction in which energy from the sun decomposes the oxygen molecule into reactive atomic oxygen, which in turn combines with molecular oxygen to form ozone.

$$O_2 + h\nu \rightleftharpoons 2O$$

$$O_2 + O \rightleftharpoons O_3$$

The ozone content of stratosphere is more or less constant. In fact, ozone is in equilibrium with oxygen. This implies that ozone is being produced from oxygen as fast as it is broken down to molecular oxygen. Ozone layer is of great importance since it absorbs the harmful UV radiations coming from the sun and prevents them from reaching to the surface of the earth.

(iii) Mesosphere

Next to stratosphere is the mesosphere which extends up to an altitude of 90 km. It is characterized by low temperature (–2° to –92°C) and very low atmospheric pressure. The region having temperature of about –92°C is the coldest region of the atmosphere. In mesosphere, the decrease in temperature is attributed to low levels of the ozone present. Ozone, as we know absorbs UV radiations from the sun. The main constituents of mesosphere are the positive ions *viz.* O_2^+, NO^+ etc.

(iv) Thermosphere

This region is above mesosphere and it extends up to a height of about 500 km. The temperature in this region rises from –92° C to 1200° C. The atmospheric gases (oxygen and nitrogen) in the region absorb the solar radiation in the far ultra violet region and undergo ionisation.

(v) Ionosphere

This is the region of atmosphere lying between the altitudes of 50 km to 100 km. In this region, there are high levels of ions like O_2^+, O^+, NO^+ and electrons; these species are formed due to solar radiations. During night, the UV radiations are not available and so the ions recombine with free electrons to give neutral species from which they originated.

(vi) Exosphere

This is upper most layer of atmosphere and is above the thermosphere. It extends up to a height of about 1600 km. This region contains hydrogen and helium and has very high temperature due to solar radiation. This region is referred to as the outer space. The division of the atmosphere into various regions is shown in a simplified way in Fig. 3.2.

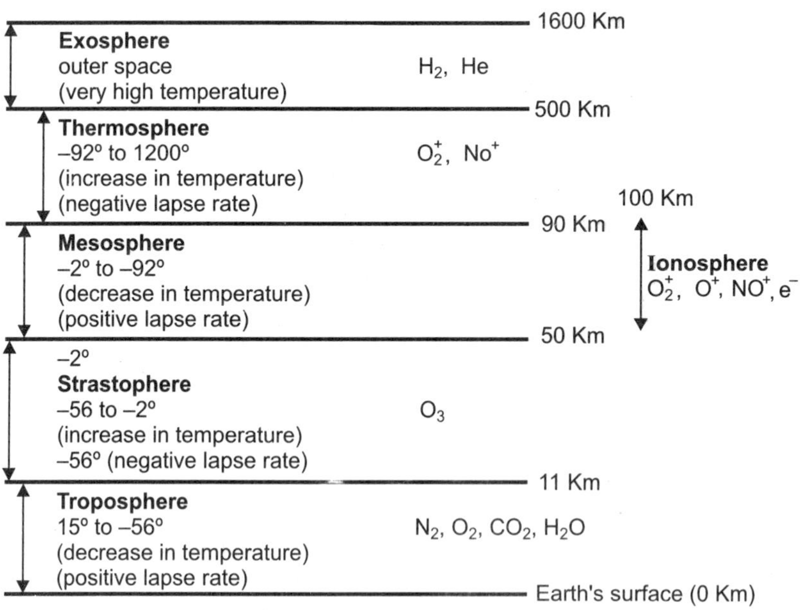

Fig 3.2: Simple Representation of Various Region of the Atmosphere

3.3 Composition of Atmosphere

The atmosphere contains a number of gases. The actual composition of dry air near sea level is given in Table 3.3. The components can be divided into three groups *viz. major components, minor components* and components present in *trace amounts*. The components are expressed in percent by volume or ppm. The value in ppm is obtained by multiplying the percent volume by 10^4.

Table 3.3 The Composition of Dry Air Near Sea Level

Components	Amount	
	% (by volume)	ppm
i. Major Components		
Nitrogen (N_2)	78.09	7,80,900
Oxygen (O_2)	20.98	2,09,400
Water vapour (H_2O)	~0.15	1,000-50,000
ii. Minor Components		
Argon (Ar)	0.934	9,340
Carbon dioxide (CO_2)	0.032	320
iii. Trace Components		
Neon (Ne)	0.00182	18.2
Helium (He)	0.000524	5.24
Methane (CH_4)	0.00018	1.8
Krypton (Kr)	0.00018	1.8
Nitrous oxide (N_2O)	0.000025	0.25
Hydrogen (H_2)	0.00005	0.5
Xenon (Xe)	0.0000087	0.087
Sulphur dioxide (SO_2)	0.0000002	0.002
Nitrogen dioxide (NO_2)	0.0000001	0.001
Ammonia (NH_3)	0.000001	0.01
Carbon monoxide (CO)	0.000012	0.12
Ozone (O_3)	0.000002	0.02
Iodine (I_2)	traces	traces

The main constituents of the atmosphere which are responsible for sustaining life on earth are nitrogen, oxygen and carbon dioxide along with water vapour. The amounts of these constituents in the atmosphere remain more or less the same (see nitrogen cycle, oxygen cycle and carbon dioxide cycle). The inert gases are present in the atmosphere in very small amounts. The oxides of nitrogen i.e. nitrous oxide and nitrogen dioxide are formed in the atmosphere by the combination of nitrogen and oxygen during lightening. Ammonia and methane are obtained in the atmosphere by bacterial decomposition of organic matter on the earth. Ozone is obtained from oxygen of the atmosphere by lighting. The burning of sulphur compounds and release of vehicular exhaust gases to the atmosphere release sulphur dioxide in the atmosphere. Carbon dioxide results by the burning of different fuels and respiration processes. Oxygen results from the photosynthetic activity of plants.

The percentage composition of the gases remains more or less constant in the atmosphere, at least up to a height of about 20 km. However, it does not hold true in case of air around large cities and industrial centres where there is an increase in carbon dioxide and a decrease in oxygen due to combustion of fuel in power plants and automobiles and also due to large population of human beings.

3.4 Radiation Balance of Earth

The earth's surface obtains nearly all its heat energy by radiation from the sun. Subsequently, the heat energy received by the earth is lost by reflection and by radiation. The net result is that the earth loses as much heat (radiation) as it received. In this way, a near steady state of the average global temperature is maintained. Over millions of years, the earth has neither progressively cooled nor warmed and the temperature of the earth has remained more or less constant.

The atmosphere plays a decisive role in maintaining the radiation balance of the earth. It is only through the atmosphere that the solar radiations reach the surface of the earth and it is also the atmosphere through which the surface of the earth radiates or reflects back the solar radiations. It is estimated that approximately 22% of the solar radiation received from the sun are reflected back by the combined effect of the clouds, snow and ice sheets on the earth surface. The reflecting capacity of the earth is called **albedo**. Of the remaining 68% of the solar energy, the atmosphere absorbs about 16-17% and 48-51% is absorbed by the surface of the earth. Finally, both the atmosphere and the earth's surface radiate back the absorbed energy as heat. It is the energy transport which plays a significant role in the earth's radiation balance. This transport of energy takes place by radiation of energy from the earth, conduction of energy through the interaction of atoms and molecules with atmosphere and convection of energy through massive air circulation. The last two factors are responsible for the loss of heat from the earth's surface via transport to clouds and consequent radiation from the clouds. The amount of heat energy radiated into the outer space is approximately 67%. Thus, the heat gained and heat lost is balanced over the global scale. On an average, the temperature of the earth is maintained around 15°C due to reabsorption of most of the outgoing radiation by water vapours and carbon dioxide. This combined effect of water vapours and carbon dioxide leads to the so called **greenhouse effect** (For more details see Chapter 4).

The heat balance varies from one region of the earth to another and this aspect is very significant. The imbalance between the heat absorbed and radiated causes atmospheric and ocean circulations to occur. This, in turn, controls the climate and weather patterns of the globe.

The earth's radiation balance is greatly affected by many human activities. Many Industrial processes are responsible for injecting a large amount of green house gases like carbon dioxide, ozone, nitrous oxide, methane, water vapours and particulate matter into the atmosphere. Deforestation, soil erosion and eruption of volcanoes also adds to the misery. The gases released into the atmosphere tend to increase the average temperature of the earth. The pollutants discharged by human activities create an imbalance in the global temperature. In fact, it has been found that over the last decades, the average global temperature has steadily risen. The discharge of pollutants in the atmosphere, if not controlled, will be disastrous for the environment.

3.5 Chemical and Photochemical Reactions in Atmosphere

The most important gases which are present in the atmosphere and undergo various types of chemical and photochemical reactions, are oxygen, nitrogen and carbon dioxide.

1. Oxygen

Oxygen originates by the process of photosynthesis and plays an important role in the troposphere and a key role in the stratosphere. In troposphere, oxygen takes part in energy producing reactions like burning of fossil fuels. Thus, methane (in natural gas) reacts with oxygen producing carbon dioxide and liberating energy.

$$CH_4 + 2O_2 \longrightarrow CO_2 + 2H_2O + 213 \text{ kcal}$$

In stratosphere, oxygen is responsible for the formation or generation of ozone. The oxygen undergoes a photochemical reaction in which energy from the sun decomposes the oxygen molecule into reactive atomic oxygen, which in turn combines with molecular oxygen to form ozone.

$$O_2 + h\nu \rightleftharpoons 2O$$
$$O_2 + O \rightleftharpoons O_3$$

The formation of ozone is essential for sustaining life on earth by preventing the harmful UV radiations of the sun from reaching the earth.

In the upper atmosphere (*viz.* mesosphere, thermosphere and ionosphere), ultraviolet radiations cause photochemical dissociation and ionisation of ozone to give rise to oxygen atoms, oxygen ion and excited oxygen atoms.

$$O_2 + h\nu \xrightarrow{135\text{-}176\ \text{nm}} \underbrace{O + O}_{\text{Oxygen atoms}}$$

$$O_3 + h\nu \xrightarrow{308\ \text{nm}} \underbrace{O^{\bullet}}_{\text{Oxygen ion}} + O_2$$

$$O + h\nu \longrightarrow \underbrace{O^{*}}_{\text{Excited oxygen atoms}} + e$$

$$O_2 + h\nu \longrightarrow O_2^{+} + e$$

The ultraviolet radiations are not available during night; so the ions recombine with free electrons to give neutral species (O_2) from which they originated.

2. Nitrogen

Nitrogen and its compounds are essential for the sustenance of life. Though nitrogen by itself is inert and un-reactive, its presence in atmosphere (to the extent of about 78%) is extremely important, since it dilutes or eliminates the harmful effects of pure oxygen (oxygen though supports life, but in pure oxygen most of the organic matter is burned. However, oxygen in about 5-6% concentration is useful for all the jobs). The origin of nitrogen in the atmosphere is believed to be from breaking down of nitrates in the soil by the *denitrification* process by some soil microorganisms.

$$4\,NO_3^- + 2\,H_2O \longrightarrow 2\,N_2 + 5\,O_2 + 4\,OH^-$$

In the presence of lightening and thunderstorm, the nitrogen of the atmosphere combines with oxygen to form nitric oxide (NO), which in turn combines with oxygen to give nitrogen dioxide.

$$N_2 + O_2 \longrightarrow 2\,NO$$
$$2NO + O_2 \longrightarrow 2\,NO_2$$

Ozone can also oxidise nitric oxide to nitrogen dioxide.

$$NO + O_3 \longrightarrow NO_2 + O_2$$

Nitrogen dioxide reacts with water vapours, producing nitric acid which is washed down to the soil in rain to give the so called **acid rain** (For more details, see Chapter 4).

$$3\,NO_2 + H_2O \longrightarrow 2\,HNO_3 + NO$$

3. Nitrous Oxide

Nitrous oxide (N_2O) commonly known as laughing gas is not produced by atmospheric nitrogen, but is released into the atmosphere by a number of chemical processes which increase the N_2O levels to about 10%. The N_2O formed rises into the stratosphere and plays a role in the destruction of the ozone layer. The various steps involved in the process are as follows:

$$N_2O + O^{\bullet} \longrightarrow 2NO$$

$$N_2O + O_3 \longrightarrow NO_2^{\bullet} + O_2$$

$$NO_2^{\bullet} + O^{\bullet} \longrightarrow NO^{\bullet} + O_2$$

The overall reaction is

$$O_3 + O^{\bullet} \longrightarrow 2O_2$$

4. Carbon Dioxide

Carbon dioxide is released into the atmosphere by burning of different types of fuels and by respiration processes. The green plants absorb CO_2 gas from the atmosphere to prepare food (carbohydrates) via the process of photosynthesis.

$$6\,CO_2 + 6\,H_2O \xrightarrow[\text{chlorophyll (green plants)}]{\text{solar energy}} C_6H_{12}O_6 + 6\,CO_2$$

Carbon dioxide is one of the very important green house gases (the others are methane, nitrous oxide, chlorofluorocarbons etc). It does not support life. In fact, animals and human beings kept in CO_2 atmosphere die due to lack of oxygen. Though CO_2 is not considered to be a pollutant but in excess amount, it has adverse effects on the climate (for details see green house effect). Before CO_2 was introduced into the atmosphere, the average temperature of earth was -15^0C. The green house gases (particularly CO_2) raised the earth's temperature to present day value of 15°C. Thus, without green house gases, the planet earth would be uninhabitable. This aspect has been discussed subsequently.

5. Ozone

Ozone is not present as such in the atmosphere. It is generated in the atmosphere from oxygen. In stratosphere, oxygen undergoes photochemical reaction in the presence of rays of sun when oxygen decomposes into reactive oxygen atoms, which in turn combines with molecular oxygen to form ozone.

$$O_2 + h\nu \rightleftharpoons 2\,O$$
$$O_2 + O \rightleftharpoons O_3$$

Ozone can be beneficial or harmful depending on the circumstances. Ozone is an irritating toxic gas and when breathed, it produces extremely harmful effects like soar throat, general bronchial irritation, coughing and fatigue; much higher concentration can even kill. Ozone is also lethal to low forms of life including bacteria. In some countries (especially Europe), it is widely used as a disinfectant for drinking water supplies. On the other hand, the same gas (ozone) forms a life saving shield that protects us from the extremely harmful effects of ultraviolet solar radiation. In the atmosphere, ozone is present as the ozone layer that surrounds the earth (within the stratosphere) centred roughly at some 30 to 35 km from above the earth's surface.

In the atmosphere, absorption of ultraviolet radiation by ozone molecules within the ozone layer regenerates an oxygen atom and an oxygen molecule.

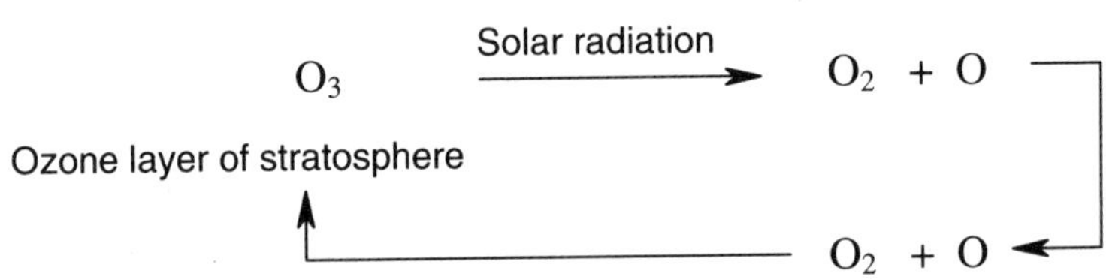

The cycling between the formation and destruction of ozone molecules results in a reasonably constant concentration of ozone that has permitted the life on earth to thrive under the invisible shield.

If chlorofluorocarbons (CFCs) are present in the atmosphere (or discharged in the atmosphere by various routes), the ozone layer is destroyed or depleted causing a number of complications. This aspect will be discussed in a subsequent section (see sec. 4.5.4).

Air Pollution

CHAPTER 4

4.1 Introduction

The composition of the atmosphere is discussed in sec. 3.3. Air is invariably impure and is always contaminated with gases like CO, NO_2, SO_2, etc. (which are poisonous in nature) and fincly divided solid and liquid particles and smog. Air becomes polluted due to the presence of the above contaminants. The presence of these contaminants in the air is called *air pollution* and the materials which pollute the air are called *air pollutants*. In fact, pollution of air (or atmosphere) is an unwanted change in the quality of the air caused by emission of gases due to the burning of fossil fuels, volcanic eruptions and discharge of particulate matter into the atmosphere by industrial activities and soil erosion.

4.2 Classification of Pollutants

The pollutants are classified into following three types:

(i) Natural Pollutants
(ii) Primary Pollutants
(iii) Secondary Pollutants

4.2.1 Natural Pollutants

The pollutants that find their way into the atmosphere as a result of natural phenomena are referred as **natural pollutants**. Some example of natural phenomena causing pollution include the following:

- Forest fires started by lightening.
- Dispersal of pollen, soil erosion and volcanic eruptions.
- Discharge of volatile organic compounds from leaves and tree.

- Decomposition or putrefaction of organic matter.
- Natural radioactivity.

The natural pollutants have been known to cause atmospheric pollution since very early time; it is perhaps as old as the earth itself. Nature deals with such pollutants in its own way. Anyhow, the concentration of natural pollutants is generally quite low and does not cause any serious problems.

4.2.2 Primary Pollutants

A **primary pollutant** is a pollutant that enters directly into the atmosphere as a result of natural or other activities. For example, when coal, oil, natural gas or wood is burnt, carbon monoxide and carbon dioxide are formed. Automobiles are responsible for contributing considerable amount of carbon monoxide. Sulphur dioxide is passed into the atmosphere by burning coal and oil containing sulphur as impurities. The sulphur of the pyrite (iron pyrite, FeS_2) that remains in coal is oxidised to sulphur dioxide.

$$4\,FeS_2 + 11\,O_2 \longrightarrow 2\,Fe_2O_3 + 8\,SO_2$$

Other primary pollutants include oxides of nitrogen, hydrocarbons and suspended particulate matter.

4.2.3 Secondary Pollutants

A **secondary pollutant** is the one that is formed by the further reaction of primary air pollutants. For example, sulphur dioxide reacts with oxygen in the atmosphere to form sulphur trioxide, which can further react with water vapours in the air to form sulphuric acid.

$$2\,SO_2 + O_2 \longrightarrow 2\,SO_3$$

$$SO_3 + H_2O \longrightarrow H_2SO_4$$

In the above case, both sulphur trioxide and sulphuric acid are secondary pollutants.

Another interesting example is that of nitric oxide (NO). It is formed by combination of the atmospheric nitrogen with oxygen of the atmosphere at very high temperature or in presence of electric sparks due to lightening etc. It is also obtained by the combustion of fuels in the internal combustion engines; which is a convenient route for the formation of nitric oxide ($N_2 + O_2 \rightarrow 2\,NO$). Thus, nitric oxide enters the atmosphere as a primary pollutant, where it further combines with atmospheric oxygen to form nitrogen dioxide NO_2). The nitrogen dioxide thus formed, may react with water of the atmosphere to form nitric acid.

$$NO + O \longrightarrow NO_2$$

$$2\,NO_2 + H_2O \longrightarrow HNO_2 + HNO_3$$

In the above example, both nitrogen dioxide and nitric acid are secondary air pollutants. Some other examples of secondary air pollutants include ozone, hydrogen peroxide, nitrates and sulphate salts (formed in the soil).

Thus, it can be summarised that the primary pollutants (e.g., CO, CO_2, SO_2, NO, hydrocarbons, particulate matter) enter into the atmosphere as a result of natural activities, exhausts from automobiles and from power generating units which use coal as the source of energy. These primary pollutants are converted into secondary pollutants (e.g. NO_2, SO_2, HNO_3, H_2SO_4, H_2O_2, O_3 etc.) by reacting with atmospheric oxygen and water vapours present in the atmosphere in the presence of solar radiation.

4.3 Major Air Pollutants

The major air pollutants are those which are produced in significant amounts and are responsible for a variety of health and environmental problems. Table 4.1 below describes various air pollutants.

Table 4.1 Different types of air pollutants

Types of Pollutants	Examples
Gaseous pollutants	Oxides of carbon (CO, CO_2) Oxides of nitrogen (NO, NO_2) Oxides of sulphur (SO_2, SO_3)
Particulate pollutants	Smoke, dust particles
Smog and photochemical smog	A colloidal dispersion of gaseous pollutants like SO_2, NO_2 etc. and water droplets with particulate matter
Metallic pollutants	Metallic particles like Pb, Hg, As, Cd, Be, Ni, V, Fe, Sn, and Al etc. released in the air by various industrial processes
Radiation	Released in the atmosphere by various nuclear devices etc.

4.3.1 Oxides of Carbon

Carbon forms two oxides, *viz.* carbon monoxide and carbon dioxide. The former (carbon monoxide) is one of the most harmful pollutants. The later (i.e. carbon dioxide is harmless but in larger amounts, it may lead to a number of problems which will be discussed later.

4.3.1.1 Carbon Monoxide (CO)

Carbon monoxide is a colourless, odourless, tasteless and non-irritating gas. It is sparingly soluble in water and is highly toxic (poisonous) in nature. It reduces the oxygen carrying capacity of blood and can be fatal at concentrations exceeding 100 ppm. It is known as a silent killer. Major sources of carbon monoxide in the atmosphere are as follows:

(i) It is produced by the incomplete combustion of all carbon containing fuels. The incomplete combustion means combustion taking place in presence of insufficient quantity of oxygen. Such combustion takes place in automobile engines (internal combustion engines).

(ii) Industrial furnaces (e.g. blast furnace) which are used at high temperature (e.g. in the manufacture of iron).

(iii) Cigarette smoke also contains carbon monoxide. Though a single cigarette emits extremely small amount of carbon monoxide, but surely the number of people smoking a number of cigarettes every day discharge considerable amount of carbon monoxide into the atmosphere.

(iv) Carbon monoxide is also a constituent of some industrial gases. For example, coal gas, obtained by distillation of coal contains 5-10% carbon monoxide. Also, water gas (obtained by passing steam over red hot coke, $H_2O + C \rightarrow CO + H_2$) contains 30-40% carbon monoxide.

(v) Besides the above sources of carbon monoxide which are due to human activities, it is also produced from some natural sources, chief amongst these are:

- Volcanic eruptions
- Forest fires
- Methane (an important and major constituent of natural gas and produced in marshy places) on oxidation under controlled conditions gives carbon monoxide.
- Decomposition or degradation of chlorophyll, the green pigment of plants.
- Marine algae and marine siphonophore produce carbon monoxide.

Carbon Monoxide Poisoning

It is known that during breathing, the haemoglobin (Hb) present in the red blood cells of blood combines with the oxygen reversibly to form oxy-haemoglobin (HbO_2).

$$Hb + O_2 \rightleftharpoons HbO_2$$

As a result of blood passing through the arteries to the tissues, the pressure of oxygen decreases and the oxygen present in the complex HbO_2 is set free, where it is utilised by the body cells.

In case, the air (we breathe) contains carbon monoxide, the oxygen present in HbO_2 is replaced by CO, since haemoglobin (Hb) has more affinity for CO than for O_2; this results in the formation of carboxy-haemoglobin (HbCO).

$$HbO_2 + CO \longrightarrow HbCO + O_2$$

Due to the formation of HbCO, the quantity of oxygen available to the body cells get reduced. In other words, carbon monoxide reduces the oxygen carrying capacity of the blood and this leads to oxygen starvation (anoxia). The deficiency of oxygen produces headache, dizziness, chocking, cardiac and pulmonary complications leading to paralysis and death.

It is well known that in winter (cold climate), when people sleep inside their homes using coal for heating, a number of cases of death have been reported due to carbon monoxide poisoning. Also, in cold climate, children are left in closed cars with heating on. There are a number of reports of children dying due to carbon monoxide poisoning (the carbon monoxide is emitted by the combustion of fuel; this CO finds way in the hot air of the heating system).

In case, carbon monoxide poisoning is detected well in time (i.e. before the person dies), it can be treated by exposing the victim to oxygen under a high pressure (2-2.5 atm). By this process, the CO present in HbCO is replaced by O_2 and the victim may be revived.

$$HbCO + O_2 \rightleftharpoons HbO_2 + CO$$

$$HbO_2 \longrightarrow Hb + O_2$$

Monitoring of Carbon Monoxide

The amount or the concentration of carbon monoxide in the air is generally determined by non dispersive infrared analysis. This analysis depends on the fact that carbon monoxide (which possesses a dipole moment), absorbs infrared radiation at certain specific wavelengths. The amount of infrared radiation absorbed by carbon monoxide gas is directly proportional to its concentration. This method is used if the level of carbon monoxide is up to a level of 150 ppm. In case the concentration of the gas is near 10 ppm, it can be estimated by gas chromatography, which makes use of a flame ionisation detector. However, if the concentration of carbon monoxide is less than 10 ppm, it is estimated by reducing with hydrogen gas over a nickel catalyst at 360^0C and the produced methane is estimated by flame ionisation detector.

$$CO + 3H_2 \xrightarrow[360^\circ C]{Ni} CH_4 + H_2O$$

4.3.1.2 Carbon Dioxide (CO_2)

Carbon dioxide is a colourless, odourless gas and is about 1.5 times heavier than air. Though it is non-poisonous, it does not support life, but animals and human beings if kept in carbon dioxide atmosphere die not due to CO_2 but due to lack of oxygen. Small amount of CO_2 is considered relatively innocuous and so is not considered to be a pollutant, but its excess quantity in the atmosphere has adverse effect on the climate (see *green house effect* and *global warming*), hence it is considered to be a pollutant in large amount. Major sources of carbon dioxide in the atmosphere are as follows:

(i) It is one of the end products of burning of fossil fuels like coal, natural gas and petroleum etc. On an average, 6000 million tonnes of CO_2 enters the atmosphere by such activities every year.

(ii) The cultivation of soil also releases large amount of CO_2, which is produced by bacteria. This activity releases more than 2000 million tonnes of CO_2 into the atmosphere every year.

(iii) A large number of industrial processes contribute to the release of large amount of CO_2 into the atmosphere. These include manufacture of lime, alcohol and other fermentation units (for the manufacture of antibiotics etc).

$$\underset{\text{Lime stone}}{CaCO_3} \xrightarrow{1250^\circ} \underset{\text{Lime}}{CaO} + CO_2$$

$$\underset{\text{Glucose}}{C_6H_{12}O_6} \xrightarrow{\text{Zymase}} 2\,C_2H_5OH + 2\,CO_2$$

Such industrial processes contribute more than 10,000 million tonnes of CO_2 to the atmosphere annually.

(iv) Some other activities like decay of dead organisms and respiration by living organisms also release CO_2 into the atmosphere. Volcanic eruptions and forest fires also contribute to the release of CO_2.

The major consumers of CO_2 in the atmosphere are given below:

a. Green Plants

These absorb CO_2 gas from the atmosphere in presence of sunlight and chlorophyll to produce their food (carbohydrates). This process is known as **Photosynthesis**.

$$n\,CO_2 + n\,H_2O \xrightarrow[\text{Chlorophyll}]{\text{Sunlight}} \underset{\text{Carbohydrates}}{(CH_2O)_n} + n\,O_2$$

b. Oceans

The oceans dissolve CO_2 gas to form carbonate rocks. In this way, both the plants and oceans keep a balance of CO_2 in the atmosphere. Out of these two, green plants are more important. It is due to this that in the interest of survival of the mankind, it is essential to plant more trees and ensure that no attempt be made for cutting or felling of the trees.

Effects of CO_2 in the Atmosphere

The main effects of carbon dioxide in the atmosphere are its contribution in the green house effect, global warming, acid rain, and maintaining the pH of rain water at about 5.6 (see sec. 4.5).

4.3.2 Oxides of Nitrogen

Nitrogen combines with oxygen to form a number of oxides, *viz.* nitric oxide (NO), nitrogen dioxide (NO_2), nitrous oxide (N_2O), dinitrogen trioxide (N_2O_3), dinitrogen tetroxide (N_2O_4) and dinitrogen pentoxide (N_2O_5). However, from the point of view of environmental pollution, the first three oxides *viz.* nitric oxide, nitrogen dioxide and nitrous oxide are important and are discussed here. Nitrous oxide is also a greenhouse gas (see green-house effect). Major sources of oxides of nitrogen in the atmosphere are as follows:

(i) During thunderstorm and lightening, atmospheric nitrogen and oxygen combine to form nitric oxide, which gets converted into nitrogen dioxide.

$$N_2 + O_2 \xrightarrow[\text{discharge}]{\text{Electric}} 2\,NO$$

$$2\,NO + O_2 \longrightarrow 2\,NO_2$$

(ii) The exhaust gases released by the automobile engines contain NO and NO_2 besides other gases.

(iii) In fossil fuel based power plants, the combustion temperature is high and under these conditions, nitrogen and oxygen combine to give oxides of nitrogen.

(iv) Nitrogen dioxide is also obtained as a by-product during the manufacture of explosives.

(v) The bulk of oxides of nitrogen in the atmosphere originate from bacterial decay of organic matter by soil micro-organisms on the surface of the earth.

(vi) Nitrous oxide is produced by the action of aerobic and anaerobic bacteria in the soil on nitrogen-based fertilizers.

Function of Oxides of Nitrogen in the Atmosphere

Nitrogen and oxygen combine in the atmosphere at high temperature to give nitric oxide, which is converted into nitrogen dioxide by aerial oxidation.

$$N_2 + O_2 \xrightarrow{\text{High Temperature}} 2\,NO$$

$$2\,NO + O_2 \longrightarrow 2\,NO_2$$

Nitric oxide can also be oxidised with ozone to give nitrogen dioxide.

$$NO + O_3 \longrightarrow NO_2 + O_2$$

Though nitrous oxide (which is produced by the action of bacteria in the soil on nitrogen based fertilizers) is unreactive in the lower atmosphere; in the stratosphere, it gives nitric oxide and nitrogen dioxide on photolysis.

$$N_2O + h\nu \longrightarrow NO + N$$

Nitrous oxide may also give nitrogen and oxygen on photolysis.

$$N_2O + h\nu \longrightarrow N_2 + O$$

The free oxygen atom may react with nitrous oxide producing nitric oxide, which is subsequently oxidised to nitrogen dioxide.

$$N_2O + O \longrightarrow 2\,NO$$

$$NO + O \longrightarrow NO_2$$

In the atmosphere, nitrogen dioxide reacts with water vapour producing nitric acid, which is washed down as acid rain (see acid rain, sec. 4.5.1).

$$3\,NO_2 + H_2O \longrightarrow 2\,HNO_3 + NO$$

Alternatively, nitric acid may combine with ammonia (present in the atmosphere) giving ammonium nitrate.

$$NH_3 + HNO_3 \longrightarrow NH_4NO_3$$

The acid rain (containing HNO_3) on coming in contact with marble ($CaCO_3$) of the monuments gives CO_2 gas and calcium nitrate.

$$CaCO_3 + 2\,HNO_3 \longrightarrow Ca(NO_3)_2 + CO_2 + H_2O$$

In the atmosphere, nitric oxide destroys the ozone (which absorbs harmful ultraviolet radiations coming from the sun.)

$$NO + O_3 \longrightarrow NO_2 + O_2$$
$$NO_2 + O_3 \longrightarrow NO + 2\,O_2$$

Thus, we see that both nitric oxide and nitrogen dioxide destroy ozone layer. This aspect has been discussed subsequently (see *ozone hole*, sec. 4.5.4).

Being corrosive, nitrogen dioxide attacks skin and corrodes teeth and causes loss of appetite. It is harmful for plants and causes extensive leaf-drops. It has been found that inhalation of NO_2 (15-30 ppm concentration) for about 2 hours, causes damage to lungs, heart, liver and kidney.

When associated with hydrocarbons and sunlight, oxides of nitrogen lead to the formation of photochemical smog (see sec. 4.5.5)

Monitoring of Oxides of Nitrogen

Method 1

In the atmosphere, both nitric oxide and nitrous oxide are finally converted into nitrogen dioxide. So the monitoring of the oxides of nitrogens is done by monitoring nitrogen dioxide. The procedure involves, bubbling the contaminated air (containing NO_2) into a solution of sodium hydroxide for a period of 24 hr. The nitrogen dioxide in the contaminated (or polluted) air is converted into sodium nitrite and sodium nitrate.

$$2\,NO_2 + 2\,NaOH \longrightarrow NaNO_2 + NaNO_3 + H_2O$$

The above solution is added to a mixture of hydrogen peroxide, sulphanilic acid, hydrochloric acid and *N*-(1-naphthyl) ethylenediamine hydrochloride. The sodium nitrite formed in the reaction diazotises sulphanilic acid. This procedure is carried out at low temperature (0-5°). The formed diazonium salt couples with *N*-(1-naphthyl)ethylenediamine dihydrochloride to give the coupled compound. The reactions involved are as follows:

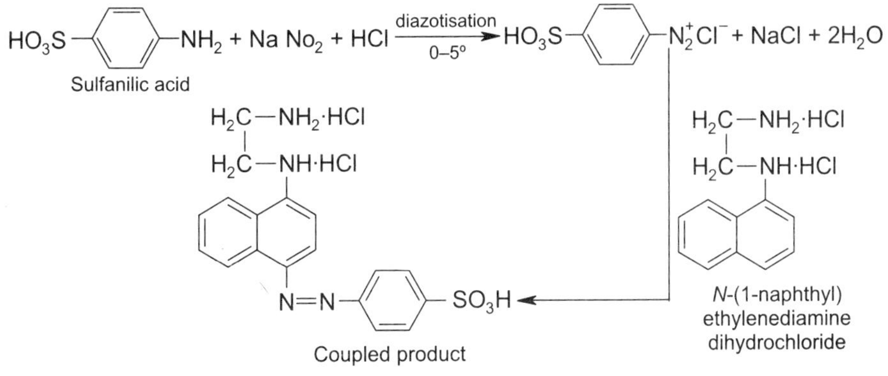

The coupled product absorbs at 540 nm. Its concentration is determined by Lambert-Beer's law. One mole of coupled salt corresponds to one mole of sodium nitrite, which in turn corresponds to one mole to nitrogen dioxide.

Method 2

In an alternative procedure, the oxides of nitrogen in the atmosphere are made to react with ozone to yield electronically excited species ($NO^{\bullet}_2$).

$$NO + O_3 \longrightarrow NO_2^{\bullet} + O_2$$
$$NO_2 + O_3 \longrightarrow NO + 2O_2$$

$NO_2^{\bullet}$ being unstable quickly returns to the ground state, forming NO_2 and emitting light in the range of 600-3000 nm.

$$NO_2^{\bullet} \longrightarrow NO_2 + h\nu$$

The intensity of light is directly proportional to the concentration of $NO_2^{\bullet}$ and hence NO molecules. The intensity of light, measured with the help of photo-multipliers gives a measure of the concentration of NO molecules.

4.3.3 Oxides of Sulphur

Sulphur forms two oxides *viz.* sulphur dioxide (a primary pollutant) and sulphur trioxide (a secondary pollutant). The first oxide *viz.* SO_2 is important from the point of view of atmospheric pollution.

4.3.3.1 Sulphur Dioxide (SO_2)

It is the most common harmful gaseous pollutant. It is a colourless gas with suffocating smell. It irritates the respiratory system of animals and humans and damages lungs. Major sources of SO_2 in the atmosphere are as follows:

(i) Sulphur dioxide is released in the atmosphere by volcanic eruptions.

(ii) Both coal and oils contain sulphur as the impurity and on burning release SO_2 into the atmosphere.

(iii) It is also produced when sulphide ores like pyrites (FeS_2), copper pyrites (CuS), copper glance (Cu_2S), zinc blende (ZnS), galena (PbS) etc. are roasted in air during the metallurgy of respective metals.

(iv) Large amount of SO_2 are released in the atmosphere from power plants, which are coal and oil based, and also from oil refineries. Even plants manufacturing sulphuric acid release SO_2 into the atmosphere.

(v) Hydrogen sulphide which is discharged into the air by natural processes is converted into SO_2 by oxidation in the atmosphere (see sec. 4.3.4).

Functions of Oxides of Sulphur in the Atmosphere

In the atmosphere, sulphur dioxide reacts with oxygen to form sulphur trioxide (SO_3) by photolytic and catalytic oxidation processes.

$$SO_2 + \tfrac{1}{2}O_2 \xrightarrow{h\upsilon} SO_3$$

$$SO_2 + \tfrac{1}{2}O_2 \xrightarrow{\text{Particulates}} SO_3$$

Sulphur dioxide also reacts photo-chemically with ozone producing sulphur trioxide.

$$SO_2 + O_3 \xrightarrow{h\upsilon} SO_3 + O_2$$

Sulphur trioxide formed in the atmosphere reacts with water forming sulphuric acid.

$$SO_3 + H_2O \longrightarrow H_2SO_4$$

The sulphuric acid formed comes down to the earth as *acid rain* (see acid rain, in sec. 4.5.1). The acid rain damages the crops and may react with ammonium salts and sodium chloride to produce sulphates.

$$H_2SO_4 + 2\,NH_4Cl \longrightarrow (NH_4)_2SO_4 + 2\,HCl$$

$$H_2SO_4 + 2\,NaCl \longrightarrow Na_2SO_4 + 2\,HCl$$

Sulphuric acid also damages the marble monuments (as in the case of nitric acid).

$$H_2SO_4 + CaCO_3 \longrightarrow CaSO_4 + H_2O + CO_2$$

At least three major environmental episodes occurred in 1930, 1948, 1952 and 1962 due to the smog conditions produced by SO_2 (see sec. 4.5.5).

Prolonged exposure of plants to even small concentration of SO_2 results in decolorization of the leaves. This is because the production of chlorophyll is hampered. The colour of fabrics, leather, paper and paints fade in presence of SO_2.

Monitoring the Presence of Sulphur Dioxide in the Atmosphere

The concentration of sulphur dioxide in the air can be determined by bubbling the test sample of air through a dilute aqueous solution of sodium tetrachloromercurate (II), $Na_2[HgCl_4]$ (obtained by the action of $HgCl_2$ on NaCl solution) [$HgCl_2 + 2\,NaCl \longrightarrow Na_2[HgCl_4]$] containing sulphamic acid (which destroys any nitrogen oxides present in the air). Sulphur dioxide in the sample of air reacts quantitatively with sodium tetrachloromercurate (II) and is converted into a stable complex, sodium dichlorosulphitomercurate (II), $Na_2[HgCl_2(SO_3)]$.

$$Na_2[HgCl_4] + SO_2 + H_2O \longrightarrow Na_2[HgCl_2(SO_3)] + 2\,HCl$$

The above complex is treated with a mixture of *p*-rosaniline dye, formaldehyde and phosphoric acid. The *p*-rosaniline dye is converted to *p*-rosaniline methyl sulphonic acid.

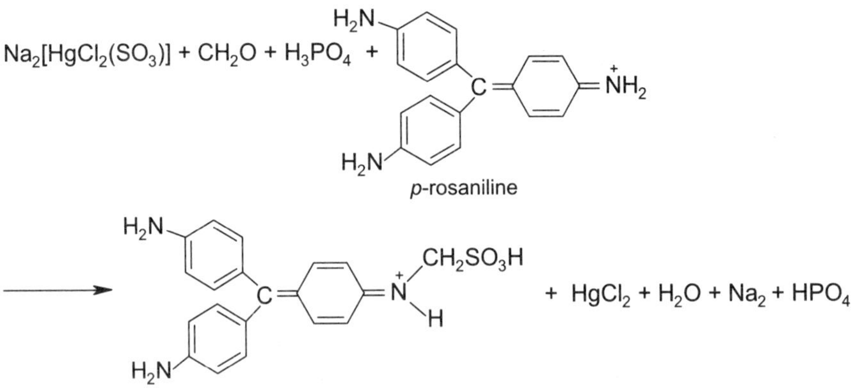

p-rosaniline methylsulphonic acid
(red-violet colour)

In the above procedure, phosphoric acid liberates SO_2 from the complex and maintains the pH of the reaction at 1.0; it also ties-up any heavy metal ions that may be present in the environment along with SO_2.

The *p*-rosanilinemethylsulphonic acid has a red-violet colour and at pH 1.0, it absorbs at 575 nm. Its concentration can be determined calorimetrically using Lambert-Beer's law. This concentration is proportional to the concentration of SO_2 in the atmosphere.

4.3.4 Hydrogen Sulphide (H_2S)

It is a colourless gas having smell of rotten eggs. It is extremely toxic and is harmful even at low concentrations. Mild exposure to H_2S can cause giddiness. It is also harmful to plants and retards their growth. An air-pollution episode due to H_2S was reported in November 1950, when an accident in a refinery complex in Roza Rica released large amount of H_2S gas on the ground level. The sources which emit oxides of sulphur (see sec. 4.3.3) also release hydrogen sulphide into the atmosphere.

(i) Natural decay of animal and vegetable matter
(ii) Volcanic eruptions
(iii) Some industrial processes like paper mills, oil refineries, natural gas plants and chemical manufacturing plants using sulphur inject H_2S to the atmosphere.

(iv) It also enters into the atmosphere by the reduction of sulphates and organo sulphur compounds.

In the atmosphere, H_2S is oxidised to SO_2 by atomic oxygen or by molecular oxygen or by ozone.

$$H_2S + O \longrightarrow HS + OH$$

$$HS + OH + O_2 \longrightarrow SO_2 + H_2O$$

$$H_2S + {}^3/_2O_2 \longrightarrow SO_2 + H_2O$$

$$H_2S + O_3 \longrightarrow SO_2 + H_2O$$

4.3.5 Chlorine (Cl_2)

Chlorine is a greenish-yellow coloured gas having disagreeable suffocating odour and is harmful even in low concentrations. It is a poisonous and toxic gas and causes irritation to mucus membrane. Exposure to chlorine may cause death. Chlorine is considered to be a pollutant. It is mostly released in the atmosphere by industrial processes from plastic industries, water treatment units and units engaged in the manufacture of caustic soda.

In the atmosphere, chlorine dissolves in moisture to give hydrochloric acid and hypochlorous acid, which comes down to the earth in the form of acid rain.

$$H_2O + Cl_2 \longrightarrow HCl + HClO$$

4.3.6 Ozone (O_3)

Ozone is a pale blue gas having pungent odour and is about 1.5 times heavier than air. Depending on the circumstances and the location, ozone can be beneficial or harmful. Thus, in the atmosphere near the surface of earth - in the air we breathe – ozone is an irritating, toxic gas. Even in low atmospheric concentration (about 1 ppm), formed near sparking, machinery, electrical generators and some types of photocopiers, it can lead to sore throats, bronchial irritation, coughing and fatigue. Higher concentration of ozone can be fatal. Ozone is also lethal to lower forms of life, including bacteria. Due to this reason, ozone is used as a disinfectant for water supplies (like chlorine) in some countries.

Ozone is produced in the upper part of the atmosphere (stratosphere) by the action of sunrays on oxygen.

$$3O_2 \xrightarrow[\text{(energy)}]{\text{Sun-rays}} 2O_3$$

In stratosphere, ozone forms a life saving shield that protects life from the catastrophic effects of extremely harmful ultraviolet solar radiation. Ozone also undergoes dissociation to give oxygen in the atmosphere.

$$O_3 \longrightarrow O_2 + O$$

During night, ultra-violet radiations are not available. This keeps a balance of the ozone layer. The oxides of nitrogen present in the atmosphere also decompose ozone into oxygen.

$$NO + O_3 \longrightarrow NO_2 + O$$

$$NO_2 + O \longrightarrow NO + O_2$$

Ozone layer is also depleted by chlorofluorocarbons. This aspect will be discussed in a subsequent section. Depletion of oxygen layer produces ozone hole (see sec. 4.5.4).

4.3.7 Hydrocarbons

Hydrocarbons are compounds made up of carbon and hydrogen. These are the most important constituents of petroleum, natural gas and LPG. These are also present in the exhaust gases of the automobiles. Hydrocarbons have adverse effects on human beings and some are carcinogenic in nature.

Methane (CH_4) is the major naturally occurring hydrocarbon emitted in the atmosphere. It is present to the extent of 90-95% in natural gas. It is also produced by bacteria during anaerobic decomposition of organic matter in soil, water and sediments.

Methane is one of the green house gas (see sec. 4.5.3) and thus, it contributes to the global warming. The hydrocarbons, particularly methane along with oxides of nitrogen contributes to the formation of photochemical smog (see sec. 4.5.5).

4.3.8 Particulates

The finely divided suspended solid particles in the atmosphere are called *particulates*. The diameter of the particulates lies in the range 0.001 – 100 μm. These cause severe health problems. Some examples particulates are as follows:

(i) Smoke emitted into air by the combustion of fuels like coal and oil in homes and factories. Power generation plants which use coke are the greatest source of smoke. The smoke basically contains particles of carbon, which remain suspended in air.
(ii) Cement dust emitted from cement factories
(iii) Insecticide dust
(iv) Coal dust
(v) Fly ash generated from power units

(vi) Milled flour

(vii) Pollen

(viii) Exhaust gases of automobiles also contain carbon particles.

The particulates are subdivided into two types, *viz. primary particulates* and *secondary particulates*.

Primary particulates are usually 1-20 μm in diameter and are directly injected into the air. Examples of this type are pollens, volcanic debris, coal dust, cement dust, fly ash and milled flour.

Secondary particulates are produced in the atmosphere as a result of reactions of primary particulates with contaminant gases. Examples of this type include vehicular emission, combustion processes and industries involving metallurgical operations. The secondary particulates are responsible for **photochemical smog**.

Some of the important sources of particulates are given as follows:

(i) **Meteorites :** These enter into the atmosphere from the outer space. On passing through the atmosphere, most of the meteorites burn up or disintegrate. The bigger particles, however, reach the earth surface in the form of meteorite dust.

(ii) **Volcanoes :** Volcanic eruptions inject enormous amounts of particulate matter into the atmosphere. The debris from volcanic eruptions can travel several thousand kilometres from the site of eruptions and damage the ecosystems of far-flung places.

(iii) **Pollens and spores :** These are injected into the atmosphere by vegetation. Being very light, these find their way in both troposphere and stratosphere.

(iv) **Forest trees :** In rural areas, trees discharge terpene vapours (the most important is α-pinene) which causes a blue haze in the atmosphere due to dispersion of sun light. The terpenes are easily oxidised by ozone to give peroxy radicals which are constituents of photochemical smog.

The frequent occurrence of forest fires is also a source of particulates. It is well known that the forest fire in 1997 in Indonesia and in May 1998 in Mexico created a number of problems due to injection of particulates into the atmosphere. These particulates consist of polynuclear hydrocarbons (for example, naphthalene, phenanthrene, anthracene, pyrene etc.) and salts of calcium, iron, zinc and magnesium.

(v) A major source of particulates in the atmosphere is mining industry, e.g. mining of coal, asbestos, and mica and ore of metals (see sec. 4.4.3).

Adverse Effects of Particulates

Particulates produce a number of adverse effects which are as follows:

(i) Particulates play a major role in the formation of photochemical smog which is known to cause a number of problems (see sec. 4.5.5)

(ii) Particulates aggravate the deleterious effects of other air pollutants. The particulates sometimes travel across the national boundaries and so the harmful effects are felt upto thousands of kilometres from their place of origin.

(iii) Women exposed to fine particulate matter (having diameter less than 10 μm) have been found to deliver children having small heads and bodies. Such children are slow to learn and have an increased risk of developing cancer. It is believed that particulates which are coated with polynuclear hydrocarbons cause irreversible damage to the DNA of the growing fetus.

(iv) Visibility is considerable reduced.

(v) Mercury if present as particulate, produces heaviness, headache, nervousness, fatigue and a number of other problems. Prolonged exposure causes the breakdown of CNS.

(vi) The exhaust gases from automobiles discharge account to the extent of about 1% of the total mass of particulates in the atmosphere. These particulates comprise of soot, lead compounds, crank case oil and hydrocarbons (besides other gaseous pollutants like CO, CO_2, oxides of nitrogen and sulphur etc).

(vii) The main components of various combustion processes are SO_2 and NO_2. These are converted into sulphates and nitrates via the formation of sulphuric acid and nitric acid. Incomplete combustion of fossil fuels produces soot particles. A number of combustion processes produce fly ash which is a mixture of oxides of silicon, aluminium, calcium and phosphorus.

(viii) Metallurgical processes emit fine metallic dust and oxides of metals which provide catalytic surface for a number of unwanted reactions in the atmosphere.

(ix) The presence of lead as particulate is extremely harmful. It is a cumulative poison and keeps on accumulating in the tissues of the human body and in the leaf tissues of plants. Accumulation of lead in human tissues causes malfunctioning of red blood cells leading to anaemia. Lead is known to damage organs like liver, kidneys and intestines and affects the CNS.

(x) Cadmium pollution may arise from the metallurgical processes of zinc, and copper. It causes lung irritation, vomiting and hypertension.

Following table lists some more metallic pollutants and their effects on human beings.

Table 4.2 Sources of some metallic pollutants and their effects

S.No.	Metallic pollutant	Major sources	Harmful effects
i.	Zinc (Zn)	Zinc refineries, galvanizing processes, brass manufacture, metal plating, and plumbing.	Zinc fumes have corrosive effect on skin and can cause irritation and damage mucous membranes.
ii.	Chromium (Cr)	Metallurgical and chemical industries, processes using chromate compounds, cement and asbestos units	Toxic to body tissues, can cause irritation, dermatitis, ulceration of skin, perforation of nasal septum. Carcinogenic action suspected.
iii.	Arsenic (As)	Arsenic containing fungicides, pesticides and herbicides, metal smelters, by-product of mining activities, chemical wastes.	Inhalation, ingestion or absorption through skin can cause mild bronchitis, nasal irritation or dermatitis. Carcinogenic activity is also suspected. Attack -SH groups of enzymes, coagulate proteins.
iv.	Beryllium (Be)	Coal, nuclear power and space industries, production of fluorescent lamps, motor fuels and other industrial use.	Damage to skin and mucous membranes, pulmonary damage, perhaps carcinogenic.
v.	Boron (B)	Boron producing units, production and use of petroleum fuel and additives, burning coal and industrial wastes, detergent formulations.	Ingestion, inhalation as dust causes irritation and inflammation. Boron hydrides can damage CNS and may result in death.
vi.	Manganese (Mn)	Ferromanganese production, organo-manganese fuel additive, welding rods, incineration of manganese containing substances.	Absorption, ingestion, inhalation, or skin contact may cause manganic pneumonia.
vii.	Nickel (Ni)	Metallurgical industries using nickel, combustion of fuels containing nickel additives, burning of coal and oil, electroplating units using nickel salts, incineration of nickel containing substances, vanaspati manufacture.	Respiratory disorder, dermatitis, cancer of lungs and sinus.
viii.	Vanadium (V)	Vanadium refining, production of vanadium containing alloy's power plants, burning of oil rich in vanadium.	Gastro-intestinal and respiratory disorders inhibition of synthesis of cholesterol, heart disease and cancer in case of chronic exposures.
ix.	Selenium (Se)	Burning of fuels and residual oils, fumes and gases from refinery wastes, incineration of paper and other wastes natural sources.	Irritation of gastro-intestinal and respiratory tracts, irritation of eyes, nose and throat. Damage to lungs, liver and kidneys.

4.4 Sources of Air Pollution

The air pollution is caused by carbon monoxide (4.3.1.1), carbon dioxide (4.3.1.2), oxides of nitrogen (4.3.2), oxides of sulphur (4.3.3), hydrogen sulphide (4.3.4), chlorine (4.3.5), ozone (4.3.6), hydrocarbons (4.3.7) and particulates (4.3.8) which have been already described in various sections mentioned in the parentheses. The present section describes some of the other major sources of air pollution. These include pollution from

- Industries
- Petroleum
- Mining Industries
- Power Plants
- Automobiles
- Agrochemical based industries

4.4.1 Pollutants from Industry

It has already been stated that industrial units dealing with the manufacture of explosives (e.g. various nitro compounds) using concentrated nitric acid and concentrated sulphuric acid release thousands of tonnes of oxides of nitrogen (sec. 4.3.2) and oxides of sulphur (sec. 4.3.3) into the atmosphere. Industries dealing with fermentation technology (e.g. in the manufacture of antibiotics and other chemicals like alcohol) release thousands of tonnes of carbon dioxide (sec. 4.3.1.2) into the atmosphere. Caustic soda manufacturing units are responsible for the release of chlorine (sec. 4.3.5) in the atmosphere. Various metallurgical processes are responsible for introducing carbon monoxide, carbon dioxide, hydrogen sulphide and metal particulates into the atmosphere. The following discussion includes some very special industries dealing with highly specialized products.

A. Polymers and Plastics

A **polymer** is a macromolecule and has very high molecular mass. Polymers are formed by the joining of a large number of repeating units called *monomer* by the process called **polymerisation**. A plastic is also a polymer (but all polymers are not plastics), which can be moulded into various shapes. The term *plastic* is used to include many synthetic polymeric materials like rubbers, fabrics, resins etc. Polymers (including plastics) form the backbone of modern civilization. In fact, in all walks of life, starting from household utensils, cloths, furniture etc., to automobiles to spacecrafts and biochemical and surgical operations, polymers are extensively used. The synthetic polymers have become so indispensable that their per capita production is associated with increased standard of living.

It is a common knowledge that polymers (including plastics) are not biodegradable. On the contrary, biopolymers (such as starch, cellulose, proteins and nucleic acids etc.) which control various life processes, are biodegradable i.e., they disintegrate by themselves during a certain interval of time. The synthetic polymers, a major portion of which are used as throw away containers and packing materials do not disintegrate by themselves, i.e., they are not biodegradable and they do not degrade over a period of time. This property makes the polymer stay intact for a much longer period - an advantage for use in household items etc. This non-biodegradability of polymers and plastics has caused a serious waste disposal problem for the mankind. With the ever increasing use of plastics, the time is not far away when the entire civilization will be buried under a pile of plastic debris. The waste plastic causes acute environmental problems in all the segments i.e., the earth, the water bodies and the air. In view of the above, endless efforts have been and are being made by researches all over the world to evolve such plastics which are biodegradable. All these attempts have led to the discovery of certain biodegradable polymers. However, such polymers are costly.

The most important reason for the polymers and plastics being non-biodegradable under natural conditions is their hydrophobicity (water-repelling capacity). In case, the hydrophobic character is improved – the micro-organisms will be able to degrade the plastics, though this would mean shorter span of life for them. The best is to achieve a balance between the advantages arising out of the durability of plastics and the disadvantages arising out of their non-biodegradability in environment. The objective could be achieved by adding some additives, which after been introduced in the final product, make these biodegradable.

n-Pentane, trichlorofluoromethane (Cl_3CF) and toluene 2,4-diisocyanate are commonly used foaming agents to make low density plastics. For example, polyurethane is formed by incorporating liquid pentane under pressure. When the mixture is extruded from an outlet, the pentane volatilises into the atmosphere, leaving polyurethane full of small holes. Chemical foaming agents produce CO_2 by a chemical reaction. It involves the reaction of an isocyanate with water.

$$RNCO + H_2O \longrightarrow RNH_2 + CO_2$$

Plasticizers like *di*-(2-ethylhexyl)phthalate, *di*-(2-ethylhexyl)adipates and polychlorinated biphenyls are added to make plastics more flexible. These plasticizers get incorporated between the polymer chains, weakening the forces of attraction between them and so they increase their flexibility.

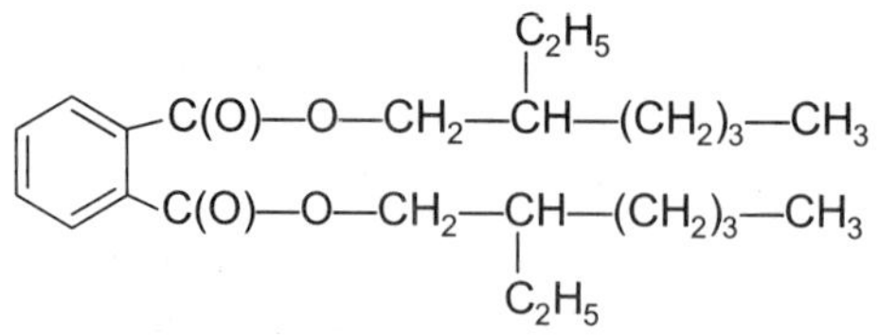

di-(2-ethylhexyl) phthalate

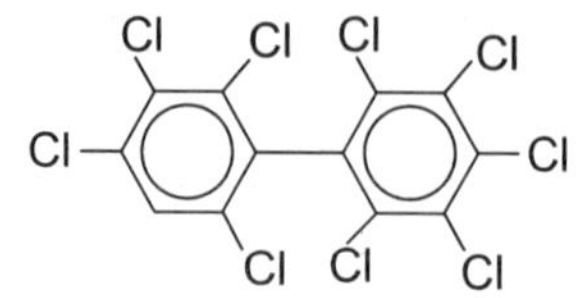

Polychlorinated biphenyl

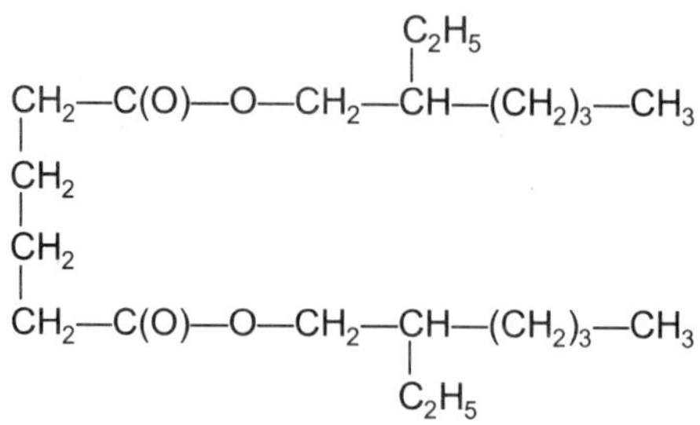

di-(2-ethylhexyl) adipate

Antioxidants like 2-hydroxybenzophenone are added so as to protect the plastics from harmful solar radiations. In the absence of antioxidants, the finished products may degrade on being exposed to sunlight for longer time.

Hazards Caused by Polymers and Plastics

Most of the hazards are associated with the manufacturing processes and disposal of the waste materials.

i. Human Health

Workers employed in polymer and plastic industries suffer from a number of health hazards.

(a) 2-Chloro-1, 3-butadiene (chloroprene) used for the production of synthetic rubber (neoprene) can cause liver damage.

(b) *Bis*-(chloromethyl)ether, used as a reagent to affect cross linking in polymers is suspected to cause lung cancer.

(c) Toluene-2,4-diisocyanate used to foam polyurethane is responsible for acute respiratory problems.

(d) Vinyl chloride used for the manufacture of polyvinyl chloride polymer results in a rare form of liver cancer (called *angiocarcinoma*).

(e) The phthalates used as plasticisers cause testicular cancer and lung damage.

(f) Polychlorinated biphenyls are released in the atmosphere by burning waste plastics in incinerators. These chemicals are washed down into the aquatic systems. Being fat soluble, they get accumulated into aquatic organisms. When human beings consume contaminated fish, they suffer from an ailment called yusho disease; the symptoms are headache, fatigue, pains in the joints, anaemia. Concentrations above 1·2 ppm can cause blindness.

(g) The polymer and plastic waste materials dumped on the land cause major problems of their disposal. The non-biodegradable nature of these materials makes them a permanent waste. In case, the waste material is incinerated, the problems associated with air pollution arise. The gases evolved by incineration are carbon dioxide (responsible for green house effect), oxides of nitrogen (responsible for acid rain) and carbon monoxide - an extremely poisonous gas. Incineration also releases particulate matter into the atmosphere. If asbestos has been used as filler during the manufacture of plastics or polymers, on incineration of plastic, asbestos is injected in the air which causes several health hazards.

Checking Polymer and Plastic Pollution

Being non - biodegradable, the problem of pollution caused by polymers and plastic can be checked to a certain extent by the following procedures:

i. Recycling of Plastics

Pollution load caused by plastics can be reduced to a certain extent by recycling the used and waste plastic items. Recycling is carried out by the following steps:

(i) Removal of non-plastic materials from plastic ones.
(ii) The plastic material is granulated and washed to remove dirt and paper labels.
(iii) The heavier substances like stones and metals are removed by passing the washed material through a series of hydraulic tanks.
(iv) The granules are dried. Air is blown over it under pressure to remove residual fibrous material.
(v) Finally, the clean granules are moulded into finished products.

The technology has been successfully used for polyvinyl chloride containers and also for recycling of polymers (e.g. polyester, polyethylene, terephthalate etc.).

ii. Conversion of Polymers into Monomers

The process for the conversion of scrap polyester, polyethyleneterephthalate (PET) into monomers is known as the **Petrette process** (Eur. Patent, 484, 963, May 13, 1992; US patent 5, 391, 263, February 21, 1995 and US Patent 5, 504, 122 April 1996). The flow sheet of the Patrette process for recovery of monomers from PET scrap is given below (Scheme 1).

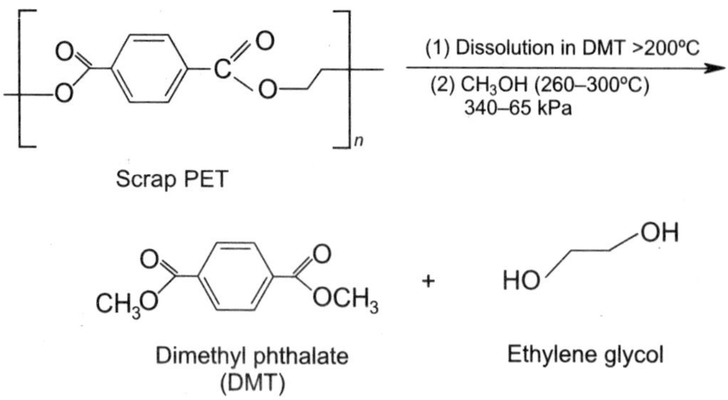

Scheme 1 Conversion of PET scrap into monomers

The monomers (DMT and ethylene glycol) are purified and polymerised again to give PET.

(iii) The best and foolproof method is to make polymers which are biodegradable. One typical illustration is the manufacture of Thermal Polyaspartate polymer (TAP), a biodegradable alternative to polyacrylate polymer (PAC). The Donlar's synthesis of TPA consists of heating aspartic acid followed by hydrolysing the formed succinimide polymer with aqueous base (Scheme 2).

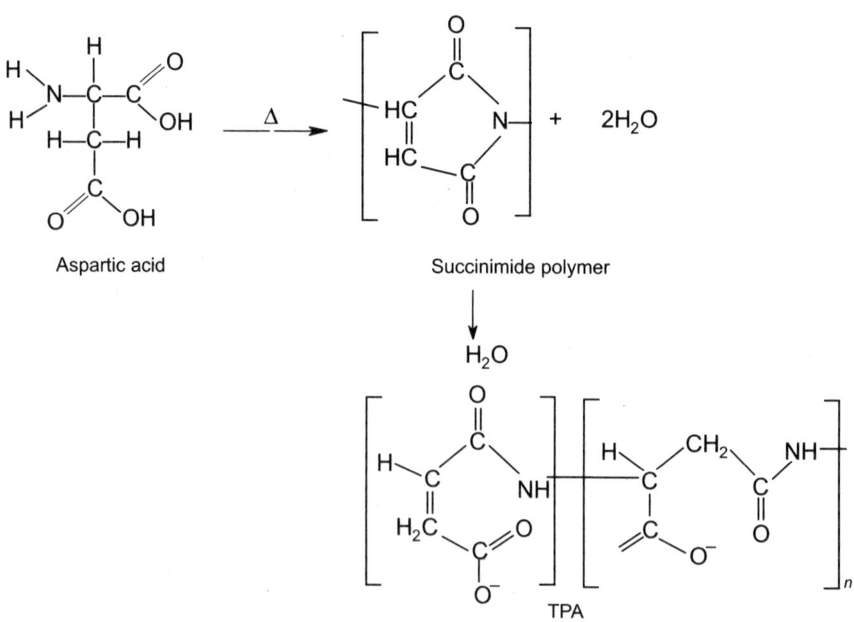

Scheme 2 Synthesis of TPA

B. Asbestos

The name *asbestos* is given to a group of silicate minerals which can be separated into fibres and used to make materials that will not burn. The name asbestos comes from a Greek word meaning *inextinguishable* or *unquenchable*. It is the only naturally occurring fibrous mineral. It finds a variety of uses and is valued as an industrial material. Due to its fibrous nature, asbestos is an environmental hazard. Asbestos fibres are readily absorbed on the surface of lung tissues; this results in a disease called *asbestosis* and is also responsible for development of lung cancer. In fact, industrial workers who have worked in asbestos industries for about 20 years or so are susceptible to be affected by asbestosis as well as cancer.

Asbestos has been in use for more than 3000 years in commercial, public and other facilities. It is used because of its incombustibility and effectiveness as a reinforcing agent when combined with other materials like cement or plastic. Its resistance to chemical and biological attack is of great value. Finally, asbestos is cheap and easily available. Some important applications of asbestos are as follows:

- The fibres of asbestos can be spun into yarn, which can be reinforced with other fibres, e.g. nylon. Such yarn is used for making clothing for fire fighters and special gloves.
- Thermal insulators
- Electrical insulators
- For making brake lining of automobiles
- Building materials

Sources of Asbestos in the Environment

Asbestos present in some rocks and minerals which is not exploited may find its way into the atmosphere by natural phenomena such as weathering, landslides etc. Workers engaged in mining, milling and fabrication processes are also exposed to asbestos. When the building materials containing asbestos deteriorate, the asbestos particles are released into the atmosphere. The airborne asbestos is finally ingested through inhalation. Likewise, asbestos is also released into the atmosphere by used brake lining of vehicles.

Monitoring of Asbestos in the Atmosphere

The concentration of asbestos in the atmosphere is determined as given below:

(i) The airborne asbestos particles are colloidal in nature and bear electrostatic charge. The first step involves the discharge or removal of the charge. For this, a sample of air (about 1000 m^3) is impinged on the plate of a electrostatic precipitator.

(ii) The particles of asbestos settled on the plate, are washed into about 100 mL of a suitable non-polar solvent which dissolves unwanted organic particulates.

(iii) The undissolved asbestos particles are filtered.

(iv) The number of asbestos particles is finally counted under an electron microscope. The concentration is reported as number of asbestos particles per mL of air.

Harmful Effects of Asbestos

Exposure to asbestos fibres has the following effects:

i. Asbestosis

Incorporation of asbestos fibres in the lung tissues causes a lung disease known as **asbestosis**. Once inhaled, the asbestos particles get embedded in the lung tissues for long periods (20-30 years) and the victim's outer lining of the lungs gets thickened; thereby, decreasing the oxygen-carbon dioxide exchange capacity and the elasticity of the lungs. The victim becomes short of breath and eventually dies of heart failure. It is found that workers engaged in asbestos industries (e.g. mining, cleaning and weaving) are more prone to suffer from asbestosis. This disease is incurable.

ii. Cancer

Workers in industries dealing with the use of asbestos such as, in building, construction, mining, manufacture of brakes and repair of clutches of automobiles face 50% chances of dying from cancer. Since cancer symptoms do not appear (or arise) until 20-30 years after being exposed to asbestos, various agencies could not recognise the dangers posed by the use of asbestos. When inhaled, the fine particles of asbestos dust are lodged in the lungs and damage the cells therein. Prolonged heavy exposure increases a person's risk of lung cancer and *mesothelioma*, a rare cancer of thin membrane that line chest and abdomen. Some developed countries have banned the use of asbestos in any form. However, some developing and under developed nations have still not banned the use of asbestos completely.

iii. Iron Metabolism

Certain varieties of asbestos contain appreciable amount of iron (25-26%). Though iron is essential for life, its excessive built up is harmful. In the body, iron forms chelate with citrate ion; these chelates enter the cells and ferrous iron reacts with oxygen producing molecular oxygen radicals ($O_2^{\bullet}$). These radicals create havoc; these oxidise DNA and proteins and cause damage to proteins. This leads to carcinogenesis.

Checking Asbestos Pollution

(i) Asbestos is most harmful in the form of dust. It is not possible to prevent atmospheric pollution, but one can take adequate precautions so that one does not inhale asbestos dust as far as possible. For example, workers who deal with asbestos must wear masks in order to prevent inhalation.

(ii) Asbestos sheets, if used must be coated with dust-proof membranes, such as paints or plastics or keeping the sheets wet while handling; this reduces the hazard. One should not be exposed to asbestos dust for long.

(iii) Asbestos dust arising out of cleaning brake lining should be cleaned by vacuum extraction procedure or with a damp rag.

(iv) Children are more susceptible to asbestos pollution than adults. This is especially important because in old buildings of schools, the roofs and tiles are made from asbestos. It is best to find such buildings and the places in which asbestos is present and these must be coated with polyvinyl chloride spray.

(v) As far as possible, the use of asbestos should be avoided. In its place, clothing made from nylon and wool blends (these provide heat and fire protection) should be used. For insulation, ceramic fibres can be used as a substitute for asbestos. Moulding of phenol-formaldehyde can be used in place of asbestos in electrical insulation materials. Also, glass fibre reinforced cement is a good substitute for asbestos based building materials.

C. Tobacco

The use of tobacco is responsible for 3 out of every 10 cancer deaths. It is known to be a major factor for cancer of the lung, larynx, oral cavity, esophagus, urinary bladder, kidney and prostrate. Tobacco smoke has been found to contain over 2500 chemical substances. It contains carbon monoxide, carbon dioxide, ammonia, nitrosamines, nitrogen oxides, hydrogen cyanide, sulphur compounds, nitrites, ketones, alcohols and acrolin. The 'tars' contain carcinogenic hydrocarbons, which include nitrosomines, benzo(á)pyrene, anthracene, acridines, quinoline, benzene, naphthol, naphthalene, cresols and insecticides (DDT) as well as radioactive compounds like potassium-40 and radium-226. Some of these compounds are known to be risk factors for human cancer.

Cigarette smoking has been shown to be a major risk factor in men and women especially in middle age for cardiovascular diseases. Chronic lung diseases related to smoking are second to heart diseases as a cause of disability. Smoking adversely affects fertility in females. Children born to smoking mothers are more prone to various abnormalities.

The cigarette manufacturing industries all over the world make an unimaginable number of cigarettes all of which are smoked. One cannot even imagine the amount of total smoke released into the atmosphere with all the harmful substances (chemicals). In fact, smoking is a major factor contributing to atmospheric pollution.

The most important question that arises is how to control pollution caused by smoking. This has to be a voluntary effort of all the people who smoke to quit smoking. Can one imagine a day when cigarette smoking is considerably reduced so that the cigarette-manufacturing companies find it non-profitable to manufacture cigarettes? The answer is probably 'NO'. The reason is that the society we live in, it is not possible to have a law to ban smoking or the manufacturing of cigarettes.

4.4.2 Pollution by Petroleum

Petroleum is naturally occurring oil, mostly found in underground reservoirs. It is formed by the decay of the remains of animals and plants. It is also found below the oceans and is believed to be formed by decay of marine plants and animals. In its crude form, petroleum is a mixture of hundreds of chemicals, chiefly hydrocarbons. It is separated into various products such as cooking gas, naphtha, petrol, diesel, kerosene and wax by fractional distillation in refineries. The pollution of air or atmosphere occurs at different stages. Some of these are:

- During drilling, some of low boiling fractions escape in the atmosphere and spill on the land. In case, drilling from below the surface of oceans, the petroleum oil may be spilled on the surface of water and may be swept by air currents to distant places.
- During transportation of crude petroleum oil, there is spilling or leakage of petroleum. This particularly occurs in oceans during transportation by huge tankers (see chapter 7).
- During fractional distillation in the refineries, a number of fractions, e.g. gas, hydrocarbon etc. find their way into the atmosphere due to leaks or poor recovery.

4.4.3 Mining Industries

Mining is the process that is used to take minerals out of the earth. Coal, asbestos, copper, mica, precious stones and many other materials are mined. Following mining industries contribute to the pollution of the atmosphere.

(i) **Coal mining :** This discharges large amount of particulate matter in the form of coal dust into the atmosphere. Workers in coal mining are prone to be incapacitated by black lung disease.

(ii) **Mica mining :** Mica is a group of silicate minerals having layered structure. It contains mixed silicates of elements sodium, potassium, calcium, aluminium, magnesium, iron and lithium. It is mined from pegmatite rocks. India is the largest producer of mica in the world contributing more than 75% of the world's consumption. In India, mica is found in Bihar, Andhra Pradesh and Rajasthan. Besides, it is also found in Rocky Mountains of USA and in Canada. Mica is widely used as an electrical insulator and in electrical capacitors. During mining and processing, mica is discharged into the atmosphere in the form of dust. This dust, on inhalation, causes a number of health hazards. The workers in the mining operations are the worst affected.

(iii) **Asbestos mining :** See sec. 4.4.1. subsection B.

(iv) **Mining of ores :** An *ore* is naturally occurring mineral which is obtained by mining and is the raw material for the extraction or isolation of metals on a commercial scale. Most of the ores are compounds (in impure form) of metals with oxygen, sulphur, silicon or phosphorus. The pollution of air or atmosphere occurs at different stages. Some of these are as follows:

- During mining by drilling etc. fine particles of the ores are discharged into the atmosphere.
- During extraction or isolation of metals - the ores are subjected to various processes like concentration, roasting, smelting etc. In all these processes, gaseous pollutants like carbon monoxide, carbon dioxide, sulphur dioxide are introduced into the atmosphere. Some metals which are comparatively of low melting point vaporise and pollute the atmosphere.

4.4.4 Power Plants

Power plants are used for generating power or electricity. In most of the conventional power plants, the source of energy is coal. Coal fired power plants are a source of serious environmental problems. These inject the gaseous pollutants like carbon monoxide, carbon dioxide, sulphur dioxide (if coal contains sulphur impurities) into the atmosphere. Besides gaseous pollutants, such power plants inject particulates in the form of fine carbon particles and fly ash (the ash left after coal is heated). A number of methods have been used to reduce the pollution caused (see sec. 4.4).

With the advancement of knowledge nuclear energy is used to produce electricity at a much cheaper rate. However, in this case, elaborate arrangements for generation and control have to be used. There are a number of problems associated with the production of nuclear energy.

- There is deleterious effects on humans due to the release of ionising radiations.
- There is considerable problem in the disposal of the spent fuel.
- There may be accidents which may have disastrous effects on mankind.

These aspects have been discussed in detail in Chapter 1 under *Sources of Energy* and Chapter 10.

4.4.5 Automobiles

Automobiles mostly run on petrol or diesel oil. The exhaust gases from the automobiles inject a number of gaseous pollutants like carbon monoxide, carbon dioxide, oxides of nitrogen and sulphur dioxide. In fact, automobiles are major contributor of atmospheric pollutants. In case of automobiles, the pollutants are removed by passing the exhaust gases through a catalytic converter.

The aeroplanes use huge amounts of gasoline and are the major contributors of exhaust gases in the higher atmosphere.

4.4.6 Agrochemical Based Industries

The agrochemicals are mostly used in agriculture. The agrochemicals include fertilizers, *insecticides* (or *pesticides*), *fungicides* and *herbicides etc.*

4.4.6.1 Fertilizers

Fertilizers are materials which, if added to soil, result in better growth of crops. Their function is to supply those nutrients to the soil which are necessary for the growth of plants and have been removed by growing of the crops. The nutrients are of three types as given below:

- ***Primary nutrients*** – These are needed in large amounts. Their examples are nitrogen, phosphorus and potassium.
- ***Secondary nutrients*** – These are required in moderate amounts. Their examples are calcium, magnesium and sulphur.
- ***Micronutrients*** – These are required in extremely small amounts. Their examples are boron, chlorine, copper, iron, manganese, molybdenum and zinc.

The fertilizers are broadly classified into three types. These are as follows:

- Nitrogen based fertilizers like sodium nitrate ($NaNO_3$), ammonium sulphate $(NH_4)_2SO_4$ and urea (NH_2CONH_2).
- Phosphate based fertilizers like super phosphate $[Ca_2H_2(PO_4)_2.CaSO_4.2H_2O]$, Triple super phosphate $[Ca_2H_2(PO_4)_2.CaHPO_4]$.

- Potassium based fertilizers like potassium chloride (KCl) and potassium sulphate (K_2SO_4).

Effects of Fertilizers on Environment

Excessive use of fertilizers is harmful to the living organisms. Some of the problems associated with the excessive use of fertilizers are as follows:

i. Soil

The nitrogen fertilizers particularly sodium nitrate and ammonium sulphate tend to increase the acidity of the soil. In highly acidic soils, the bacterial population decreases and the population of fungi increase. Pathogenic fungi, such as *plasmodiophora brassicate* are found in acidic habitats.

Though fertilizers replace the inorganic nutrients in the soil, the soil's organic matter (which decay to produce humus which is necessary for plant growth) is not replaced. Also, the soil organisms use organic matter as a source of energy. Thus, dependence only on fertilizers may lead to a change in the physical, chemical and biotic characteristics of the soil.

In acidic conditions of the soil, some nutrient cations like Ca^{2+}, Mg^{2+}, are leached away from the soil and soluble aluminium ions (which are toxic to plants) are formed.

Excessive use of fertilizers results in the excessive growth of weeds and herbs.

ii. Human Health

Due to excessive amounts of nitrogen fertilizers, concentration of nitrate ions increases in leafy vegetables. Consumption of such leafy vegetables results in bioaccumulation of nitrates in the bodies. Nitrate by itself is not harmful but in human body, it is reduced to nitrite which is a probable factor for the occurrence of cancer.

Excessive amounts of nitrates may also lead to **blue baby syndrome** or methaemoglobinaemia in small children. The nitrates in the body are reduced to nitrites which react with haemoglobin (in the blood stream) and reduce the oxygen carrying capacity of the blood. This condition is referred to as blue baby syndrome. The symptoms include diarrhoea and vomiting and the child's complexion become stale blue, which may be fatal.

iii. Eutrophication

Eutrophication is a major problem associated with the use of fertilizers. This aspect will be discussed subsequently (see Chapter 6 on *Water Pollution*).

Checking Fertilizer Pollution

Nitrogen based fertilizers are essential for the growth of plants. However, problems arise due to discharge of nitrates to the ecosystem. The nitrates are produced as a result of microbial oxidation of nitrogen-based fertilizers. Most of the problems can be overcome by:

- Rotation of crops on scientific basis.
- Using fertilizers in accordance with professional advice. Only recommended dose should be used. It is safer to use less amount of fertilizer than required.
- Using organic manure as far as possible.

4.4.6.2 Pesticides

Pesticides are the chemicals, which are used for eliminating pests. These are, in a way, essential for increasing the yields of the crops. However, according to the findings of WHO about 7,50,000 people are poisoned by pesticides every year, resulting in about 14,000 deaths. Developing countries account for 30% of the pesticide consumption but share more than 60% of the casualties. The term, pesticides includes insecticides, fungicides and herbicides.

4.4.6.2.1 Insecticides

Insecticides are used to eliminate the insects, which eat away the crops. Use of insecticides is helpful for

- Controlling the insect pests which reduce the yield of the crops.
- Stopping the spoilage of food crops (eg. grains) during storage by controlling the insects.
- Besides protecting the crops, the insecticides also protect the civilization by controlling the spread of diseases like sleeping sickness, bubonic plague and malaria.
- Controlling household insects like cockroaches, houseflies and mosquitoes.

Types of Insecticides

Broadly speaking, the synthetic insecticides are of the following four types:

a. Organochlorines

These have been in use since 1950. One of the most common and best known insecticides is DDT [2,2-*bis*(*p*-chlorophenyl)-1,1,1-trichloroethane.]

Other examples include aldrin and dieldrin.

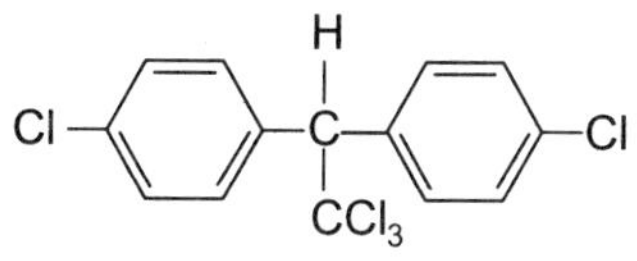

DDT

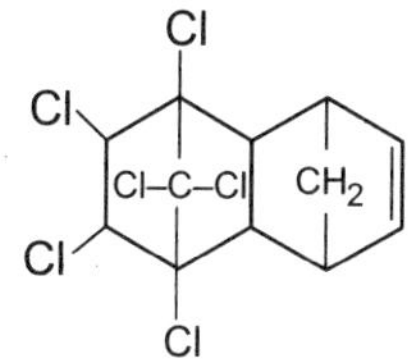

Aldrin

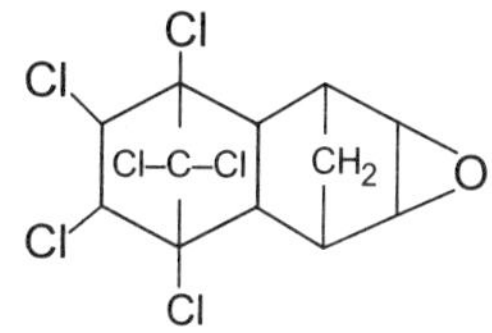

Dieldrin

These insecticides are persistent in the environment and so show biological activity for long periods of time. However, these insecticides not only kill unwanted insects (responsible for destroying the crops) and are responsible for prevention of diseases but also destroy the non-target organisms (organisms which are helpful for the crops). Also, these insecticides keep on accumulating in the environment and cause pollution (in other words, they are not biodegradable). They also get accumulated in the vegetables, fruits and even water. They are harmful for human beings and have been phased out.

b. Organophosphates

Organophosphate insecticides have very low persistence in the atmosphere and are biodegradable. Being toxic in nature, the organophosphates kill not only the non-target insects, but are also harmful for humans and the wild life. They are unstable and so have to be applied at frequent intervals. Typical examples include parathion and diazinon, demeton and *bis* (*di*-ethylphosphoric) anhydride.

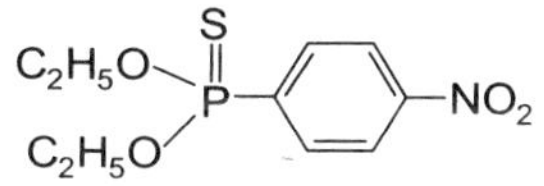

Parathion

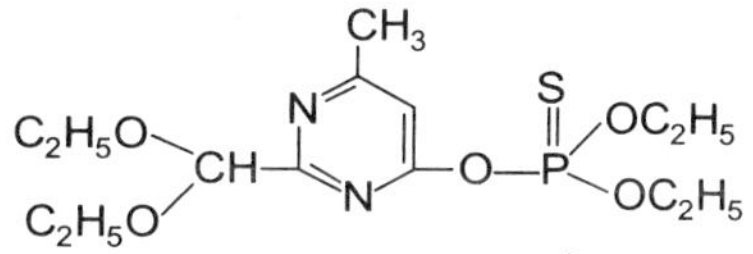

Diazinon

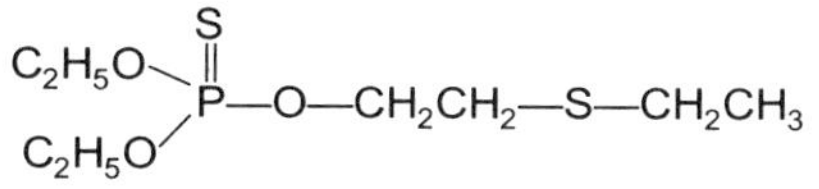

Demeton

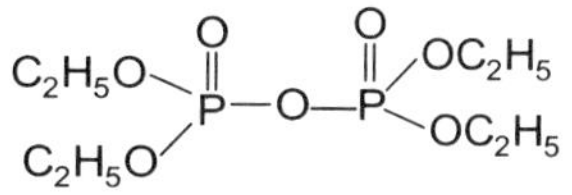

bis-(diethylphosphoric)anhydride

c. Carbamates

These insecticides are less toxic than organophosphates and less presistant than organochlorines. Some examples include carbaryl and methiocarb.

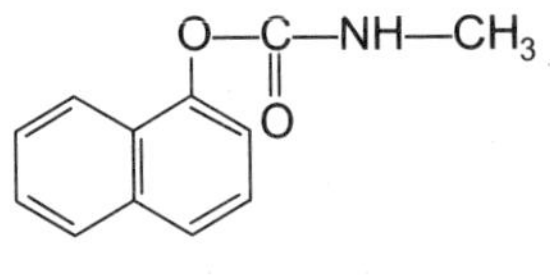

Carbaryl

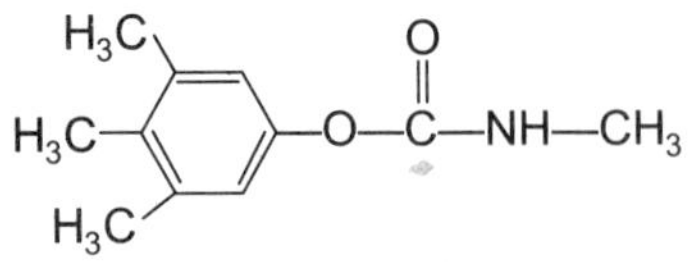

Methiocarb

d. Pyrethroids

These are the recent group of insecticides and are the structural analogs of pyrethrum, the naturally occurring insect repellent present in chrysenthemum. These are neither persistent nor toxic, though they are costly. Still about 33% of the world's insecticides are pyrethroids. Some examples are pyrethrum and deltamethrin.

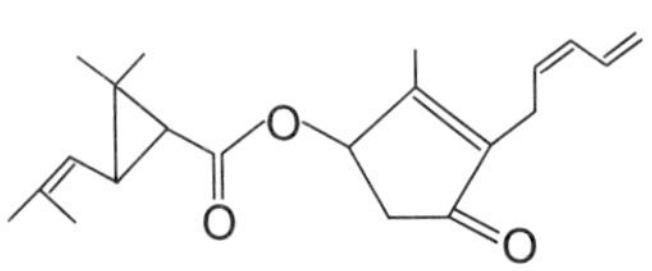

Pryrethum

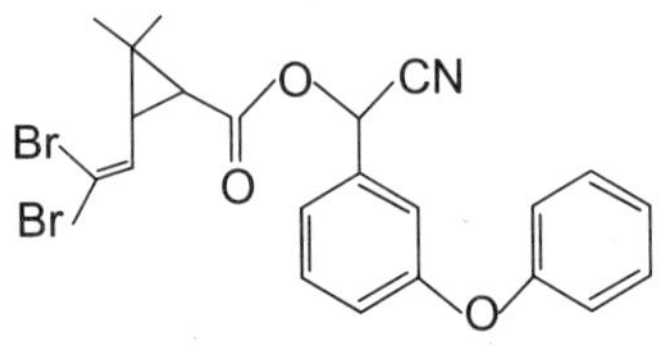

Deltamethrin

Effects of Insecticides on the Environment

Excessive use of insecticides has affected the environment in the following ways:

(i) Insecticides get accumulated in the fatty tissues of various organisms including the humans via the plants and vegetables etc. Accumulation in food chain has disastorus effects.

(ii) The target insects develop resistance against insecticides after a period of time. In this way, the effectiveness of the insecticides decrease considerably.

(iii) Due to accumulation of insecticides (e.g. DDT) in the fatty tissues of birds, their capacity to reproduce declines considerably resulting in the decline of their population.

(iv) In case the water in the rivers and oceans surface is contaminated with insecticides (which results with the help of rain water), the fish are the worst affected.

(v) In industries manufacturing insecticides, the workers are affected by slow poisoning effects of the insecticides.

(vi) Accidents in manufacturing units are responsible for the environmental pollution. The worst incident happened in Bhopal (Madhya Pradesh,

India) on Dec. 2, 1984, when methyl isocyanate gas used for the manufacture of insecticides got leaked which killed and disabled lakhs of people. This aspect will be discussed in chapter 15.

Since insecticides are essential for agricultural production, it is not possible to ban the manufacture or use of such products. The best option is to look for alternative insecticides, which will kill the pests but are not detrimental to the environment.

The following two types of insecticides have been used to control insects:

i. Hormonal Insecticides

The hormonal insecticides kill the insects via upsetting its normal metabolic processes. A hormone (called a Juvenile hormone) shows insecticidal properties. This hormone is initially present in insect larva but its production stops before undergoing metamorphosis (i.e., when it grows into adult insect). In case, the larva is continuously exposed by the Juvenile hormone, the larva cannot mature – the result is that its reproduction stops.

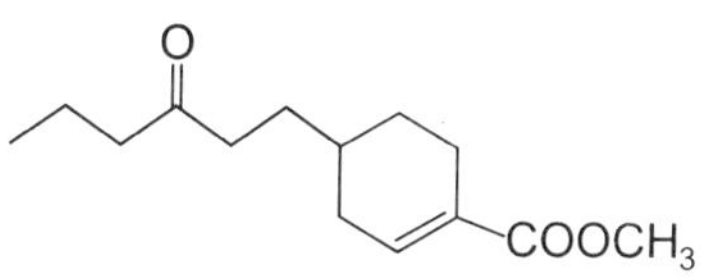

a Juvenil hormone

The hormonal insecticides are effective only at a certain stage (the stage before metamorphosis occurs). These insecticides do not let the larva develop into adults.

ii. Sex Attractants

A group of compounds, known as pheromones have the capacity to attract the male insects from a very long distance. If these sex attractants (in extremely low concentration) are kept in a place, then male insects are attracted towards it from a radius of about 2 miles. Once assembled in a small area, they can be killed by any means.

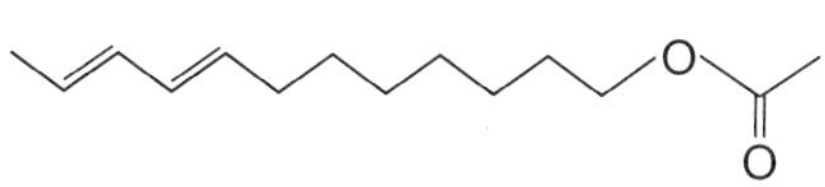

a pheromone

4.4.6.2.2 Fungicides

Fungicides are chemicals which when applied to plants or their seeds and bulbs, inhibit or prevent fungal growth or fungal diseases. Infact, use of fungicides ensures vegetation and crops free from any disease. Though fungicides are somewhat harmful to the environment, their benefits (in producing disease free crops) outweigh the risks.

What are Fungi?

Fungi are plants without chlorophyll. So, they cannot make their food by photosynthesis like other green plants. Fungi live as parasites at the expense of living organism. Fungus stay in only a particular portion of a plant, compared to insects which can move about on the plant surface. Fungicides play a vital role in controlling the growth of fungi.

The main uses of fungicides are:

- Protection of commercial crops from diseases. In the absence of fungicides, the plants may rot (due to fungi) and do not grow as expected.
- The germination of seeds for making seedlings is successful, if seeds are placed in a solution of fungicide for about 20-30 minutes and then put in the soil for germination in the usual way.
- The fungi affected plants / crops are rendered free of fungi by spraying fungicides on them.
- To preserve wooden articles.
- The main use of fungicides is to act as growth inhibitors for the fungus.

Broadly speaking fungicides are of two types:

i. Systemic Fungicides

These fungicides, when applied to the plant, rapidly distribute itself in various plant tissues. Some examples are as follows:

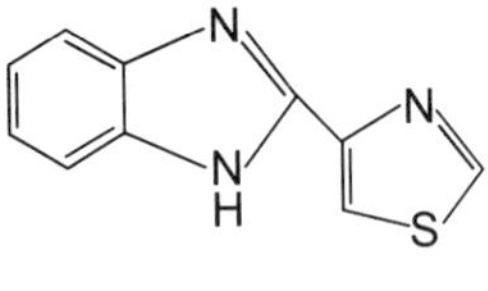

Thiobendazole

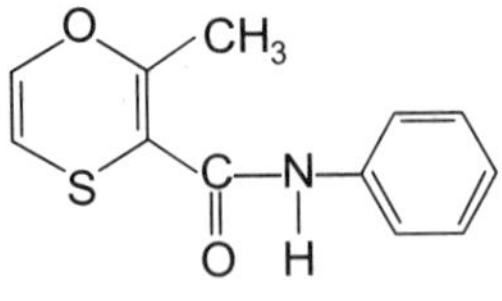

Carboxin

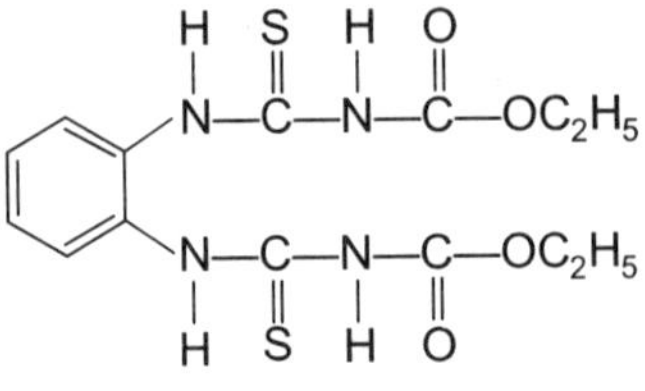

Thiophanate

ii. Non-systemic Fungicides

These protect only that part of the plant where it is applied. Some examples are given below:

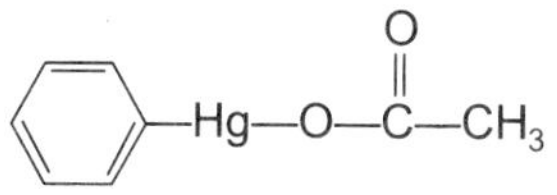

Phenyl mercury acetate

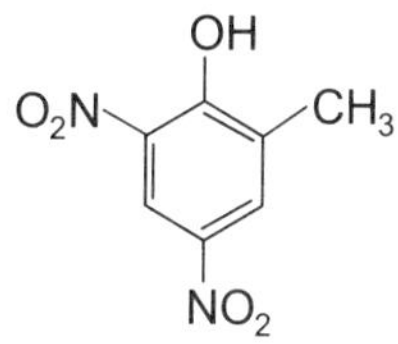

4,6-dinitro *o*-cresol

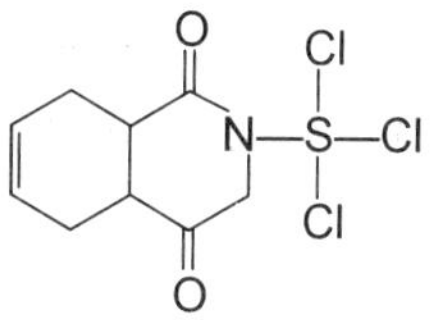

Captain

Effects of Fungicides on Environment

Though fungicides are a boon for the agriculture, they have the following implications on the environment:

- In the soil, both bacteria and fungi compete for food. When fungi are eliminated by the use of fungicides, the bacterial population increases. Thus, though the plants are protected from fungal attack but they become susceptible to bacterial diseases.
- The fungicides in the soil increase the number of harmful bacteria and decrease the population of useful types. For example, the application of 4,6-dinitro-*o*-cresol and captan prevent root nodule bacteria from initiating symbiotic nitrogen fixing process.
- Certain fungicides are toxic to soil arthropods. For example, captan at a higher dose reduces the population of the useful soil invertebrates like springtails, collembola, earthworm and millipedes.
- Mercurial fungicides, e.g., phenyl mercury cyanide used in seed treatment, cause reduction of the seed eating birds like bobwhite quaal and pigeons. These insecticides also impair the egg hatching in pheasants resulting in reduction of population of such species.
- Certain fungicides e.g. captan inhibit the growth of fresh water algae even in very low concentrations.
- Continued use of fungicides makes the pests (that are desired to be eliminated) resistant to them.
- Mercurial fungicides are also responsible for human poisoning and deaths. This happens when ingestion of flour and wheat seeds treated with organo-mercury fungicides leads to mercury poisoning. Such poisonings have been recorded in Iraq in 1956, 1960 and 1972, in Guatemala in 1965 and in Pakistan in 1969.

Redressal of the Effects of Fungicides

The environmental effects of pesticides can be minimised by using them selectively in low concentrations. The problem of resistance of the pests can be taken care of by the development of new fungicides, which should be more effective, less toxic and more environmental friendly. One such fungicide is a derivative of β-methoxyacrylic acid.

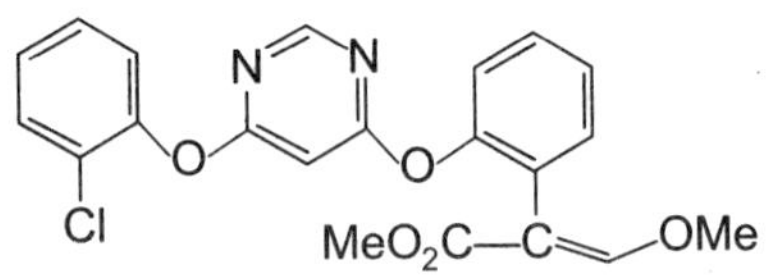

It is likely to be very useful anti-fungal agent, since it is effective against a wide range of fungal pathogens.

4.4.2.6.3 Herbicides

Herbicides are chemicals that kill unwanted weeds or interfere with their growth. Like fungicides, herbicides are instrumental for increasing the yields of disease free crops (inspite of their environmental effects).

Weeds are unwanted plants which grow at places where they are not required, eg. in agriculture crops and deplete the nutrients which are meant for the betterment of the crops. Herbicides basically control the growth of weeds. The main uses of herbicides are as follows:

- To remove weeds from agricultural fields; the weeds retard the growth of the main crops.
- To remove unwanted growth from the lawns, gardens, orchards, pastures etc.

Herbicides are of following two types:

i. Contact Herbicides

These herbicides are quick acting and kill the weeds by direct contact. Some examples include the following.

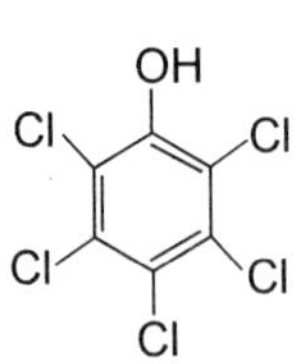

Pentachlorophenol

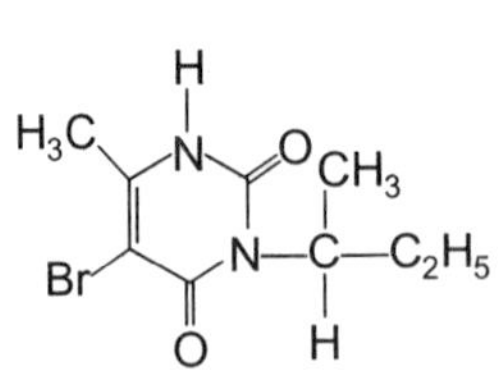

Bromacil

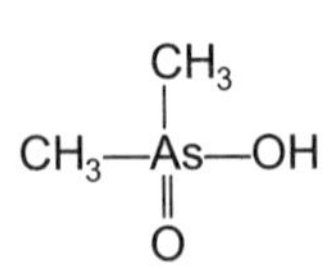

Cacodylic acid

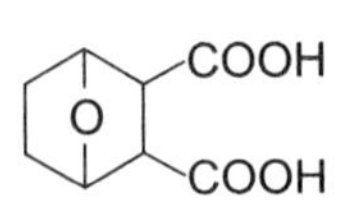

Endothal

Contact herbicides are ineffective against those weeds whose roots produce new shoots at a fast rate.

ii. Systemic Herbicides

These diffuse within the plant's body and are very useful. Examples include the following:

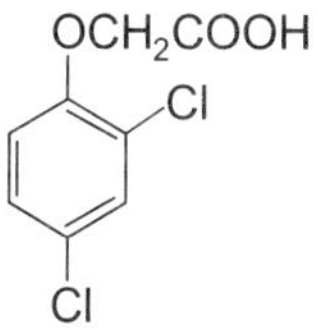
2, 3-D

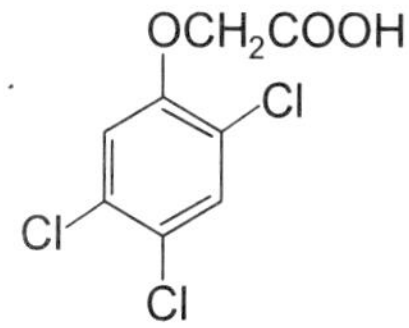
2,4,5-T

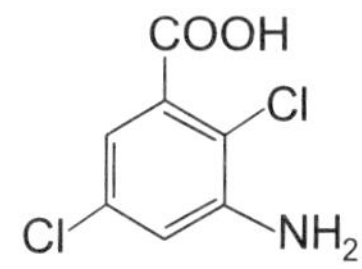
Chloramben

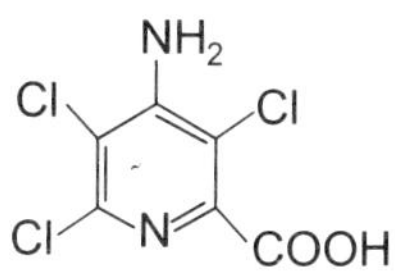
Pictoram

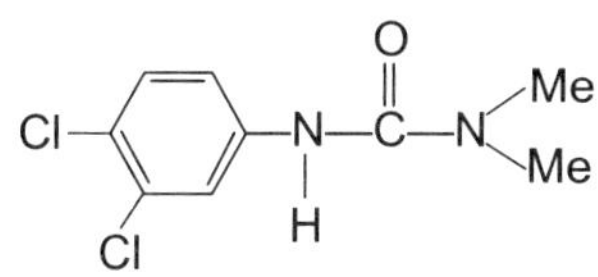
Diuron

Systematic herbicides are effective at low concentrations.

Herbicides act as growth inhibitors for the weeds. For example, the contact herbicide, pentachlorophenol acts as mitotic poison. Some herbicides (eg. diuron) act as a photosynthetic inhibitor. Finally some herbicides (e.g. pictoram) act as a respiratory inhibitor.

Effects of Herbicides on Environment

Like fungicides, herbicides are extremely helpful for agricultural products, but they do have following effects:

- Residues of herbicides in soil increase fungi count resulting in fungal diseases.
- Herbicides like 2,4-D reduce the population of springtals and collembola.
- Herbicides eg. 2.4-D and 2,4,5-T when sprayed on the trees etc. are harmful to millards, quails and partridges.
- Herbicides are hazardous to live stock which eat weeds exposed to sublethal doses of 2,4-D and 2,4,5-T.
- The production of freshwater algae is reduced due to 2,4-D and 2,4,5-T.
- Herbicides are extremely harmful to the health and survival of human race. On being exposed to 2,4-D and 2,4,5-T which are used to defoliate forest areas (as in case of Vietnam war) produce birth defects eg, children being born with deformed or missing limbs etc. This is attributed to damage of male sperm cells.

The weeds developed resistance to the conventional herbicides by their continued application. The only way to solve this problem is to develop new and novel herbicides. One such herbicide is δ-aminolevulinic acid. It uses solar energy to produce toxic components that kill the weeds. On exposure to sunlight, δ-aminolevulinic acid in converted into magnesium tetrapyrrole, which is a precursor to chlorophyll biosynthesis.

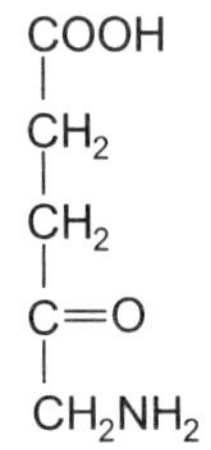

δ-aminolevulinic acid

The magnesium tetrapyrrole absorbs light energy and undergoes photochemical reaction producing singlet oxygen (a highly reactive form of oxygen). The liberated free radical kills the weeds. The left over δ-aminolevulinic acid or magnesium–tetrapyrrole is biodegradable and so does not cause any problems.

4.5 Effects of Air Pollution

The effects of pollutants (in the air or atmosphere), *viz.* carbon monoxide (4.3.1.1), carbon dioxide (4.3.1.2), oxides of nitrogen (4.3.2), oxides of sulphur (4.3.3), hydrogen sulphide (4.3.4), chlorine (4.3.5), ozone (4.3.6), hydrocarbons (4.3.7) and particulates (4.3.8) have already been discussed in the sections given in the parenthesis. In addition to the above, various pollutants are responsible for the following.

- Acid Rain
- Green House Effect
- Global Warming
- Depletion of Ozone
- Formation of Smog

4.5.1 Acid Rain

The formation of nitric acid and sulphuric acid as secondary pollutants (sec. 4.2.3) in the atmosphere leads to acid rain. It includes all types of rain and snow whose excessive acidity causes environmental problems like destruction of vegetation and marine life and the corrosion and etching of buildings that are exposed to the atmosphere.

Chemically speaking, all rain is acidic with or without air pollution. This is due to the natural presence of carbon dioxide in the atmosphere which dissolves in rain drops or rain water (even moisture present in the atmosphere does the same function) to form carbonic acid.

$$CO_2 + H_2O \longrightarrow H_2CO_3$$

Due to the above reaction, carbon dioxide can dissolve in water until the solution becomes saturated. This results in the rainwater attaining an acidic pH of 5.6. Due to this, the purest form of rain (in the absence of any of the air pollutants) reaches the earth as an acidic solution of pH 5.6. In view of the above, the acid rain is described as rain whose pH is lower than 5.6.

How is acid rain caused?

(i) **Due to the presence of oxides of nitrogen in the atmosphere** During thunderstorm and lightening, atmospheric nitrogen and oxygen combine to form nitric oxide, which gets converted to nitrogen dioxide, which reacts with water to form nitric acid and is washed down as acid rain (see also sec. 4.3.2).

$$N_2 + O_2 \xrightarrow[\text{discharge}]{\text{Electric}} 2\,NO$$

$$2\,NO + O_2 \longrightarrow 2\,NO_2$$

$$3\,NO_2 + H_2O \longrightarrow 2\,HNO_3 + NO$$

The oxides of nitrogen enter into the atmosphere from other sources also (sec. 4.3.2).

(ii) **Due to the presence of sulphur dioxide in the atmosphere** Sulphur dioxide is released into the atmosphere from a number of sources (see sec. 4.3.3.). In the atmosphere, sulphur dioxide reacts with atmospheric oxygen to form sulphur trioxide by photolytic and catalytic oxidation processes. The sulphur trioxide thus formed, reacts with water forming sulphuric acid which comes down on earth as acid rain.

$$SO_2 + \tfrac{1}{2}O_2 \xrightarrow{h\upsilon} SO_3$$

$$SO_3 + H_2O \longrightarrow H_2SO_4$$

(iii) The presence of hydrogen sulphide (sec. 4.3.4) and chlorine (sec. 4.3.5) in the atmosphere also contributes to the formation of acid rain.

The acid rain is not confined to the regions where the harmful pollutants (like oxides of nitrogen and sulphur) are emitted. Since winds carry atmospheric contaminants across national borders, the acid rain may fall hundreds of miles away from the pollution sources. Acid rain, in fact, has become an international issue. It is the responsibility of all countries to make sure that the pollutants are not discharged into the atmosphere; these should be checked at the initial source only.

Harmful Effects of Acid Rain

Acid rain causes extensive damage to the environment by affecting the following:

i. Vegetation

The acid rain makes the soil acidic. This adversely affects the plants and animals. The acid rain contains H^+, SO_4^{2-} and NO_3^- ions which when added to soil, leach the nutrients from the soil. The acid rain also damages the leaves of plants and trees. This is responsible for damaging forests and other vegetation.

ii. Fertility of Soil

The activity of symbiotic nitrogen fixing bacteria, present in the nodules of leguminous family, is inhibited. This is responsible for destroying or reducing the fertility of the soil.

iii. Aquatic Life

Acid rain renders the river or even ocean waters acidic, thereby, adversely affecting marine animals. Changes in pH of fresh water affect the reproduction and survival of many species of fish.

iv. Building and Monuments

Acid rain causes extensive damage to buildings and monuments made from marble, limestone, slate and mortar. Limestone which is very common building material, is attacked readily.

$$CaCO_3 + H_2SO_4 \longrightarrow CaSO_4 + H_2O + CO_2$$

Most of the monuments are made from marble which is mostly $CaCO_3$ and so is attacked by acid rain.

The damage caused to the buildings and monuments by acid rain is irreversible. A most glaring example is slow degradation of Taj Mahal in Agra (UP, India). The pollutants in and near Agra arise from iron foundries, rubber manufacture, brick kilns, and oil refinery at Mathura.

v. Ecological Balance

Acid rain is responsible for wiping out many bacterial and blue green algae; thereby disrupting the whole ecological balance.

vi. Human Health

Acid rain has been found to be very dangerous to human health. Acidic conditions can play havoc within human nervous system, respiratory system and digestive system by making the person an easy prey to various neurological diseases.

It is of utmost importance to control the acid rain. Short-term control of acid deposition (soil) can be achieved by using lime.

4.5.2 Green House Effect

It is the phenomenon in which the earth's atmosphere traps the heat form the sun and prevents it from escaping into outer space. Without this effect, the average surface temperature of the earth would be about 33^0 C lower than it is today. The green house effect received its name because the earth's atmosphere acts much like the glass or plastic roof and walls of a greenhouse. Sunlight enters a greenhouse through the transparent glass or plastic and heat the plants but the heat emitted by the plants in the form of infrared radiation cannot pass through the glass or plastic roof and walls. As a result, the temperature inside increases.

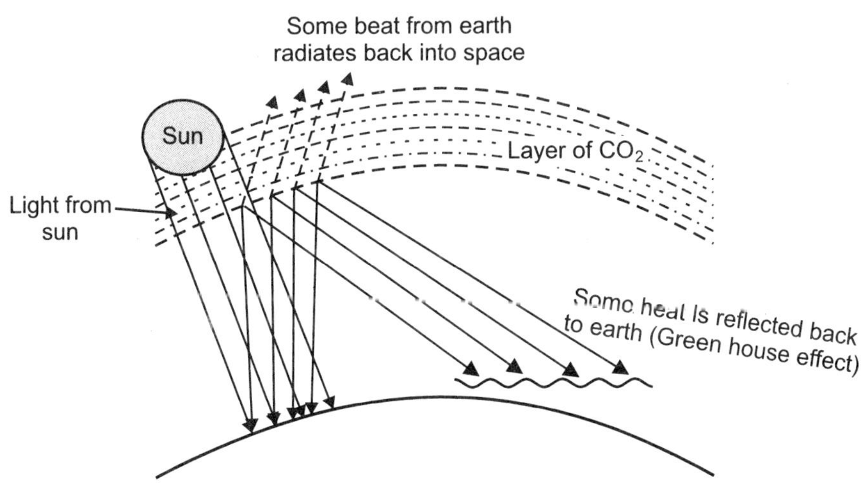

Fig. 4.1: Green House Effect

In the atmosphere, there is a protective layer of O_3 gas at a height between 15-60 km (the thickest layer of ozone exists at a height of 23 km from the surface of the earth and a blanket of CO_2 gas exists in the lower part of the atmosphere (i.e. below 15 km). When sunlight consisting of ultra-violet rays, visible light and infra-red, fall on the top of the atmosphere, the harmful ultraviolet rays are absorbed by the O_3 layer and so these do not reach the earth's surface. However, the visible light and infra-red rays pass though the CO_2 layer and fall on the earth. The result is that the earth gets heated due to heating effect of the infra-red radiations. As the earth becomes hot, it starts emitting heat (infra-red rays) back to the atmosphere.

The emitted infrared rays are absorbed by the CO_2 layer in the atmosphere. In this way, CO_2 in the atmosphere does not allow the infra-red radiation reflected by the earth's surface to go out of the atmosphere. We can say that the layer of CO_2 gas in the atmosphere traps all the infra-red radiations (heat rays) coming from the earth's surface. These trapped infra-red rays heat the earth's atmosphere. The heating up of the earth due to the trapping of infra-red radiations (reflected from the earth's surface) by CO_2 layer in the atmosphere is called **'green-house effect'**. Thus, the temperature of the earth is raised. The rise in the temperature produced by the 'green-house effect' in the earth's atmosphere depends on the amount (proportion) of CO_2 gas in the atmosphere.

Besides CO_2, which is known as the green house gas, there are some other green houses gases like methane (CH_4), chlorofluorocarbons and nitrous oxide which have the same effect. We, thus, see that the presence of excess of CO_2 in the atmosphere brings about climate changes, see global warming given below.

4.5.3 Global Warming

An increase in the average temperature of the earth's atmosphere due to green house effect can have far reaching effects on the climate and consequently on the key life support systems of the planet. Since the last century, human activities, mainly burning of fossil fuel, has raised the atmospheric concentration of green house gases, which may intensify the green house effect and result in an increase in the average global temperature. Since global climate including rainfall, storms, wind patterns, ocean currents and sea level is intimately related to global heat flows and temperature patterns, there is a general expectation that the earth's climate may be significantly modified in the next 50 years. This would, in turn, alter the earth's delicate ecological balance.

Between the years 1878-1990, the temperature has increased from 0.45°C to 0.15°C. During 1990-2025, it may be further rise by 1°C and between 2025-2100 there may be rise of 3°C.

Besides what has been stated above, the rise in temperature will melt all the glaciers (snow-mountains), flooding the low-lying areas of the earth. An increase in global temperature is also likely to increase the incidence of infectious diseases such as malaria, dengue, sleeping sickness and yellow fever.

Table 4.3 gives some major greenhouse gases and their sources.

Table 4.3 Sources of major greenhouse gases

Gas	Sources
Carbon dioxide, CO_2	1. Fossil fuel combustion (coal, oil, natural gas) 2. Deforestation and land use changes
Methane, CH_4	1. Enteric fermentation in cattle and insects 2. Biomass burning and garbage land fills 3. Coal mines and natural gas leaks 4. Rice paddies 5. Swamps and tundra
Chlorofluorocarbons	1. Aerosols (spray propellants) CFCs 2. Refrigeration and air-conditioning 3. Plastic foams 4. Industrial solvents 5. Sterilants for medical supplies
Nitrous Oxide, N_2O	1. Fertilizer use 2. Fossil fuel combustion 3. Biomass burning 4. Changing land use

4.5.4 Depletion of Ozone (Ozone Hole)

The ozone layer protects life on earth by absorbing the high energy ultraviolet radiations from the sun reaching the earth. These UV radiations can cause skin cancer and also damage the crop plants. In 1987, the thinning of the ozone layer was first observed over the Antarctics, which is known as **Ozone hole**. The thinning or the destruction of the ozone layer or the depletion of the ozone layer may occur due to the following:

(i) The oxides of nitrogen present in the atmosphere decompose ozone into oxygen:

$$NO + O_3 \longrightarrow NO_2 + O_2$$

$$NO_2 + O \longrightarrow NO + O_2$$

We, thus, see that the oxides of nitrogen (which are formed during thunderstorm and lightening by combination of atmospheric nitrogen and oxygen) destroy or deplete the ozone layer. It is believed that the above process destroys more than 60% of ozone in the atmosphere.

Various nuclear tests conducted in various parts of the world are also responsible for the destruction of ozone layer. Since nuclear tests produce such a high temperature that the atmospheric nitrogen and oxygen combine forming nitric oxide, which as stated above, is responsible for destroying ozone.

(ii) Depletion of ozone layer by chlorofluorocarbons (CFC's) - As the name indicates, chlorofluorocarbons contain carbon, chlorine and fluorine. The CFC's are non-flammable and non-toxic. They are used as refrigerants (coolants for refrigerators and air conditioners), for making polystyrene, for fire-fighting (special types of CFC's, known as halons are used as fire extinguishers; halons are bromine containing derivatives of chlorofluorocarbons), as propellant for aerosol sprays and as blowing agents for foam plastics.

Though unreactive, the CFC's are a major factor for ozone depletion. Being inert, CFC's do not easily degrade in the troposphere and so they find their way into the stratosphere which contains the ozone layer. In stratosphere, the CFC's are photo-chemically dissociated into reactive chlorine atoms. The latter catalyse the depletion of stratospheric ozone as shown below:

$$Cl^{\bullet} + O_3 \longrightarrow ClO^{\bullet} + O_2$$
$$ClO^{\bullet} + ClO^{\bullet} \longrightarrow (ClO)_2$$
$$(ClO)_2 + h\upsilon \longrightarrow Cl^{\bullet} + ClOO^{\bullet}$$
$$ClOO^{\bullet} \longrightarrow Cl^{\bullet} + O_2$$
$$Cl^{\bullet} + O_3 \longrightarrow ClO^{\bullet} + O_2$$

The net reaction is $2O_3 + h\upsilon \longrightarrow 3O_2$

It is estimated that for every reactive chlorine atom generated in the stratosphere 1,00,000 molecules of ozone are depleted. The depletion of ozone by chlorofluorocarbons can be schematically represented below in Fig. 4.2.

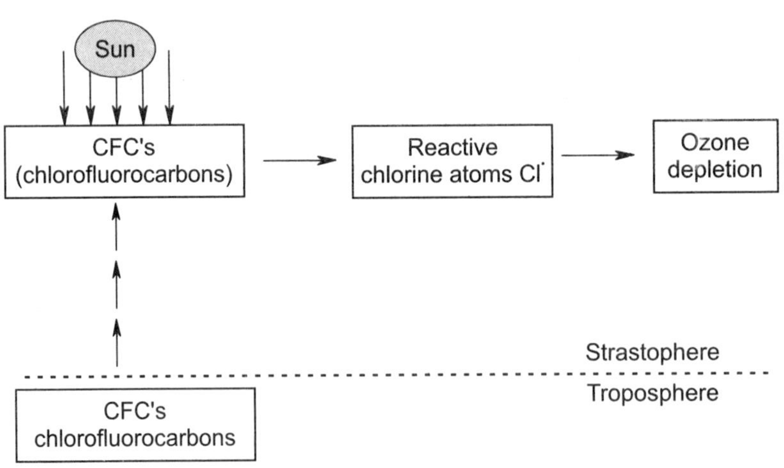

Fig. 4.2: Ozone Depletion by CFC's

Concerned about the ozone hole (which permits harmful UV radiations to reach the earth), a protocol was signed at Montreal, Canada in 1987 in which representative of a number of developed countries reached an agreement to limit the production of CFC's. The target set was to reduce the use of these harmful CFC's by 50% by 1998. It was found that the ozone depletion caused by CFC's was much more alarming. In view of this, the Montreal Convention was followed by a conference held in London in 1990 in which many more countries participated. The aim of the Montreal Protocol was affirmed and it was decided to speed up the phasing out of CFC's.

Prevention or Minimisation of Ozone Depletion

As already discussed ozone depletion is effected by oxides of nitrogen and CFC's. The oxides of nitrogen find their way into the atmosphere by the combination of atmospheric nitrogen and oxygen particularly during thunderstorm and lightening. There is virtually no way to prevent this mode of ozone depletion. However, using alternative substitutes can control the depletion of ozone by CFC's.

Due to inert nature and enormous stability of CFC's, they stay in the atmosphere for very long time. For example, the two most commonly used CFC's viz. CFC-11 (CCl_3F) and CFC-12 (CCl_2F_2) stay in the atmosphere for 75 and 110 years, respectively. This implies that even if the use of CFC's is stopped, the already discharged CFC's in the atmosphere will continue to deplete the ozone for a considerable length of time. In view of this, it is more or less essential to phase out the use of CFC's and their substitutes must be discovered. The substitutes should have zero potential of ozone depletion and their green house potential should also be very low. It has already been stated that CFC's are the constituents of green house gases and contribute to global warming.

The following CFC's are most likely to replace the usual CFC's. These are HCFC-123 ($CHCl_2CF_3$) and HFC-134 a (CH_2FCF_3). These have very low ozone depletion potential and their green house potentials are also quite low. However, their degradation products are toxic in nature.

It should be mentioned at this stage that certain related chemicals like carbon tetrachloride (CCl_4) and trichloroethane (CH_3CCl_3) also have ozone depleting potential. Use of these chemicals also has to be phased out. At present, there are no substitutes for these (trichloroethane is used as a solvent and carbon tetrachloride is used as a raw material for the manufacture of CFC's and pesticides and as a fire extinguisher).

Checking the Depletion of Ozone

It has been reported (Science, 1991, 254, 1191-1194) that the depletion of ozone could be prevented or at least minimized by injecting alkanes (e.g. ethane, propane etc.) in the stratosphere (which mainly contains ozone). The alkanes will react with ozone distroying chlorine atoms (from CFC's) and immobilize them.

4.5.5 Smog

Smog is a mixture of smoke and fog. The name smog is derived from combination of smoke and fog.

$$\text{Smoke} + \text{fog} \longrightarrow \text{smog}$$

Smog is of two types:

i. Classical smog

ii. Photochemical smog

i. Classical Smog

It is commonly known simply as smog and is produced from gases like SO_2 and NO_2 present in air (produced by the burning of fossil fuels like coal and petrol or gasoline in homes, industries and automobiles). These gases get mixed with moisture in the tiny droplets of fog producing sulphuric acid and nitric acid, respectively.

$$SO_2 + \frac{1}{2}O_2 \longrightarrow SO_3$$

$$SO_3 + H_2O \longrightarrow H_2SO_3$$

$$H_2SO_3 + \frac{1}{2}O_2 \longrightarrow H_2SO_4$$

$$2\,NO_2 + H_2O \longrightarrow HNO_2 + HNO_3$$

$$HNO_2 + \frac{1}{2}O \longrightarrow HNO_3$$

The resultant product mixture consisting of sulphuric acid and nitric acid get condensed on the solid particulate matter of smoke or dust present in the air to give colloidal dispersion called smog or classical smog.

Smog is usually formed during winter season and is present in the lower atmosphere. It decreases visibility and affects the human health. It is responsible for breathing problems like bronchitis and asthma and may contribute to heart problems. It produces irritation to eyes, nose and throat. It also reduces visibility and creates problems in road and air traffic.

It is well known that in December 1952, a dense cloud of smog was formed over London city and remained for about 5 days and resulted in death of 4000-5000 people, and created health problems to numerous other people. The cause of death was pneumonia, bronchitis and other respiratory problems. This smog, commonly called **London Smog** formed after the introduction of coal as a fuel, which produced both sulphur dioxide and smoke. The London smog is also called **Reducing Smog**.

ii. Photochemical Smog

This type of smog is formed by the combination of smoke, dust and fog with air pollutants particularly oxides of nitrogen and hydrocarbons. These pollutants (*viz.* hydrocarbons and oxides of nitrogen) originate form automobile exhaust, though they are also released in the atmosphere by other industrial activities. Some hydrocarbons like isoprene and α-pinene are also emitted from trees.

The formation of photochemical smog can be explained in the following way (see Chart 1). The various steps involved are:

(a) Hydrocarbons (which are reactive containing a C=C group) from automobiles interact with ozone to form a hydrocarbon free radical. ($RCH_2^\bullet$). These hydrocarbon free radicals may also be formed by the reaction of hydrocarbons with atomic oxygen [evolved from NO_2 by UV radiation form the sun, NO_2 + UV radiation $\rightarrow$ NO + O].

(b) The free radical, $RCH_2^\bullet$, reacts rapidly with oxygen to form another free radical, $RCH_2^\bullet O_2$.

(c) The second free radical $RCH_2O^\bullet$, in turn, reacts with nitric oxide present in the atmosphere (either from N and O in presence of sun light or by irradiation of NO_2) to produce the third free radical $RCH_2O^\bullet$ (the NO is converted into NO_2).

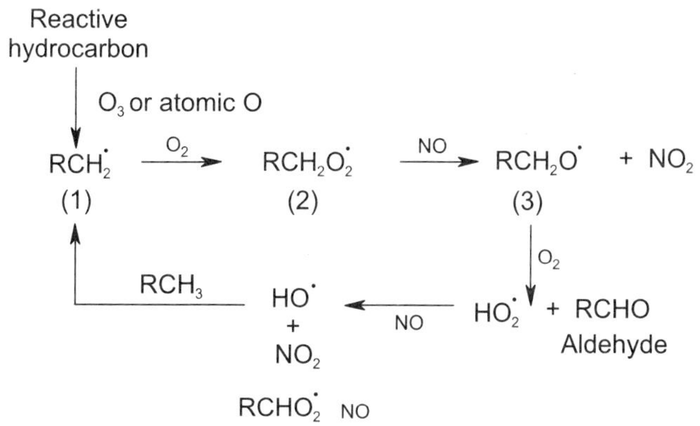

Chart 1. Steps involved in the formation of photochemical smog

(d) The third free radical ($RCH_2O^\bullet$) subsequently reacts with oxygen yielding an aldehyde (RCHO) and generating a hydroperoxyl radical ($HO_2^\bullet$).

(e) The hydroperoxyl radical ($HO_2^\bullet$) reacts with nitric oxide to give NO_2 and generates hydroxyl radical ($HO^\bullet$).

(f) The hydroxyl radical ($HO^\bullet$) being extremely reactive reacts with a hydrocarbon (RCH_3) regenerating the hydrocarbon free radical ($RCH_2^\bullet$).

The above cycle is repeated a number of times leading to rapid build up of photochemical smog products.

(g) The aldehyde (RCHO) formed in step (d) reacts with hydroxyl radical ($HO^\bullet$) producing an acyl radical ($RCO^\bullet$), which in turn reacts with oxygen to give peroxyacyl radical ($RCOO_2^\bullet$). The formed peroxyacyl radical finally reacts with nitrogen dioxide (NO_2) to generate peroxyacetylnitrate (PAN). The steps involved in the last sequence of reactions from the aldehydes are shown in Chart 2.

$$\underset{\text{Aldehyde}}{RCHO} \xrightarrow{HO^\bullet} \underset{\text{Acyl radical}}{RCO^\bullet} \xrightarrow{O_2} \underset{\text{Peroxy acyl radical}}{RCOO_2^\bullet} \xrightarrow{NO_2} \underset{\text{Peroxyacetylnitrate (PAN)}}{R{-}COO_2NO_2}$$

Chart 2. The conversion of aldehydes to PAN

It is the peroxyacetylnitrate (PAN), a constituent of smoke, which produces irritation in the eyes. The components of smog, e.g., ozone and PAN, affect the respiratory tract of human beings. It also causes nose and throat irritation and can lead to several problems (disease) of the eyes, lungs and heart. The photochemical smog also affects the growth of plants and causes damage to the vegetation.

Originally, the photochemical smog was observed at Los Angeles and was called **Los Angeles Smog**. Due to this visibility was drastically dropped and the residents complained of eye irritation. Subsequently, it was found that this type of smog is not typical of Los Angeles and was found to be formed in other regions also.

The formation of photochemical smog can be suppressed

(i) By preventing the release of oxides of nitrogen and hydrocarbons in the atmosphere by motor vehicles by using catalytic converters.

(ii) Some objective can be achieved by using a better quality of oil so that there is least or no emission of the oxides of nitrogen and hydrocarbon. This is possible by the use of so called **Reformed gasoline**, in which about 10% of the aromatic hydrocarbons are replaced by fuel oxygenates, such as methyl tertiary butyl ether. The additives make the fuel burn more completely and efficiently. With this, the emission of volatile organic hydrocarbons is suppressed. Also, the gasoline containing oxygenates burns at a lower temperature resulting in the reduction of the emission of nitrogen oxides.

(iii) With the advancement of scientific knowledge, it is now possible to fabricate automobile engines so that there are no unburnt hydrocarbons left. This technology has been employed in most of the advanced countries.

4.6 Control of Air Pollution

The air pollutants pose a great danger to one and all. The harmful effects of various pollutants *viz.*, carbon monoxide (sec. 4.3.1.1), oxides of nitrogen (sec. 4.3.2), oxides of sulphur (sec. 4.3.3), hydrogen sulphide (sec. 4.3.4), chlorine (sec. 4.3.5), ozone (sec. 4.3.6), hydrocarbons (sec. 4.3.7) and particulates (sec. 4.3.8) have been discussed at appropriate places mentioned above. Air pollution can best be described as a **'TIME BOMB'** which may explode any time, at any place causing severe, undetermined problems. This is amply clear from the various air pollution episodes (accidents due to natural or man made causes) which have occurred since very early times. Some of the air pollution episodes are given below.

(i) The first episode of severe air-pollution was recorded in 1930 in **Meuse River Valley of Belgium**. Due to thermal inversion, SO_2 gas was trapped at 38 ppm level. Due to this episode, about 60 people died and several cattle were killed.

(ii) In 1994, Los Angeles in California experienced a serious type of air pollution caused by **photochemical smog**. It was characterised by reduced visibility, eye irritation and plant damages (for details see sec. 4.5.5).

(iii) In 1948, in **Donora**, Pennsylvania, USA 40% of the population was affected causing 20 deaths. The reason was poisoning by SO_2 at 2 ppm level.

(iv) In 1952, December 5-9, heavy fog conditions prevailed for 5 days at a stretch in London and the toll was 4000-5000 people. Thousands of people were affected. The peak SO_2 concentration was 1.3 ppm and smoke 4 mg m^{-3}. Similar smog conditions recurred in December 1962 though the deaths were around 700.

(v) In November 1950, the **Paza Rica** air pollution episode resulted from an industrial accident of a refinery in which about 25 people died and many more became ill. The reason was the release of huge amount of H_2S

from the refinery which spread over the ground around the refinery and seeped into nearby homes.

(vi) In July 1976, an explosion took place in a chemical plant (at **Seveso, Italy**) manufacturing a herbicide (*viz.* 2,4,5-trichlorophenoxy acetate). A white cloud of TCDC (2,3,7,8-tetrachlorobenzo-10-dioxin) came out contaminating buildings, ground and soil. After three weeks, the Italian Government evacuated about 800 people from the affected area. About 200 people suffered from skin diseases and also from liver ailments. The after effect was birth of deformed and premature babies.

(vii) On December 3, 1984, an accident occurs in a Union Carbide factory manufacturing a carbamate pesticides located at Bhopal, Madhya Pradesh, India. The reactions involved in the manufacture of the carbamate are given below:

$$\underset{\text{Methyl amine}}{CH_3NH_2} + \underset{\text{Phosgene}}{COCl_2} \longrightarrow \underset{\text{Methyl isocyanate (MIC)}}{CH_3{-}N{=}C{=}O} + 2HCl$$

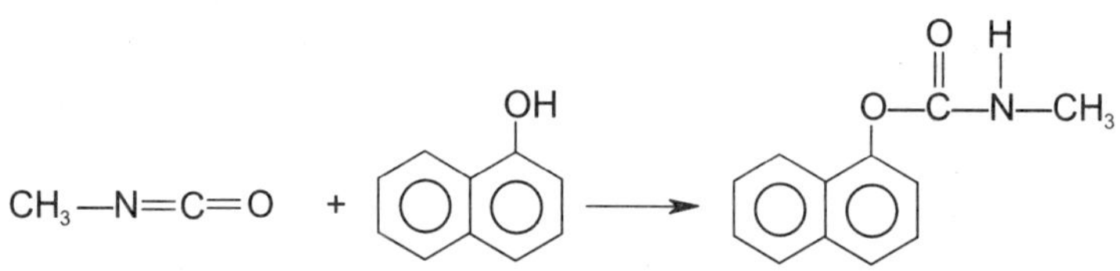

MIC α-naphthol 1-naphthyl-N-methyl carbamide (carbaryl)

The starting material methyl isocyanate (CH_3-N=C=O) (MIC), was stored at low temperature in storage tanks. On December 3, 1984, a massive leak of MIC occurred in the storage tank and caused havoc. About 22,000 people residing in the area died and a very large number (about 1,20,000) were disabled for rest of their lives.

This incident is now commonly known as the Bhopal Disaster, and Bhopal where the episode occurred is known as the **city of death**. Also see sec. 15.2.

(viii) On 26 April 1986, a worst nuclear catastrophe occurred at Chernobyl in the Ukraine area of the Soviet Union. It is believed to be due to human error in controlling rods.

The accident resulted in the reactor core temperature more than 2000^{0}C. Fuel and radioactive remains shot into air hitting the surrounding areas. The effect was felt over most of Europe. The accident resulted in the death of more than 2000 people. Soil, water and vegetation over 100 sq km area around Chernobyl was damaged. The neighbouring countries such as Poland were severely affected. Even after cleaning up operation and dumping tons of sand, clay etc. to bury the molten core, the radiation continuously emitted – more than 3,00,000 times the permissible dose.

(ix) A nuclear reactor in Three Mile Island, in USA made news due to an accident that took place due to a complex operational failure due to human error.

In view of the disastrous effects of atmospheric pollution, it is most important to make sure that suitable control measures should be taken to prevent the atmospheric pollution. However, in case of nuclear catastrophe, adequate precautions and steps should be taken so that such catastrophes do not occur again. Given below are some control measures for various pollutants.

A. Oxides of Sulphur

The presence of sulphur dioxide in the atmosphere is very hazardous. It is very necessary to check its emission into the atmosphere. The possible approaches are the desulphurisation of the fuel (coal is the major source of SO_2) and removal of SO_2 from flue gases.

i. Desulphurisation of Coal

Coal contains sulphur either in inorganic form (as iron pyrites) or inorganically bound form, in approximately equal ratio. The inorganic sulphur is removed by subjecting grounded coal to hydraulic washing. Iron pyrites, being heavier settle down while the sulphur free coal floats on the surface and is collected. The organically bound sulphur is removed by passing hydrogen gas over the fuel in presence of catalyst (mixed cobalt molybdenum oxide). By this process, the organically bound sulphur is converted into hydrogen sulphide, which is absorbed in a solution of diethanolamine, subsequently H_2S is oxidised to sulphur.

$$H_2S + [O] \longrightarrow H_2O + S$$

The organic sulphur can also be removed from coal by a microbial method using a strain designated as IGT-57. However, this method does not remove inorganic sulphur and is too slow to be of utility.

ii. Desulphurisation of Flue Gases

In this procedure, powdered coal and powdered lime stone are heated, when the following reactions take place.

$$\underset{\text{Limestone}}{CaCO_3} \xrightarrow{\Delta} CaO + CO_2$$

$$CaO + SO_2 \longrightarrow CaSO_3$$

$$CaO + SO_2 + O \longrightarrow CaSO_4$$

The formed calcium sulphite and calcium sulphate are removed by scrubbing with a slurry of lime water [obtained by hydrolysis of lime, $CaO + H_2O \longrightarrow Ca(OH)_2$].

The method is economical but the disposal of waste $CaSO_4$ in large amount creates problems.

Alternatively, SO_2 in flue gases is reduced by mixing it with carbon monoxide (from another polluting unit) over a catalyst.

$$2\,CO + SO_2 \xrightarrow{\text{Catalyst}} 2\,CO_2 + S$$

A number of catalysts have been used. The disadvantages of the catalytic method are the cost and deactivation of the catalyst.

Alternatively, SO_2 in the exhaust gases is removed by using scrubbers. The scrubbers force the gases through a slurry of calcium carbonate, or magnesium hydroxide. The SO_2 dissolves in water and reacts with $CaCO_3$ or $Mg(OH)_2$ to form a salt. If $Mg(OH)_2$ is used, the $MgSO_3$ that is formed which can be isolated and heated to regenerate SO_2, which can be collected and used as a raw material for other industrial processes.

$$SO_2 + Mg(OH)_2 \longrightarrow MgSO_3 + H_2O \text{ (removal of } SO_2\text{)}$$

$$MgSO_3 \xrightarrow{\Delta} SO_2 + MgO \text{ (regeneration of } SO_2\text{)}$$

Fig.4.3 below gives a schematic diagram of a scrubber.

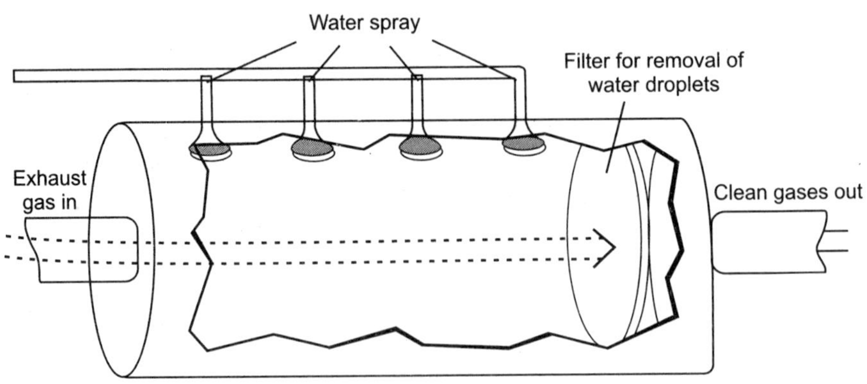

Fig. 4.3: A scrubber (schematic diagram)

In both the above methods, the oxides of nitrogen which are invariably present in flue gases are not removed. Both oxides of nitrogen and sulphur can be removed using a single technology as described below.

B. Oxides of Nitrogen

It has already been stated that the important sources of oxides of nitrogen in the atmosphere are fossil fuel based power plants and automobile exhaust - both of which require control procedures.

i. **Checking the emission of oxides of nitrogen (and simultaneous checking emission of oxides of sulphur) from coal.** This is affected by the so called **coal conversion technology.**

The procedure involves heating powdered coal with sodium hydroxide solution under moderate pressure. This removes sulphur along with many trace elements from coal. The coal obtained is free of sulphur and will not produce oxides of sulphur on combustion. The sulphur free coal reacts with steam to produce water gas (a mixture of carbon monoxide and hydrogen);

$$\text{Coal} + \text{steam} \rightarrow CO + H_2$$

The water gas is sprayed with water and burned. The presence of water lowers the combustion temperature and so nitrogen and oxygen do not combine to form oxides of nitrogen.

Alternatively, **fluidised-bed combustion** technique is used. In this method, a bed of ash or sand is placed on a grate at the bottom of a boiler. Air is blown upwards through the grate and simultaneously crushed solid fuel is fed into the boiler. The fuel heats up the particles of ash or sand. The heat is picked up by the coils in which water is fed at the bottom and steam comes out from the top (see Fig.4.4). Subsequently, limestone is added to the mixture in the boiler. Limestone combines with the evolved SO_2 from the fuel.

$$2\,CaCO_3 \xrightarrow{\Delta} 2\,CaO + 2\,CO_2$$

$$2\,CaO + 2\,SO_2 + O_2 \longrightarrow 2\,CaSO_4$$

The system operates at 800-1000^{0}C. At this low temperature, oxides of nitrogen are not formed. So, the flue gases are free from both the oxides of sulphur and nitrogen.

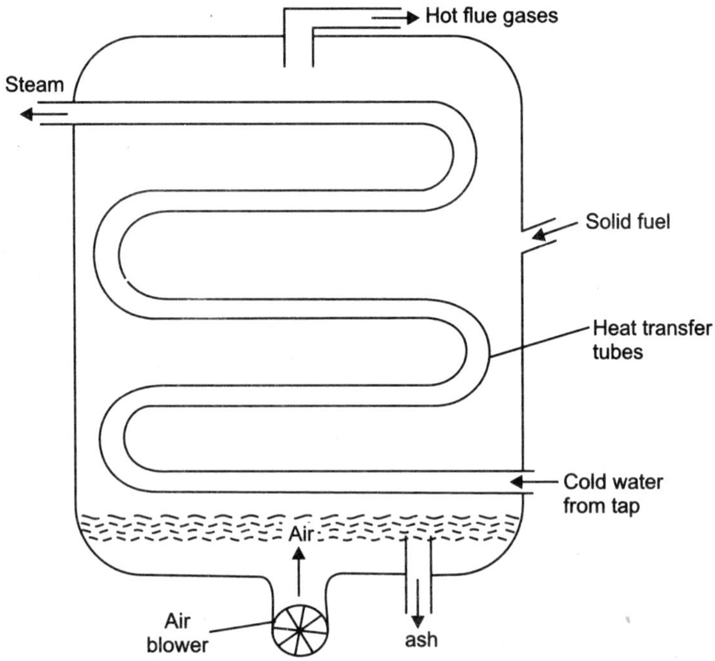

Fig. 4.4: Fluidized bed combustion

ii. Checking the emission of nitrogen (and simultaneous carbon monoxide and unburnt hydrocarbons) from automobile exhaust.

The automobile exhaust has become a major source of atmospheric pollution (oxides of nitrogen and carbon monoxide) since the number of automobiles has increased enormously. The automobiles exhaust pollutants are controlled by installing catalytic converters in the exhaust system, so that the exhaust gases are first passed through a catalytic converter before being released into the atmosphere as harmless gases like CO_2, H_2O and N_2. The only limitation of this procedure is that the gasoline or petrol should be unleaded since lead poisons the catalyst. It may be appropriate to state that anti-knocking compounds like lead tetraethyl are now not added to the petrol. Instead, the composition of the petrol is so adjusted (by the addition of some branched chain hydrocarbons) so as to have the desired octane rating. The catalytic converter contains a mixture of platinum, palladium and rhodium supported on a honeycomb bed (made of ceramic or metal). Various reactions taking place in the catalytic converter are:

$$2\,CO + O_2 \xrightarrow{\text{Catalyst}} 2\,CO_2 \qquad \text{(i)}$$

$$HC + O_2 \xrightarrow{\text{Catalyst}} CO_2 + H_2O \quad \text{(ii)}$$
Hydrocarbons

$$2\,NO + 2CO \xrightarrow{\text{Catalyst}} N_2 + 2\,CO_2 \quad \text{(iii)}$$

$$2\,NO + CO \xrightarrow{\text{Catalyst}} N_2O + CO_2 \quad \text{(iv)}$$

$$2\,N_2O \xrightarrow{\text{Catalyst}} 2\,N_2 + O_2 \quad \text{(v)}$$

$$NO + HC \xrightarrow{\text{Catalyst}} N_2 + CO_2 + H_2O \quad \text{(vi)}$$

As seen, in the catalytic converter

(a) CO gets converted into CO_2 (eq. (i))

(b) Hydrocarbons are oxidised into CO_2 and H_2O (eq. (ii))

(c) Nitric oxide reacts with CO to give N_2 and CO_2 (eq. (iii))

(d) Nitric oxide may also be partially reduced to N_2O (eq. (iv)) and the formed N_2O decomposes into N_2 and O_2 (eq. (v))

(e) Nitric oxide and hydrocarbons are converted into N_2, CO_2 and H_2O (eq. (vi))

C. Carbon Monoxide

Carbon monoxide is the most harmful pollutant (being poisonous in nature) in the atmosphere. Its main sources are incomplete combustion of all carbon containing fuels particularly in automobile engines and industrial furnaces and cigarette smoking. It is removed from the source in the following way.

(i) From automobile exhaust gases– it is removed (along with oxides of nitrogen and unburnt hydrocarbons) by installing catalytic converter in the exhaust system.

(ii) Accumulation of carbon monoxide in any space (like in submarines and space flights) it is removed by providing catalytic burner to oxidise CO to CO_2.

(iii) Cigarette smoking should be done only in specific area with full ventilation. However, it is not possible to control CO from the cigarette smoke. The best way is to quit smoking.

Control of pollution from polymers and plastics (sec. 4.4.1.A), asbestos (sec. 4.4.1.B) are given at appropriate places.

D. Particulates

The common sources of particulate emission into the atmosphere are the powerhouses operating on coal. The smoke that comes out of the chimneys contains fine particles of carbon which remain suspended in air. These particulates are removed by installation of an electrostatic precipitator in the chimney. By this process when a high electrostatic potential is applied across the chimney, the carbon particles coming out of the chimney are attracted by charged electrodes. This results in the formation of soot which gets deposited in the chimney which is cleaned from time to time to remove soot. The particuıates coming in the exhaust of automobiles are removed by catalytic converter.

A schematic diagram of an electrostatic precipitator is shown in Fig.4.5. As shown, the exhaust gases from power houses and other industrial houses (which used coal or other fuels as a source of energy) are passed between two charged vertical plates or electrodes. The particles pick up electrons supplied by the negative electrodes and move to the positive ore. Thus, the liquid particles (if present as in aerosols, which are small particles of liquid and solids dispersed in smoke) accumulate and flow to the bottom of the collector. The solid particles drop off with agitation.

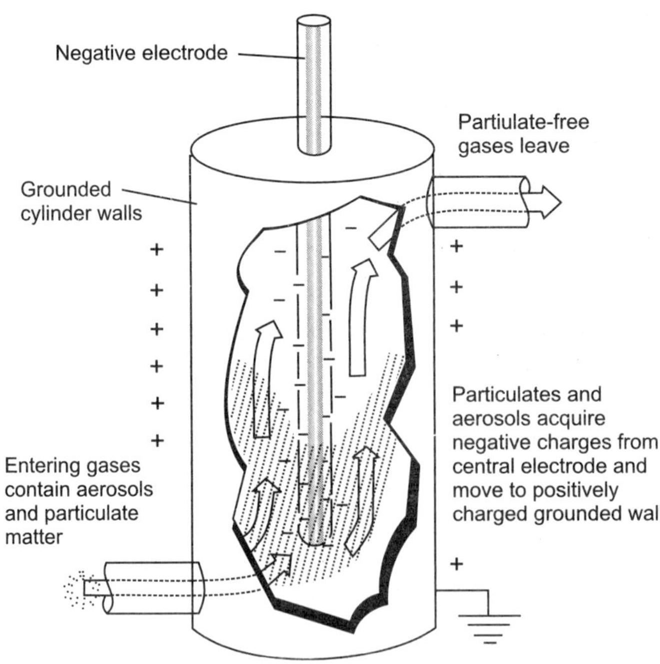

Fig. 4.5: An Electrostatic Precipitator

In some of the industries, the filters (called the industrial filters) are used to remove particulates. These filters work like the filters used in air conditioners or vacuum cleaner filters. The exhaust gases are passed through them, leaving particulates and aerosols behind the filter. A typical industrial filter is shown in the Fig. 4.6.

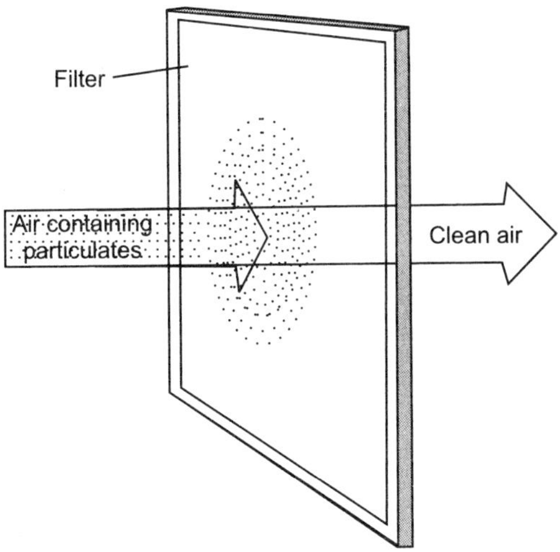

Fig. 4.6: An Industrial Filter

E. Use of Alternative Source of Energy

Following are some of the alternative sources of energy:

(i) Use of hydroelectric power (if feasible) in place of power generation by coal.

(ii) Use of nuclear power. This has its own problems of pollution, disposal of the waste and nuclear accidents.

(iii) Use of alternative energy sources for automobiles. Of these natural gas (CNG) is particularly appealing because of least potential for air pollution. In fact, in Delhi, it is now mandatory for taxis and buses to use CNG.

Electric cars powered by batteries may seem to be completely pollution free alternative. However, their use might only shift environmental concern from one source or location to another. The electricity requirement for charging all the transportation batteries would have to be met by existing or new electric power generators which by itself may be of environmental concern.

Water

CHAPTER 5

5.1 Introduction

Water is the most essential component of life. It is a unique liquid in the sense that without it life is impossible. On a global scale, there is abundance of water. The main problem is the availability of water in the right place at the right time and in the right form. The survival of mankind has always depended on water. It is well known that in earlier times, civilizations flourished on riverbeds. Also, riverbeds were responsible for wiping out the populations due to floods.

5.2 Availability and Sources

Water can be found in liquid, solid or gaseous form at a number of places. Depending on the specific location, the residence time may vary for a few days to thousands of years (Table 5.1). As already mentioned (sec. 2.5.2), more than 99% of Earth's water in its natural state, is not available or is unsuitable for human use. Thus, the amount of water which is available for all people, plants (including agriculture) and animals on the earth is much less than 1% of the total amount.

Table 5.1 Global water supply

Location (source)	Percentage of total water	Approx. average residence time
Oceans	97.2	Infinity
Atmosphere	0.001	9 days
Rivers and Streams	0.0001	2 weeks
Ground water (up to depth of 0.8 km)	0.31	Hundreds to thousands of years
Lakes (fresh water)	0.009	Tens of years
Ice caps and glaciers	2.15	Tens of thousands of years

Source: US Geological Survey.

Water is the only natural resource which is used in tremendous quantities. However, the availability of water at any particular point depends on several factors in the hydrologic cycle including rates of evaporation, precipitation and transportation. With the enormous population explosion and rapid increase in agricultural procedures and industrialization, the demand for water has considerably increased. In view of this, it becomes absolutely essential to look for other sources of water supply. Two such possibilities are the use of ground water and desalination of seawater.

5.2.1 Ground Water

Ground water normally refers to the water below the water table where saturated conditions exist (up to about 800 meters depth). It constitutes about 0.31 percent of the total global water and is expected to be available for thousands of years. In some advanced countries like USA, about half the population depends on ground water for their water requirements. In India and other developing countries, ground water is the main source of water required for human consumption, agriculture and other needs. It is generally believed that ground water is pure and safe to drink. This is not true. In fact, ground water can easily be polluted. In view of this, it is necessary to take appropriate steps so that the ground water does not get polluted and if polluted it should be processed so that the harmful pollutants are removed. This aspect is dealt in Chapter 6.

Too much withdrawal of ground water creates another problem. The level of water in the earth goes down since replenishment of ground water takes a very long time.

5.2.2 Desalination of Sea Water

The oceans, as we know, account for about 97.2 percent of the total global water. However, it is saline in nature and contains dissolved salts to the extent of about 3.5% (A cubic metre of sea water as on average contains about 40 kg of salts). The seawater cannot be used for human consumption or far agricultural purposes. Before seawater can be used, it has to be desalinated. A number of processes are used in different parts of the globe where seawater is desalinated and rendered fit for human consumption. These processes that remove salt require large amount of energy resulting in a very high cost of water.

A process that can be used for desalination of seawater is *reverse osmosis*. In this process, the solute (water in case of sea water) is separated from dissolved salts by making it flow through a semipermeable membrane under an applied pressure in excess of normal osmotic pressure. The dissolved salts are retained in the semipermeable membrane and clear water passes through. This process cannot be used on a large scale due to cost factors.

The most important process of desalination involves the use of solar energy. In fact, such methods, procedures and equipments have been developed. The principle involved is to focus the sun's energy on to a relatively small amount of water in a suitable container. The sun's energy can be focussed with the help of mirrors. The water evaporates or gets converted into steam; the steam is condensed into water. The procedure is basically **distillation**. Distilled water is prepared by this process.

Desalination will always remain a costly process and can be used only when alternate water sources are not available.

5.2.3 Conservation of Water

In view of considerable increase in the consumption of water over the years, it is in the interest of mankind to conserve water. *Water conservation* is the careful use and protection of water resources and it involves both the quantity of water used and its quality. Conservation is an important part of sustainable water use. Broadly speaking maximum water is used in agriculture, for domestic purposes and in industrial units. Following procedures are used for conserving or reducing the use of water and achieving the results.

i. Agriculture Use

Agriculture is the largest water user. About 80% of total fresh water is used for agriculture. Improved agricultural irrigation can reduce the consumption of water by 20-30% and this could be a tremendous saving. Following procedures can be used for this purpose:

- Reduced seepage and evaporation of water by using lined or covered canals.
- Use computer monitoring and scheduled release for maximum efficiency. This procedure is used in some of the advanced and developed countries but not in under developed or developing countries.
- Irrigate at times when evaporation is minimum; for example, at night or in the early morning.
- Use improved irrigation systems, such as sprinklers or drip irrigation. It applies water more effectively to crops.
- Avoid or minimise loss of water by run off. This can be achieved by improving the quality of soil to increase infiltration. If possible, use mulch around plants to retain water.
- Development of crops that use less water. This also helps in the conservation of water.
- Agricultural water should be priced to encourage conservation. Subsidising water only encourages its over use.
- In under developed or developing countries, the farmers should be made aware of the tremendous benefits that arise due to judicious use of water. This aspect is the most important one.

ii. Domestic Use

About 10% of the total fresh water is used for domestic purposes. It is mostly used in bathrooms and for washing of clothes and dishes. The consumption of water for domestic use can be considerably reduced by the following practices.

- Use more-efficient bathroom fixtures, such as low flow toilets that require 1.5 gal per flush rather than the standard 5-gal-per-flush toilet.
- Use low flow showerheads that deliver less but sufficient water.
- Turn off the water when not absolutely needed for washing, brushing teeth, shaving etc.
- Flush the toilet only when really necessary.
- All leaks should be fixed or rectified at the earliest.
- Using dishwashers or washing machines can reduce water consumption.
- Water lawns and plants in early morning, late afternoon or at night to reduce evaporation loss.
- Use drip irrigation and place water-holding mulch around garden plants.
- The concept of an individual 'I pay for the use of water - so why bother' is not correct. The understanding of the concept of using minimum water to conserve water resources is most important.

iii. Industrial Use

Industries use substantial amount of fresh water for their various activities. These not only use water but are also responsible for discharging the waste water containing various chemicals etc. back to the water bodies causing serious water pollution problems. Following guidelines should be followed in industries:

- Water removal for steam generation of electricity could be reduced by about 25% to 30% by using cooling towers that use less or no water.
- Curbing water withdrawals by restoring to in-plant treatment and recycling of water or by developing new equipment and processes that use less water.

iv. Use of Wastewater

An important aspect of the conservation of water includes treatment of wastewater from households and industries before it is released or discharged into the surface water (river, lake etc). If allowed to be discharged directly into water bodies, the water gets polluted and can cause numerous health hazards (see sec. 6.5). The treated water can also be used for some other purposes like agriculture etc. Wastewater treatment is usually carried out in three stages, *viz. primary treatment*, *secondary treatment* and *ultra treatment.*

In the **primary treatment**, the raw sewage water is first passed through a couple of screens to remove large floating material. The water is then passed through chamber (called the *grit chamber*) which is packed with sand and small stones; this is helpful in removing the dirt, which is disposed off. The resulting filtered water then enters a sedimentation tank, in which particulate matter settles down to form the sludge. Sometimes, it is necessary to add certain chemicals (like alum), which hasten and help the settling process. The sludge is removed and used after processing as manure. The above treatment removes about 30-40% of the pollutants in the form of suspended matter.

In the **secondary treatment**, the water from the primary treatment, *viz.*, from the sedimentation tank is taken to an aeration tank in which air is pumped in along with some sludge from the final (second) sedimentation tank. The sludge contains aerobic (oxygen rich environment) bacteria that consumes or break down organic material (pollutants) in the waste. After several hours in the aeration tank, the water is taken to the second or final sedimentation tank, in which most of the sludge settles down. The total sludge is subjected to microbial digestion; this process evolves methane gas which is used as a fuel. The water from the final sedimentation tank is then disinfected in disinfection tanks usually by chlorine; this removes disease-causing organisms. The treated water is discharged into water bodies or used for various purposes.

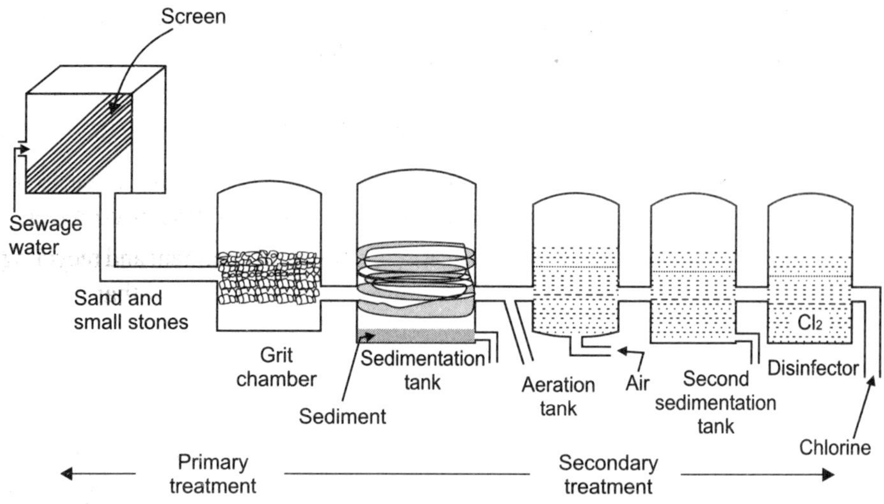

Fig. 5.1: Sewage Water Treatment

5.3 Characteristics of Water

Water is a unique liquid, without which no life can survive. This is attributed to the following characteristics.

(i) Water is regarded as a **universal solvent**. No other liquid can be compared to water as solvent. Since natural water is slightly acidic (due to dissolved carbon dioxide which gives carbonic acid, H_2CO_3 (for more details see sec. 4.5.1; pH of natural water is 5.6). It can dissolve a variety of compounds from simple salts to minerals including sodium chloride (common table salt) and calcium carbonate (calcite) in limestone rock. Water also reacts with complex organic compounds including many amino acids found in the human body.

(ii) Water possesses the highest **heat of fusion** and **heat of evaporation**, collectively known as the **latent heat,** of all known liquid substances at ordinary temperature. The latent heat of water is responsible for moderating the temperature of the biosphere. It also plays a vital role in the evaporation of water and its condensation as rain and as dew (see hydrological cycle, sec. 2.8.1.1)

(iii) Compared with other common liquids water has a high surface tension, a characteristic that is very important in many physical and biological processes involving movement of water through and into organisms.

(iv) Water is a poor thermal conductor as compared to metals.

(v) Water has a high viscosity which allows organisms to swim using simple movements.

(vi) Water protects the aquatic animals and plants from the mechanical disturbances.

(vii) Among common compounds and molecules- water is the only one whose solid form (ice) is lighter than its liquid form; (it expands by about 8% when it freezes and becomes less dense). This is responsible for the floating of ice on water. Can one imagine a situation if ice were heavier than water– it would sink to the bottom of water bodies (the biosphere would be entirely different than what it is today) and life if existed, could be greatly altered. Freezing of the water bodies from the bottoms onwards will result in complete freezing of the water bodies resulting in the death of all life in water.

(viii) Water is **transparent to sunlight**. This allows photosynthetic organisms to live below the surface of water.

(ix) Water exerts **pressure.** Organisms living at sea level experience a pressure of about 15 psi (~1 atmosphere, which is equal to 760 mm mercury). With depth, the pressure increases at the ratio of about one atmosphere for every 10 meters of descent. The physical characteristics like solubility, ionic dissociation and surface tension – all are influenced by the pressure.

5.4 Water Quality Parameters

No single parameter is possible for water. Depending on the use for which water is intended to be used, it should have the right parameters. For example, if water is to be used for cleaning purposes, no special treatment is required and no parameters are to be judged. However, if water has to be used for washing purposes (laundry etc), it should be ascertained that the water is not hard. Hard water is also harmful for use in boilers etc. as it leads to scale formation. For drinking purposes, water must be pure and free of any of the pollutants. Physical parameters and maximum contaminant level has been prescribed by different regulatory bodies. Tables 5.2 and 5.3 give standards for industrial and drinking water (human consumption) by BIS.

Table 5.2 BIS standards for water for industrial and drinking purposes

Physical parameters	BIS standard
Colour	Colourless
Odour	Odourless
Taste	Light, sour-sweet
*p*H	6-8.5
Specific conductance	300 μ mho cm^{-1}
Dissolved oxygen	4-6 ppm

Table 5.3 Maximum amount of contaminants level permissible for potable water

Contaminant	Maximum contaminated level (ppm) as BIS standard
Chlorides (Cl^-)	600
Sulphate (SO_4^{2-})	1000
Cyanide (CN^-)	0.0001
Fluoride (F^-)	3
Nitrate + Nitrite ($NO_3^- + NO_2^-$)	12
Phosphate (PO_4^{3-})	0.1
Calcium $(Ca)^{+3}$	100
Magnesium $(Mg)^{+2}$	30
Barium $(Ba)^{+2}$	1
Copper $(Cu)^{+2}$	1
Arsenic $(As)^{+3}$	0.002
Lead (Pb^{+2})	0.1
Iron (filterable)	0.3
Chromium (Cr^{+2})	0.05
Zinc (Zn^{+2})	0.05
Pesticides	0.005
Total bacterial count	$1x10^6$

The quality of water is basically judged by three parameters, *viz.* Dissolved Oxygen (DO), Biochemical Oxygen Demand (BOD) and Chemical Oxygen Demand (COD). However, following other parameters are also useful for determining the quality of water:

1. Alkalinity
2. Most Probable Number (MPN)
3. Total solids
4. Oxidation
5. Transparency
6. Silica content
7. Hardness
8. Dissolved inorganic impurities
9. Toxic metals
10. Microbial contaminants in sewage

5.4.1 Dissolved Oxygen (DO)

Dissolved oxygen is the amount of oxygen gas (O_2) that is dissolved in the water of any source. If the amount of dissolved oxygen is high, the degree of self-purification of water is high; in such cases, milder water treatment is required. However, if the amount of dissolved oxygen is less, all the impurities would not be eradicated by the self-purification phenomena and so in such cases, better purification methods are necessary. In this way, the value of dissolved oxygen in water is a useful measure of the condition of water (or the degree of pollution in water).

Oxygen in water may be directly absorbed from the atmosphere. It may also be released in water by aquatic plants during respiration or photosynthetic procedure. The level of oxygen or the amount of oxygen that can be dissolved in water depends on its temperature (see Table 5.4 below). The lower the temperature of water, the more will be the amount of dissolved oxygen. The amount of oxygen that can be dissolved in water decreases as the content of soluble salts in water increases. Due to this, the availability of oxygen is comparatively less in seawater than in river water. The solubility of oxygen in water is also directly proportional to the atmospheric pressure. Thus, a lake or pond at a higher elevation will have a lower value of dissolved oxygen than the one near the sea level.

Table 5.4 Dissolved oxygen in ppm (saturation value) in fresh and sea water

Type of Water	Temperature °C					
	0	**5**	**10**	**15**	**20**	**25**
Fresh water	14.6	12.8	11.3	10.2	9.2	8.4
Sea water	11.3	10.0	9.0	8.1	7.4	6.7

The short fall in dissolved oxygen or the degree of unsaturation is determined by subtracting measured value of dissolved oxygen at a given temperature from the saturation value of dissolved oxygen at the same temperature (by referring to table above).

The percentage saturation at a given temperature is given by

$$\% \text{ saturation} = \frac{\text{Measured dissolved oxygen}}{\text{Standard dissolved oxygen (from the table)}} \times 100$$

5.4.2 Biochemical Oxygen Demand (BOD)

Biochemical oxygen demand (BOD) is a measure of the amount of oxygen that would be needed by the micro-organisms to decompose the organic and inorganic pollution load in polluted water. The micro-organisms transform the pollutants into non-hazardous compounds. During this process, dissolved oxygen is consumed.

$$\underset{\text{Organic matter}}{CH_2O + O_2} \xrightarrow[\text{Aerobic oxidation}]{\text{Micro-organisms}} CO_2 + H_2O + \text{Micro-organisms (Bacterial cell)}$$

The inorganic pollutants may include, sulphites, sulphides, thiosulphates and ferrous iron. These pollutants are attacked by oxygen in presence of enzymes released by the micro-organisms.

For determination of BOD, the water sample is first saturated with oxygen and then incubated at constant temperature (usually 20°C) for five days. This allows sufficient time for the micro-organisms in polluted water to affect oxidation of pollutants. After 5 days, the remaining amount of dissolved oxygen is determined and the biochemical oxygen demand is obtained by substraction.

$$\text{BOD} = \begin{array}{c}\text{Saturation value of dissolved}\\ \text{oxygen per litre at 20°C}\end{array} - \begin{array}{c}\text{Amount of } O_2 \text{ dissolved per}\\ \text{litre at 20°C after 5 days}\end{array}$$

The result is called the 5-day BOD and is expressed in milligrams of the oxygen per litre of water or in ppm. So, BOD is a measure of the oxygen demand placed on a system to accomplish the complete oxidation or decomposition of suspended colloidal or dissolved organic matter. The 5-day BOD analysis is considered as an accepted standard test. Drinking water should have a 5-day BOD of less than 1.5 mg L^{-1}. The BOD of raw sewage ranges from 200-400 mg L^{-1} or 200-400 ppm, indicating considerable pollution. The major contributors to BOD are the chemical industries, pulp and paper industry and the food processing industry. The BOD water status is given in Table 5.5.

Table 5.5 BOD-water status

BOD (mg litre)$^{-1}$	Status
1	Very clean water
2	Clean water
3	Fairly clean water
5	Not used for drinking and pharmaceutical preparation
10	Contaminated water
15	Unfit for fish reproduction population
150	Sewage discharge from homes
200	Industrial waste water
350	Waste water from paper industries
1000	Food processing waste water
2000	Dairy discharge waste

BOD is not a true indication of pollution, but gives an idea of the problems arising due to waste materials in the water supplies. In fact, BOD is directly proportional to the organic matter oxidized.

5.4.3 Chemical Oxygen Demand (COD)

Like BOD, chemical oxygen demand (COD) is also a measure of the amount of oxygen required to degrade or breakdown the organic matter. In BOD, the degradation of organic matter is affected by micro-organisms but in COD, the degradation is affected by some chemical oxidising agent like potassium dichromate. It is believed that biochemical oxygen demand (BOD) does not give a correct assessment of the pollution load in water. This is because some compounds like detergents are resistant to microbial degradation and some other pollutants like cellulose are slow to get oxidised. Also the toxic compounds, if present in polluted water may poison the micro-organisms before they can act on pollutants. Lastly, the BOD determination takes too long time (5 days).

The above problem can be overcome by another analytical method called chemical oxygen demand (COD). In this case, the water sample is treated with an oxidising agent (usually potassium dichromate in acidic medium), which oxidises most of the polluting substances including those which are resistant to microbial oxidation. The unused potassium dichromate is determined by back titration with a suitable reducing agent like Mohr's salt. The amount of potassium dichromate consumed is determined by substraction.

$$\text{Amount of } K_2Cr_2O_7 \text{ consumed} = \text{Amount of } K_2Cr_2O_7 \text{ initially added} - \text{Amount of } K_2Cr_2O_7 \text{ determined by back titration}$$

The amount of oxygen used in the oxidation can be calculated from the strength of $K_2Cr_2O_7$ consumed using the following equation:

$$K_2Cr_2O_7 + 4\,H_2SO_4 \longrightarrow K_2SO_4 + Cr_2(SO_4)_3 + 4\,H_2O + 3(O)$$

The results are expressed in terms of the amount of oxygen in ppm that will be required to oxidize the contaminants. On the basis of COD status, the quality of water can be found out (see Table 5.6 below).

Table 5.6 COD Status

COD in mg L^{-1}	Status
0-5	Very clear water, used for drinking
5-20	Fairly clean
20-100	Unfit for drinking but can be used for washing and agriculture

COD determination is reliable and fast for assessment of the quality of water. However, it suffers from the drawback that aromatic hydrocarbon derivatives are not easily oxidised by most chemical oxidising agents. So, their presence cannot be judged by COD determination.

5.4.4 Alkalinity

The total amount of substances that cause increased concentration of OH^- ions either on dissociation or as a result of hydrolysis is called **alkalinity** of water. In natural water, alkalinity is due to the presence of HCO_3^-, SiO_3^{2-}, $HSiO_3^-$and CO_3^{2-}. Sometimes, the presence of salts of weak organic acids bind H^+ and thus result in increasing the concentration of OH^- ions rendering the natural water alkaline.

The alkalinity may be regarded as **bicarbonate alkalinity** (A_b), **carbonate alkalinity** (A_c) or **hydrate alkalinity** (A_h), depending upon the presence of HCO_3^-, CO_3^{2-} and OH^- ions, respectively. Carbonate hardness (see sec. 5.4.10.1) is also responsible for the alkalinity of water. Alkalinity of water is determined by titrating it with a mineral acid using methyl orange as an indicator.

5.4.5 Most Probable Number (MPN)

High population of bacteria like *E.coli* and *coliforms* is found in water polluted with organic wastes. Both E.coli and coliforms can be detected and enumerated with the help of MPN method. Coliform is present in human intestine and generally is not harmful but its presence indicates the presence of human waste in water, which may contain other harmful substances. Polluted water shows high values of MPN.

5.4.6 Total Solids

The total solid content is the amount of non-volatile substances present in water in a colloidal and dispersed state. It is expressed in milligram per kilograms (mg/kg). For determining the content of total solids in water, calcium and magnesium bicarbonates are converted into carbonates, *viz.* $CaCO_3$ and $MgCO_3$. In addition to the total solids, there are three other types of solids; these are:

(i) Fixed residue solid
(ii) Mineral residue solid
(iii) Sulphate solids

Fixed residue solid is obtained by calcining the total solids for 10-15 min at 800° C; this results in removal of organic substances due to combustion and decomposition of carbonates. The residue obtained is the fixed residue solid content.

The quantity obtained by adding all cations and anions, including CO_3^{2-} and Al_2O_3, Fe_2O_3 and SiO_2 in water, is called the **mineral residue**. If the total solids are treated with concentrated sulphuric acid, all the cations present in water are converted into sulphates, the total mass of which constitutes the **sulphate content** or **sulphate solid**.

5.4.7 Oxidation State

The contamination of water by organic substances is represented by oxidability and is expressed in milligram of oxygen required to oxidise the organic compounds contained in 1 kg of water. The oxidability is denoted as mg/kg. Oxidability can also be expressed by the amount of $KMnO_4$ (in mg/kg) used to oxidise the organic substances. Oxidability, however, does not represent the total content of organic substances in water, since complete oxidation of all organic substances does not occur under the conditions in which oxidability is determined.

5.4.8 Transparency

It indicates the concentration of suspended solids, which can be determined by the weight method. The transparency can also be determined by the **type method** (cm) of a column of water in a glass tube, through which it is still possible to discern printing or by the **cross method** which two crossed black lines 1 mm thick are visible on a white paper placed on the bottom of a glass tube.

5.4.9 Silica Content

It is the concentration of silicic acid, H_2SiO_3, expressed as silica (SiO_2), contained in water. The concentration of SiO_2 in natural water varies from 5-10 to 90 mg L^{-1}. It is an indication of the extent of mineralization of water. Raw water contains silicic acid in the ionic ($HSiO_3^-$) as well as in colloidal state. The presence of silica in feed water for boiler units causes a number of operating conditions due to formation of silicate scales of low thermal conductivity; even heavy deposits on the blades and nozzles of turbines are formed. In modern practice, silica is removed from raw water using ion exchangers.

5.4.10. Hardness

Water from all sources contain a great degree of dissolved mineral salts. Normally, water contains dissolved magnesium and calcium bicarbonates, sulphates and chlorides in considerable amounts, along with traces of different cations and anions which introduce hardness to water. Water that produces or forms an insoluble precipitate when it is boiled or when soap is added is called **hard water**. In fact, hard water does not give lather with soap, such water may not always be unfit or harmful for drinking but cannot be used in laundry work and for steam generators in boilers. Water required for pharmaceuticals preparations, dairies, laundries and for boiler feed must be soft water. Water which gives lather easily with soap is called **soft water**. Before water is used, one must ensure the nature and amount of dissolved quantities present.

Soap reacts with soluble calcium or magnesium salts and forms an insoluble white scum or precipitate with hard water; this causes a loss of soap and reduces the cleaning action. A large amount of soap is needed to produce lather with hard water.

$$\underset{\text{Sodium stearate (soap)}}{2\,C_{17}H_{35}COO^-Na^+} + \underset{\text{hard water}}{CaCl_2} \longrightarrow \underset{\text{Calcium stearate (insoluble scum)}}{(C_{17}H_{35}COO)_2Ca\downarrow} + 2\,NaCl$$

$$\underset{\text{Sodium alkanesulphonate (detergent)}}{C_{15}H_{31}SO_3^-\,Na^+} + \underset{\text{hard water}}{CaCl_2} \longrightarrow \underset{\text{Calcium alkane sulphonate}}{(C_{15}H_{31}SO_3)_2Ca\downarrow} + 2\,NaCl$$

Soft water does not contain dissolved calcium and magnesium salts. The presence of sodium compounds does not contribute to hardness in water.

Depending on the nature of the anions present in it, the hardness of water is of following two types:

(i) Temporary or carbonate hardness
(ii) Permanent or non-carbonate hardness

i. Temporary or Carbonate Hardness

When water contains calcium or magnesium bicarbonate, the hardness is called **temporary hardness** or **carbonate hardness**. Such hardness can be removed by boiling, when calcium and magnesium bicarbonates form insoluble carbonate and settle down.

$$\underset{\text{Soluble}}{Ca(HCO_3)_2} \xrightarrow{\Delta} \underset{\text{Insoluble}}{CaCO_3\downarrow} + H_2O + CO_2\uparrow$$

$$Mg(HCO_3)_2 \xrightarrow{\Delta} \underset{\text{Insoluble}}{MgCO_3\downarrow} + H_2O + CO_2\uparrow$$

Temporary hardness can also be removed by adding calculated quantities of lime. This method is called **Clark's method**. However, if too much lime is added, the water becomes again hard.

$$Ca(HCO_3)_2 + Ca(OH)_2 \longrightarrow 2\,CaCO_3\downarrow + 2\,H_2O$$

$$Mg(HCO_3)_2 + Ca(OH)_2 \longrightarrow MgCO_3 + CaCO_3 + 2\,H_2O$$

ii. Permanent or Non-carbonate Hardness

It is due to the presence of soluble salts (chlorides and sulphates) of calcium magnesium, iron or other heavy metals. Unlike temporary hardness, it cannot be removed by boiling. Carbonate hardness (temporary hardness) and non-carbonate hardness (permanent hardness) together constitute the **total hardness**. Before removing permanent hardness, the water should be boiled to ensure removal of temporary hardness.

5.4.10.1 Units of Hardness of Water

Bicarbonates, chlorides and sulphates of Ca^{2+}, Mg^{2+}, etc which give an insoluble precipitate with soap contribute to hardness. It is most convenient to express the hardness caused by these different salts in terms of equivalent hardness of a single salt. As a matter of convention, calcium carbonate (sparingly soluble in water and molecular mass =100) is chosen for this purpose and is expressed in parts per million (ppm). For this purpose, all the hardness-causing impurities are first converted in terms of their respective mass equivalents of $CaCO_3$ and the sum of these is expressed in ppm.

Thus, 100 g of $CaCO_3 \equiv$ 111 g of $CaCl_2 \equiv$ 136 g of $CaSO_4 \equiv$ 95 g of $MgCl_2$ $\equiv$ 120 g of $MgSO_4 \equiv$ 162 g of $Ca(HCO_3)_2 \equiv$ Equivalents of calcium carbonate

$$= \frac{\text{Mass of hardness producing substance} \times \text{chemical equivalent of } CaCO_3}{\text{Chemical equivalent of hardness producing substance}}$$

$$= \frac{\text{Mass of hardness producing substance} \times 50}{\text{Chemical equivalent of hardness producing substance}}$$

One chemical equivalent of $CaCO_3$ = 50 because molecular mass of $CaCO_3$ = 100

The most common units of hardness are given below.

(a) **Parts per million (ppm)** is the parts of $CaCO_3$ equivalent hardness in 10^6 parts of water.

1 ppm = 1 part of $CaCO_3$ equivalent hardness in 10^6 parts of water

$$1 \text{ ppm} = \frac{1 \text{ part of hardness}}{10^6 \text{ part of water}}$$

(b) **Milligrams per litre (mg/litre)** is the number of milligrams of $CaCO_3$ equivalent hardness present per litre of water.

1 mg/litre = 1 mg $CaCO_3$ equivalent hardness in 1 litre
but 1 litre of water weights = 1 kg = 1000 g
= 1000 × 1000 mg = 10^6 mg

1 mg/litre = I mg $CaCO_3$ equivalent per 10^6 mg of water
= 1 part $CaCO_3$ equivalent per 10^6 part of water
= 1 ppm

A common method for accurately determining the hardness of water is complexo-metric EDTA titration. It can also be determined by **Hehner's method**, where the alkalinities of water before and after boiling are determined. Some times, soap titration method is also commonly used. In this method, standard soap solution is titrated against hard water and it will not give lather until all the hardness causing metal ions are precipitated in the form of insoluble soap.

It is well known that water is used in a number of industries such as textile, sugar, paper, dyeing, laundries, bakeries and pharmaceutical industries. Following are the disadvantages of using hard water in various industries:

(i) **Textile Industry**. In textile industry, hard water causes wastage of soap when used for washing yarn, fabric etc. also the calcium and magnesium soaps get fixed on to the fibres, which do not produce good shades on dyeing.

(ii) **Sugar Industry**. Hard water causes problems in the crystallisation of sugar. Also the sugar produced may be deliquescent.

(iii) **Paper Industry**. Using hard water gives undesirable finish to the paper.

(iv) **Dyeing Industry**. Calcium and magnesium salts in hard water react with dyes forming undesirable precipitates.

(v) **Bakeries**. The presence of bacteria and fungi affect the yeast's action resulting in poor quality of the product.

(vi) **Pharmaceutical Industry**. Hard water containing magnesium and calcium salts in pharmaceutical products like drugs, injections etc. may produce undesirable and harmful effects.

Due to the above mentioned disadvantages of using hard water in industries and boilers etc., it is necessary to remove the hardness of water i.e. to make it soft; for rendering it fit for use.

5.4.10.2 Water Softening

The process of removing hardness-producing salts from water is known as **softening of water**. Following methods can be used for the softening:

i. Lime Soda Process for Water Softening

In this process, the hardness of water is first estimated and then the water is treated with calculated quantities of slaked lime [$Ca(OH)_2$] and soda ash (Na_2CO_3). Following reactions take place:

$$\left.\begin{array}{l} Ca(HCO_3)_2 + Ca(OH)_2 \longrightarrow 2CaCO_3\downarrow + 2\,H_2O \\ Mg(HCO_3)_2 + Ca(OH)_2 \longrightarrow CaCO_3\downarrow + MgCO_3\downarrow + 2\,H_2O \end{array}\right] \text{Temporary hardness removed}$$

$$\left.\begin{array}{l} MgCl_2 + Ca(OH)_2 \longrightarrow CaCl_2 + Mg(OH)_2\downarrow \\ MgSO_4 + Ca(OH)_2 \longrightarrow CaSO_4 + Mg(OH)_2\downarrow \\ Al_2(SO_4)_3 + 3\,Ca(OH)_2 \longrightarrow 2\,Al(OH)_3\downarrow + 3\,CaSO_4 \\ FeSO_4 + Ca(OH)_2 \longrightarrow Fe(OH)_2\downarrow + CaSO_4 \\ Fe(OH)_2 + H_2O + \tfrac{1}{2}O_2 \longrightarrow 2\,Fe(OH)_3\downarrow \end{array}\right] \text{Permanent hardness removed}$$

The lime treatment can be done either in cold (this process takes longer time) or in the hot (80-150°C).

ii. Permutit Process

In this process, hard water is passed through a column of zeolites. These zeolites can be either naturally occurring (e.g. natrolite, $Na_2O.Al_2O_3.SiO_2.H_2O$; and anclcini, $Na_2O.Al_2O_3.4SiO_2.3H_2O$) or synthetic (e.g, permutit, an artificial zeolite of general formula $Na_2O.Al_2O_3.nSiO_2.xH_2O$ where n = 5 – 13 and x = 3 – 4). When water is passed through a bed of permutit, the sodium ions in permutit are exchanged with other cations like Ca^{2+}, Mg^{2+} etc. Thus, sodium zeolite is converted into calcium and magnesium zeolites and water becomes free of Ca and Mg salts but richer in sodium salts. In this way, zeolite is a 'sodium exchanger'. This process removes both temporary and permanent hardness. When zeolite is exhausted (i.e. all the Na in zeolite is replaced with Ca and Mg), it can be regenerated by washing zeolite bed with a concentrated solution of brine (or sodium chloride); this process converts exhausted zeolite into the original form and can be used again. Soft water obtained by this process is most suitable for laundry purposes but not for boiler purposes. This process cannot be used for treating acidic waters, as permutit disintegrates in acidic condition.

iii. Ion Exchange Process

This procedure is useful for removing cationic and anionic impurities in water. In the process, water is passed first through an cation exchanger. The cation exchanger is a resin containing ionic sulphonic acid group. The cations, (M^+), present in impure water are exchanged with the hydrogen ion (H^+) of the resin. Thus,

$$\underset{\text{Cation exchange resin}}{\text{Resin/SO}_3^-\text{H}^+} + \underset{\text{cation}}{\text{M}^+} \longrightarrow \text{Resin/SO}_3^-\text{M}^+ + \text{H}^+$$

The water is subsequently passed through an anion exchanger resin, which contain ionic quaternary ammonium hydroxide group; the anions, X^-, present in impure water exchanges with the hydroxide ions (OH^-) of the resin. Thus,

$$\underset{\text{Anion exchange resin}}{\text{Resin/N}^+\,\text{OH}^-} + \underset{\text{anion}}{\text{X}^-} \longrightarrow \text{Resin/N}^+\,\text{X}^- + \text{OH}^-$$

The H^+ ions obtained in the cation exchange process combine with the OH^- ions formed in the anion exchange process to produce water.

$$H^+ + OH^- \longrightarrow H_2O$$

When the resins are exhausted, they are regenerated by treatment with acid (in case of exhausted cation exchange resin) or alkali (in case of exhausted an ion exchange resin).

$$\underset{\text{Cation exhausted change resin}}{Resin/SO_3^-M^+} + H^+ \longrightarrow Resin/SO_3^-H^+ + M^+$$

$$\underset{\text{Anion exhausted exchange resin}}{Resin/N^+X^-} + OH^- \longrightarrow Resin/N^+OH^- + X^-$$

This method is used to obtain demineralised water.

5.4.11 Dissolved Inorganic Impurities

As the name suggests, these substances are dissolved in water. The nature of water depends on these dissolved inorganic impurities. Following table gives the dissolved impurities that may be present in water and their effects.

Table 5.7 Dissolved inorganic impurities

Nature of cations	Impurity anion	Effects
Calcium and magnesium	HCO_3^- CO_3^{2-}	Causes alkalinity and hardness
	Cl^- SO_4^{2-}	Causes hardness
Sodium	HCO_3^-	Causes alkalinity
	CO_3^{2-}	Causes alkalinity
	Cl^-	Causes change in taste
	F^-	Affects tooth enamel
Metal and their ions	Fe	Causes hardness, imparts red colour, changes taste
	Mn	Imparts brown or black colour
	Pb	Causes lead poisoning
	As	Causes arsenic poisoning
Gases	O_2	Corrodes metals
	CO_2	Caused acidity, corrodes metals
	H_2S	Bad smell; acidity, corrosion

5.4.12 Toxic Metals in Water

Natural waters and wastewaters contain a number of trace elements, which are toxic at higher levels. Following table gives some of such elements along with their sources and effects.

Table 5.8 Some toxic elements in water

Elements	Sources	Effects and significance
Arsenic	Mining by-product, pesticides, chemical waste	Toxic, possibly carcinogenic
Cadmium	Industrial discharge, mining waste, metal plating, water pipes	Replaces zinc biochemically, causes high blood pressure, kidney damage, destruction of testicular tissue and red blood cells, toxicity to aquatic biota
Beryllium	Coal, nuclear power and space industries	Acute and chronic toxicity, possibly carcinogenic
Boron	Coal, detergent formulations, industrial wastes	Toxic to some plants
Chromium	Metal plating, cooling tower water additive, (chromate) normally found as Cr (VI) in polluted water	Essential trace elements; possibly carcinogenic as Cr (VI)
Copper	Metal plating, industrial and domestic waste, mining, mineral leaching	Essential trace elements, not very toxic to animals, toxic to plants and algae at moderate levels.
Fluorine (Fluoride ion)	Natural geological sources, industrial waste, water additive	Prevents tooth decay at about 1 mg/l, causes mottled teeth and bone damage at about 5 mg/l
Lead	Industry, mining, plumbing, coal, gasoline	Toxic (anaemia, kidney disease, nervous disorder), wild-life destroyed
Manganese	Mining industrial waste, acid mine drainage, microbial action on manganese mineral at low pH	Relatively non-toxic to animals, toxic to plants at higher levels, stains materials (both-room fixtures and clothing)
Mercury	Industrial waste, mining, natural sources	Toxic to animals
Selenium	Natural geological sources, sulphur, coal	Essential at low levels but toxic at higher levels
Zinc	Industrial waste, metal plating, plumbing	Essential in may metallo-enzymes, toxic to plants at higher levels

Besides the above, the water bodies contain a large number of pesticides which are discharged into water bodies by run offs from agricultural land, (for more details see sec. 6.4.2.4).

Inhibition of Enzyme Action

The toxic chemicals (*viz.* metal ions) which may be present, inhibit enzymes. Heavy metal ions, particularly Hg^{2+}, Pb^{2+} and Cd^{2+} attack the active sites (e.g. $-SCH_3$ and –SH in methionine and cysteine amino acids, which are part of enzyme structure).

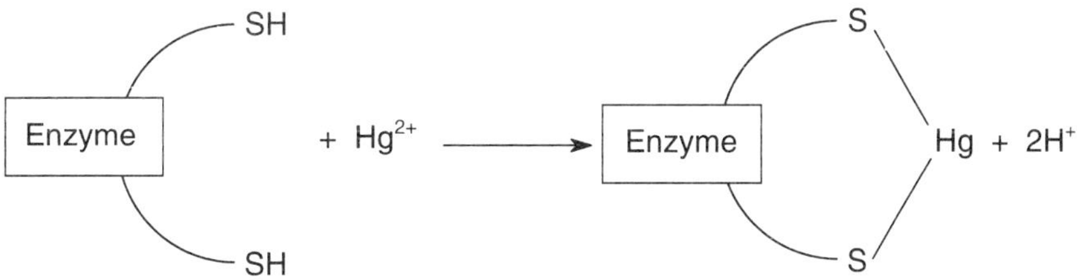

In this way, the enzyme action is inhibited. Also, in metallo-enzymes, the metal of the metallo-enzyme is replaced by another metal ion of similar size and charge. For example, Zn^{2+} in some metalloenzyme is replaced or substituted by Cd^{2+} leading to cadmium toxicity. The enzymes, which are inhibited by Cd^{2+} include adenosine triphosphate, alcohol dehydrogenase, amylase, carbonic anhydrase. Pb^{2+} inhibits acetylcholinesterase, alkaline phosphate, adenosine triphosphate, and carbonic anhydrase.

The biochemical effects of some toxic elements like arsenic, lead, cadmium and mercury form the subject matter of subsequent discussions.

Effects of Arsenic

Arsenic occurs in insecticides, fungicides and herbicides. It finds its way into agricultural land and enters the food chain. Arsenic exerts its toxic action by attacking -SH groups of an enzyme resulting in inhibition of enzyme action.

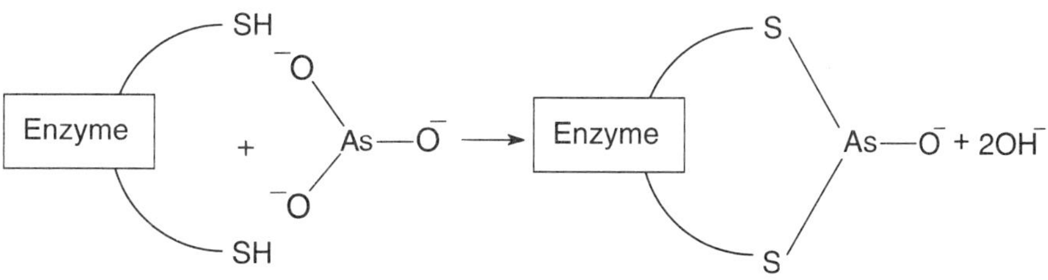

Arsenite

The enzyme, pyruvate dehydrogenase, which is responsible for generating cellular energy is deactivated by complexation with As(III). In this way, the generation of ATP is prevented.

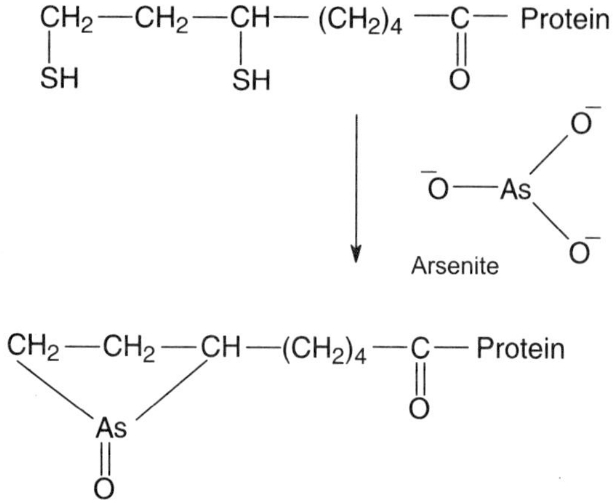

Arsenic also interferes with some biochemical processes involving phosphorous; this is due to chemical similarity of arsenic and phosphorous. Thus, ATP (adenosine triphosphate), as we know, is generated from glyceraldehyde 3-phosphate by phosphorylation (route A). However, in presence of arsenic, glyceraldehyde 3-phosphate gives 1-arseno-3-phosphoglycerate (route B). Thus, in the later route (B), in place of phosphorylation, arsenolysis occurs, the product formed is 1-arseno-3-phosphoglycerate, which prevents ATP synthesis.

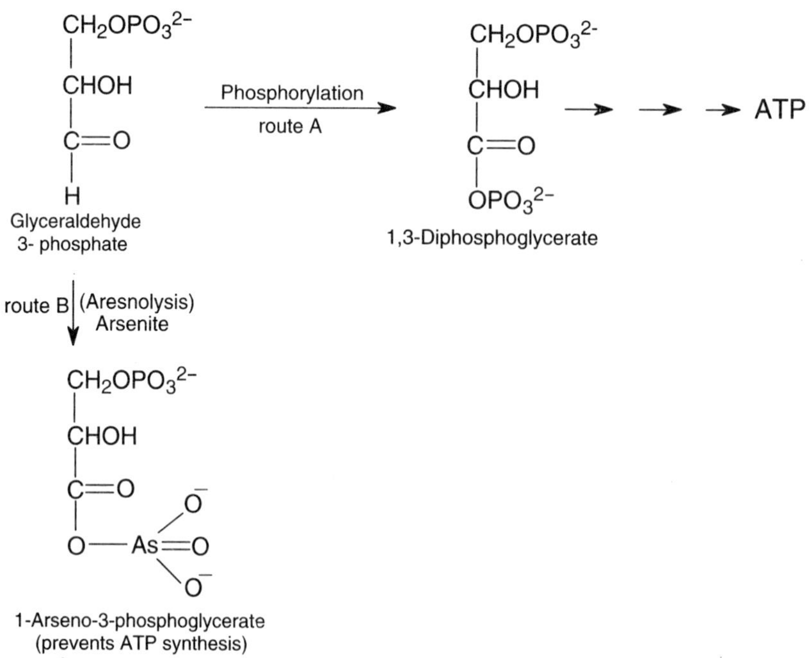

Arsenic (III) compounds in higher concentrations coagulate proteins by attacking the sulphur bonds. The antidotes for arsenic poisoning are chemicals having -SH group, which binds to As(III). An example is 2,3-dimercaptocpropanol.

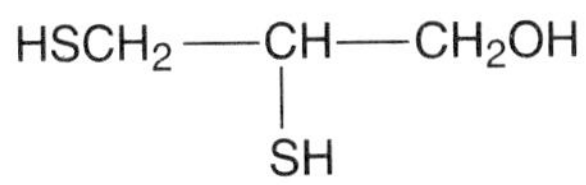

Effects of Lead

Lead occurs abundantly in nature in the form of sulphide ore. During metallurgical operations, it finds its way into water bodies. Earlier lead was added to gasoline in the form of tetra ethyl lead [$Pb(C_2H_5)_4$] along with scavengers like 1,2-dichloro- and 1,2-dibromoethane. However, the use of leaded petrol has been banned in most of the countries and so is no longer considered to be the major source of lead in the atmosphere or water bodies. Lead pipes, which are used for transporting water is a minor source of lead in water bodies, since the lead pipes are coated with a layer of carbonate. However, even the small amount of lead in water bodies enters the food chain. The biochemical effect of lead is its interference with heme synthesis leading to haematological damage. Lead inhibits the aminolevulnic acid dehydrase enzyme. Due to this, the overall effect is the disruption of the synthesis of not only of haemoglobin but also of other respiratory pigments like cytochromes. Also, lead does not permit utilization of oxygen and glucose for life sustaining energy production.

Lead, being similar to Ca^{2+}, gets accumulated in the bones. Subsequently, it is remobilised along with phosphates from the bones which exert a toxic effect.

Lead poisoning can be cured by administration of chelating agents which strongly bind Pb^{2+}. One such example is administration of calcium chelate of EDTA. Pb^{2+} displaces Ca^{2+} from the chelate and the resulting Pb^{2+} chelate is excreted in the urine.

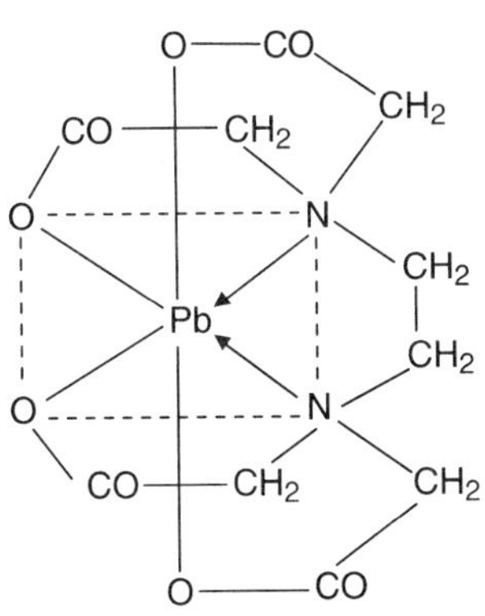

Pb-EDTA Chelate

Effects of Cadmium

Cadmium occurs in nature along with zinc in zinc blende (ZnS). Cadmium poisoning was first observed in Japan in the form of *itai itai* or 'Ouch Ouch' disease; in this disease, the bones become fragile. Higher levels of cadmium cause kidney problems, anaemia and disorders of bone marrow, hypertension, anaemia, renal dysfunction and even cancer. Most of the cadmium ingested into the body is trapped in kidneys and eliminated. However, a small fraction is bound effectively by the body proteins, metallothionein, present in kidneys. When excessive amounts of Cd^{2+} are ingested, it replaces Zn^{2+} at major enzymatic sites and thus, causes metabolic disorders.

Effects of Mercury

Mercury is one of the most well known toxic metal. The toxicity or the toxic effects of mercury came to lime light in 1953-1960 in Japan by the incidence of **Minamata disease**. At Minamata Bay in Japan, more than 100 people died and thousands were permanently paralysed by consuming fish contaminated with mercury. Even genetic defects were observed in some new born babies whose mothers had consumed contaminated fish from the Bay. The cause of the disease was contamination of water with mercury rich effluents discharged into the Bay by Chisso Chemical Company. It was discovered (1963) that the illness was due to methyl mercury poisoning, caused by eating contaminated fish from the bay. The Chisso plant used mercuric chloride catalyst, a non-toxic inorganic mercury compound in the production of acetaldehyde which was released in the industrial effluents. Subsequently, the sediments from the Minamata Bay were found to be rich in methyl mercury chloride. It was, therefore, clear that inorganic mercury underwent bio-methylation in aquatic system. It was also evident from the data obtained which showed the presence of methyl mercury chloride in concentrated form in the body of the fish.

It has been found that methylation of mercury in water bodies is brought about by a number of micro-organisms called methanogenic bacteria. This process is called bio-methylation. The bio-methylation takes place under anaerobic conditions. The process is facilitated by the enrichment of water with organic impurities which permit growth of methanogenic bacteria; the bio-methylaiton proceeds smoothly in the pH range 5.5-6.5.

The Minamata incident was followed by a more tragic incident of mercury poisoning form Iraq (1972) where about 500 people died after consuming wheat which had been sprayed with mercury containing pesticide.

Sources of Mercury in Aquatic Systems

Mercury is discharged into the aquatic system by both natural and man-made sources. Amongst the natural sources is the ore of mercury, cinnabar (HgS) in igneous rocks. On an average, about 800 tonnes of mercury are released into the environment due to weathering of rocks. The lava, produced by volcanic eruptions is also a natural source of mercury contamination of environment. Due to this activity, the mercury content of lake Ontake in Japan has shown a steady rise due to repeated volcanic eruptions of the nearby Mount Kisoontake.

The mercury content in the aquatic systems has increased considerably due to human activities resulting from rapid industrialisation and urbanisation. Some of the human activities include the following:

- Widespread use of mercurials as insecticides, fungicides, bactericides and pharmaceuticals.
- Use of mercury amalgamated with silver, tin, cadmium and copper to make dental fillings.
- The chlor-alkali process using mercury cathodes is the single largest source of mercury in natural waters.
- Even after closing down various industries using or producing mercurials, the amount of mercury already accumulated in the beds of water bodies continues to be a cause of major concern.
- Mercury vapour lamps, electrical switches, mercury batteries are the second largest source of mercury discharge in the environment.
- The impact of seed dressing is enormous. It causes widespread dispersal of mercury compounds. Also, mercury finds its way into the human food chain.

Toxic Effects of Mercury

The toxicity of mercury depends on its chemical species. Thus, elemental mercury (Hg) is non-toxic and quite inert. If swallowed, it is excreted without any serious harm. However, the vapours of mercury (it inhaled) are quite toxic. The vapours on inhalation enter the brain through blood stream causing damage to the central nervous system.

Mercurous ion (Hg_2^{2+}) forms insoluble chloride with chloride ions. Since the stomach in our body contains hydrochloric acid (chloride ions), it is not toxic. On the other hand, mercuric ion (Hg^{2+}) is quite toxic; due to its affinity for sulphur. It can attach itself to the sulphur containing amino acids of proteins. It also forms bonds with haemoglobin and serum albumin, both of which contain sulphydryl groups. Since this ion (Hg^{2+}) does not travel across biological membranes, it does not get access into biological cells.

The organomercurials particularly methyl mercury (CH_3Hg^+), which is fat soluble, are the most toxic species. The most harmful effect of RHg^+ is due to its ability to move through the placental barrier and enter foetal tissues. Attachment of Hg to cell membranes inhibits transport of sugars across the membranes and allow the passage of K (Potassium) to the membrane. In case of brain cell, this results in energy deficiency in the cells. This is the reason why babies born to mothers having mercury poisoning suffer from irreversible damage to the central Nervous System.

Bioamplification of Mercury

The organomercurial (e.g. methyl mercury chloride) is incorporated in the food chains. Being soluble in the tissue of simple organisms, these penetrate inside the bodies of simple organisms. The complex species which feed on simple organisms accumulate the organo-mercurials in their system. This process is called bioamplication. Due to the process, organo-mercurials concentrate in the tissues of aquatic organisms, particularly fish. Bioamplication depends on the following parameters of water, *viz.* *p*H, temperature, hardness, salinity and contaminants. At lower *p*H (<5), the uptake of methyl mercury chloride is favoured. However, at *p*H more than 5, methyl mercury chloride is converted into methyl mercury hydroxide (CH_3HgOH), which is insoluble in the lipid constituent of fatty tissues and so the bioamplication is reduced. The bio-amplification of methyl mercury chloride in fish increases with temperature and salinity. Bioamplication is reduced in hard water or in the presence of chelating agents like EDTA.

Effects of Mercury Poisoning

- Mercury inhibits enzyme action.
- Mercury reacts with sulphydryl groups in proteins and thus, disturbs the distribution of ions and interferes with the movement of fluids across the membrane.
- It is responsible for neurological disorders including dysphasia, ataxia, deafness and death.

Antidotes

The permissible limit of mercury in food is 0.5 ppm. The level of mercury in blood should not exceed 1.0 ppm and in drinking water, it should not exceed 0.005 ppm. Some common chemotherapeutic agents for the treatment of mercury poisoning are given below:

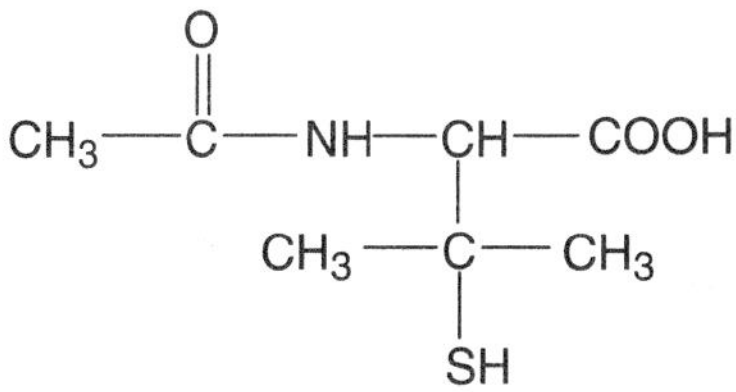

N-Acetyl-DL-penicillamine

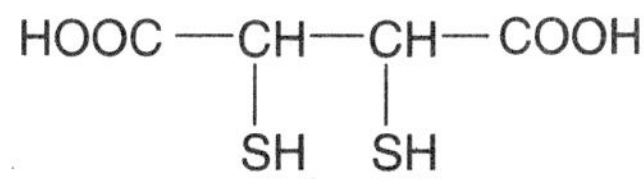

Dimercaptosuccinic acid

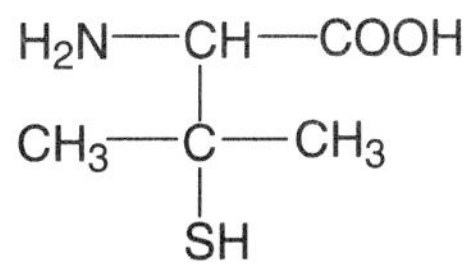

D-Penicillamine

These chemotherapeutic agents convert the intracellular methylmercury salts into a mobile complex that passes out of the body as urinary or faecal excretions.

Checking Environmental Pollution of Mercury

It has already been stated that mercury released into the environment by various agencies causes havoc. In view of this, it is essential that the environment pollution due to mercury be checked. This can be achieved by the following procedures.

- All chlor-alkali plants must be discouraged for using mercury electrodes and should be encouraged to switch over to new technology. In fact, diaphragm cells are now being used.
- All alkyl mercury pesticides must be banned and replaced by less toxic substituents.
- Mercury-amalgam dental fillings must be banned.
- Release of mercury and its compounds as waste into water bodies should be curtailed by instituting recovery processes. Advance waste treatment can also contribute to reduction of pollution caused by mercury.

Using all possible precautions and new technologies, it has been possible to reduce the mercury pollution. The main problem the world is facing with it is that the water bodies whose underlying sediments are already overburdened with mercury will continue to release the metal for a considerable length of time. Attempts to dredge up the mercury containing sediments makes the problem worse since dredging stirs up the bottom level and mercury will spread in whole of the water bodies. It has been suggested to prevent the release of mercury from the sediments by covering it with a layer of ferrous sulphide and sand. However, this method is too cumbersome considering the vast oceans.

5.4.13 Microbial Contaminants in Sewage

A number of microbes are found in sewage. These include Algae, Fungi, Protozoa, Bacteria and Viruses etc.

Algae

These are small plant organisms that make their own food by photosynthesis. They appear in green, brown or blue-green colour. The carbohydrates, proteins and oxygen produced by algae are used by other aquatic species for their survival. The presence of algae affects the value of water for water supply due to problems caused by them in taste and odour.

Fungi

They are aerobic, multicellulor, non-photosynthetic chemoheterotrophic eucaryotes. They are saprophytes and are responsible for the decomposition of carbon in the biosphere. They grow in low moisture area and in low pH environment. They are essential for the breakdown of organic matter in carbon cycle.

Protozoa

Protozoa are single celled organisms that live on dead organic matter. They act as scavengers consuming bacteria and algae. Protozoa present in waste water include amoebas, flagellates and free swimming and stalked ciliates. A number of protozoa are pathogenic.

Bacteria

A clan to extremely small relatively simple microorganisms. Many of them are disease causing, responsible for dreadful diseases such as cholera, tuberculosis, typhoid in humans and animals. They have a single cell covered with a cell wall. Inside the cell, instead of a proper nucleus, just a loop of hereditary material is present. There are bacteria which can live in oxygenated environment and also those which can live without oxygen. Most bacteria get their nourishment from dead and decaying organisms or by parasitizing animals and plant. There are, however, some bacteria which can synthesize organic compounds using chemical energy.

Viruses

An extremely small (25 to 300 nanometer in length) obligate parasite which can multiply only in a living cell of a plant or an animal. Most of the viruses are highly specific to a host.

Bacteria, viruses and protozoa etc. are pathogenic organisms found in waste water. These are usually excreted by person infected with gastrointestinal tract diseases lire typhoid and paratyphoid fever, dysentery, diarrhoea and cholera.

Water Pollution

CHAPTER 6

6.1 Introduction

Water pollution is the degradation of the quality of water that makes it unsafe or harmful to human beings, animals and aquatic life. To a certain extent, contamination of water is inevitable, since water being a universal solvent can dissolve various types of substances in it. Water pollution disturbs the normal use of water for irrigation, agriculture, industries and for human consumption. Water pollution can be caused by natural or anthropogenic processes. Natural processes include pollution of water bodies by decomposed vegetables, animals and weathered products. The anthropogenic processes include industrial, agricultural, radioactive and mining processes.

Water pollution is more dangerous than air pollution. This is because the sources of water contaminants are more diverse. There is slower rate of self purification. Contamination of water is more sensitive to propagation of life in comparison to air pollution.

6.2 Types of Water Pollutants

Water pollutants are of different types, *viz.* biological agents, chemical agents and physical agents.

6.2.1 Biological Agents

These include bacteria, parasitic fungi and protozoa. The main source of biological agents include human sewage, animal and plant wastes, decaying organic matter, industrial wastes (oil refineries, paper mill, food processing units etc.) and natural land and urban runoffs and waste water discharge from slaughter houses. Some of the contributory factors include over population, unplanned industrial and human settlements and lack of civic amenities.

The oxygen requiring bacteria feed on the biological wastes and this results in depletion of oxygen in the water bodies. Life is destroyed in the absence of oxygen. The water acquires foul odour.

6.2.2 Chemical Agents

These can be inorganic chemicals and minerals, organic chemicals and radioactive substances. The **inorganic chemical agents** may be acids, salts (nitrates, phosphates etc.), metals like lead and mercury. The main sources include natural run off from land, industrial wastes, leaded gasoline, lead smelting, agricultural chemicals run off from agricultural lands, mining, oil fields, food processing units etc. These pollutants in water are toxic to various forms of life and human beings and can cause genetic defects. The water is rendered unfit for human consumption (domestic, agricultural and industrial use). It upsets the ecosystem of water bodies and causes eutrophication.

The **organic chemicals** include agrochemicals (pesticides, herbicides), detergents, and chlorine containing compounds, oil, grease and plastics. The main sources include agriculture, forestry, pest control industries, home and industrial wastes, paper industry, bleaching process etc. Such polluted water is toxic to aquatic life and other organisms that are dependent on such water bodies. It may also cause eutrophication of water bodies.

The **radioactive** substances are released into water bodies as a result of nuclear wastes from research laboratories and hospitals. These are also released by processing of uranium ores and nuclear plants. The radioactive substances enter the food chain and are responsible for birth and genetic defects. They are also the causative agents for cancer.

6.2.3 Physical Agents

The **physical agents** include suspended solids, particulates and heat. The main source is soil erosion, run offs from the agriculture, mining and power plants. Suspended organic and mineral solids absorb toxic substances like heavy metals and pass them through food chain. Aquatic animals face problems in breathing through gills in such waters. Thermal pollution occurs when hot water enters the water body. Increase in temperature lowers the solubility of oxygen in water, there by affecting the aquatic life.

6.3 Types of Water Pollution

Water pollution can be divided into following categories depending on the source and storage of water:

(i) Ground water pollution
(ii) Surface water pollution
(iii) Sea water pollution

6.3.1 Ground Water Pollution

Water is considered to be a gift of nature. Over 98% of the fresh water on the earth lies underground in the form of ground water (see sec. 5.2.1). It is the main source of drinking water for about 50% of the population on the global scale.

Ground water gets polluted due to human activities (eg. industrial, domestic and agricultural activities). Generally, ground water contamination is irreversible. This lead to the degradation of water quality, i.e. objectionable taste, odour and excessive hardness. Following table gives the sources of water pollution along with the pollutants incorporated in water.

Table 6.1 Sources of contamination of ground water

Sources of contamination	Contaminants discharged into ground water
Domestic wastes	Pathogenic organisms, nutrients and solids from domestic waste
Industrial wastes	Toxic heavy metals along with hazardous organic and inorganic effluents
Agricultural wastes	Fertilisers, pesticides, insecticides, herbicides. Leaches from agricultural land containing nitrates, phosphates and potash pollute the ground water

Note: Ground water can also be polluted by septic tanks, refuse dumps.

The average chemical composition of ground water is given in Table 6.2. For comparison, the chemical composition of surface water (river) and sea water is also given.

Table 6.2 Chemical composition of ground water, river water and sea water

Parameters (ppm)	Ground water (well)	Surface water (river)	Sea water
Chloride	9.6	13	19×10^3
Bicarbonate	330	110	140
Calcium	92	36	400
Magnesium	34	8.0	1.35×10^3
Iron(III)	0.09	0.02	0.01
Nitrate	13	0.1	—
Potassium	1.4	1.2	380
Silica	10	1.2	6.0
Sulphate	84	22	2.65×10^3
Sodium	8.2	6.5	10.5×10^3
Total dissolved solid	434	1.65	—

Harmful Effects of Ground Water pollution

Polluted ground water is a major cause for the outbreak of epidemics and chronic diseases like typhoid, jaundice, dysentery, diarrhoea, tuberculosis and hepatitis. The use of polluted ground water for irrigating damages the crops and decrease the yield of agricultural products. Polluted ground water affects fertility by killing bacteria and soil micro-organisms. A typical case of ground water pollution was reported on Sept 18, 1988 by the 'Hindu' Newspaper. It reported the contamination of ground water in Medak district of Andhra Pradesh in India from different chemical industries which released poisonous chemicals. The ground water become greenish in color and emitted foul smell. It caused dysentery, vomiting, jaundice and diarrhoea and a number of deaths were reported. Some other harmful effects will form the subject matter of subsequent sections.

In view of the above it is extremely necessary to protect ground water from getting polluted. This can be achieved by taking appropriate steps for predisposal treatment of toxic industrial effluents and municipal discharges. Also, location of wells for drinking water supplies should be very carefully decided.

6.3.2 Surface Water Pollution

Surface water in the form of rivers, streams and lakes is present to the extent of only 0.0091 percent of the total global water (see sec. 5.2). All segments of our society (urban, rural, industrial, agricultural) are responsible for surface water pollution. Various processes and materials may pollute fresh water bodies. Some sources of surface water pollution are given in Table 6.3 below:

Table 6.3 Sources of surface water pollution

Sources of contamination	Contaminants discharged into ground water
Agricultural run offs	Agrochemicals (pesticides, herbicides, fungicides)
Accidental spill of chemicals	Different chemicals
Leaks from surface, storage tanks or pipelines	Gasoline, oil
Run off from industrial sites like factories, refineries, mines etc.	Solvents, chemicals
Radioactive material processing plants	Radioactive materials

The chemical composition of surface water (river water) is given in Table 6.2. The maximum permissible limits of effluents discharged into surface water are given in Table 6.4 below:

Table 6.4 Permissible Limits of Various Characteristics of Effluents

Characteristics of effluents	Maximum permissible limit
pH	5-9
BOD	20 mg/L
Suspended solids	30 mg/L
Particulates	15 mg/L
Cyananide	0.1 ppm
Phenols	0.5 ppm
Sulphate	1 ppm
Toxic metals	0.5 ppm

Depending on the extent of pollution in surface water, it is harmful for life processes and sustenance of life. The harmful effects are similar to that of polluted ground water (see sec. 6.3.1). As far as possible, efforts should be made so that the surface water is free from following contaminants:

- Compounds imparting colour, odour, taste and turbidity eg. oils, greases, phenols, toxic materials.
- Substances which float on the surface as oil, scums etc.

- Toxic radionuclides
- Materials which are harmful for aquatic flora and fauna.
- Thermal effluents should not be discharged in the surface water.
- Chlorinated compounds

Lakes all over the world contribute to the surface water. The largest fresh water lake is Lake Superior in North America covering about 31,800 sq km. The deepest lake is Lake Baikal in Siberia, which is about 1940 m deep. Besides these, other well-known lakes are Dal and Nagin lakes in Kashmir and Nainital (India) and lake Michighan in USA. Due to pollution caused by various industrial, agricultural and chemical run off some of the lakes are in the process of dying.

6.3.3 Sea Water Pollution

Oceans are the major source of water in the world. Water bodies cover more than 70% of the earth's surface. Due to population explosion and depletion of land resources, the food materials (e.g. fish etc.) are being used more and more. In this respect, human beings are responsible for destroying the aquatic environment either by mismanaging or by over exploitation. Oil pollution in the sea is the main factor which poses serious threat to the marine ecosystem and fisheries of the world. This aspect of marine pollution is discussed in chapter 7. Besides these, seawater is polluted by the discharge of various types of wastes from diverse human activities.

6.4 Sources of Water Pollution

Water is the only natural resource available in plenty and is essential for the survival of all living organisms. However, clean (or potable) water has become a precious commodity since its quality is degraded by various sources of pollution. Some of these sources are:

Sewage and Domestic waste
Industrial Effluents
Agricultural discharge /run off
Soaps and detergents
Toxic or heavy metals
Thermal Pollution
Petroleum oil Spills

6.4.1 Sewage and Domestic Waste

The word sewage is generally used for combined liquid wastes discharged from all domestic (toilets etc.) and industrial sources. It consists of mixture of all liquid domestic wastes, including faecal matter, urine, soap and detergents. The total spent water (including sewage) of the community flowing through a sewer is referred to as **municipal waste**.

It is believed that about 75% of water pollution is caused by sewage and domestic wastes. There is uncontrolled dumping of wastes from rural areas, town and cities into water bodies. Due to population explosion, tremendous quantities of sewage are generated. Most of the sewage or municipal waste receive no treatment, particularly in most of the underdeveloped countries and some developed countries of the world.

Characteristics of Sewage

The characteristics of sewage include both physical and chemical characteristics.

Physical Characteristics

(i) **Appearance :** The sewage water normally has a high turbidity and appears greyish. Some times, the colour becomes black due to partial decomposition.

(ii) **Odour :** Sewage after some time develops a characteristic rotten egg odour of hydrogen sulphide. This is due to the putrefaction.

(iii) **Temperature :** The temperature of sewage is normally a few degrees higher that the supply water. The temperature of sewage has an effect on sedimentation and biological activity.

Chemical Characteristics

The chemical characteristics of sewage can be of two types *viz. inorganic* or *organic*.

(i) **Inorganic:** The common mineral matter comprises of nitrates, phosphates, sulphates, carbonates and chloride of calcium, magnesium, sodium, potassium and iron. Besides, some trace quantities of toxic metals such as copper, chromium, zinc, manganese, lead and nickel may also be present.

(ii) **Organic :** The common organic matter includes fatty acids, esters, amino acids, amides, amino sugars, proteins and their decomposition products. These constituents affect the *p*H and the odour.

6.4.2 Industrial Effluents

Industrial wastewater results from all manufacturing industries. These discharges are contaminated with materials used in the plant (starting materials), manufactured end products, intermediate products and by-products. All these materials are responsible for pollution of water bodies. The types of pollutants depend on the various industrial processes.

The industrial pollutants include suspended matter, chlorides, surfactants etc. The most troublesome are sulphates, nitrates, ammonium salts, salts of heavy metals and oils etc. These affect the colour, taste and odour of water in water bodies. These pollutants affect the dissolved oxygen (DO), biochemical oxygen demand (BOD) and chemical oxygen demand (COD) of the water.

Various pollutants from different industries are discussed below:

6.4.2.1 Textile Industry

Textile industry involves various operations and is among the major water users. The wastes from textile industry are variable in character. Many compounds produced by other industries are used in various stages. A composite waste from cotton textile unit comprises starches, dextrins, gums, glucose, waxes, pectins, alcohols, fatty acids, acetic acid, soaps, detergents, sodium hydroxide, carbonates, sulphides, sulphates, chlorides, dyes and pigments, dye carrier like phenols and benzoic acid, peroxides and chloride bleaching compounds.

Organic pollutants (in waste water) such as dyes, starches and detergents undergo chemical changes by using the dissolved oxygen from the receiving water bodies and thus destroy fish life. High concentrations of soluble inorganic salts make the water in the water bodies unsuitable for industrial and municipal use. Metals such as chromium and zinc are toxic to aquatic life. All these pollutants affect the colour and odour of water.

As per environmental laws, it is the responsibility of the textile industries to remove the wastes before the water is discharged into water bodies. It is, of course, best to reuse the processed water in the same industry. This procedure also results in saving the materials which can be used again in various processes.

6.4.2.2 Paper and Pulp Industry

The demand of paper and products obtained from it, have considerably increased. New pulping and bleaching processes have been developed. However, the main pollution associated with paper and pulp industry is the pollution resulting from the discharge of waste materials into the water bodies. The products obtained from paper are fine papers, coarse papers, tissues, specialities papers and cardboards. All these are made from pulp, which in turn is obtained from wood and the supporting and conducting tissues of the trees and bamboos.

The discharge of spent water from pulp and paper industries (into water bodies) contains two types of pollutants; *viz. suspended matter* and *dissolved organic substances*. The suspended matter consists of dispersed solids such as coarse and fine particles of bark, wood slivers and slit. The dissolved organic substances include lignins, tannins, carbohydrates, salts of fatty acids and other organics. Besides above, the wastes contain caustic alkalis (used for pulping), filler and coating materials (eg. clays, calcium carbonate and titanium dioxide), inks, sizing and hydrolysis products. The chemicals used in sizing, coating and bleaching processes form extremely harmful constituents of the waste.

The suspended matter in the wastes can form bottom deposits in receiving water bodies. These are harmful to aquatic life and on decomposition, produce odoriferous components. The water in the water bodies becomes opalescent and so the penetration of sunlight into water is considerably reduced, resulting in reduced photosynthetic processes.

The dissolved organics in the discharged water are slowly acted upon by micro-organisms and converted into simple harmless compounds. However, in this process, oxygen dissolved in water is consumed and this results in the death of some of the aquatic organisms.

A commonly used pulp-bleaching agent is chlorine. Though, it is known for its excellent bleaching ability but when discharged into aquatic systems, it reacts with other industrial effluents to produce a number of chlorinated products; the most harmful of these is tetrachlorodibenzo-*p*-dioxin.

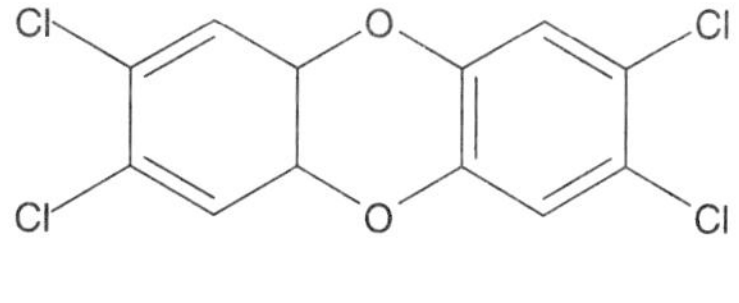

Tetrachlorodibenzo-*p*-dioxin

The pulp and paper industry discharge is rich in sodium salts of resin acids and higher mercaptans, which are toxic to fish and other aquatic organisms.

The pollution caused by paper and pulp industry is minimised by the following procedures:

(i) Removal of particulate pollution : This is affected by allowing the particles to settle down (under the influence of gravity) in settling tanks. In case the suspended particles are light, coagulants such as alum or silica are added to expedite the settlement processes. The particulate settled material is removed from the bottom of the settling tanks and from this material, the fillers, pigments and coating materials might be isolated and reused.

(ii) **Removal of dissolved organic matter :** The effluents after removal of suspended matter is treated with micro-organisms and oxygen in specially designed tanks called the stabilization basins. In these basins, mechanical stirrers stir the liquid material and air is bubbled in. By this process, the organic effluents are biologically oxidised into harmless substances.

(iii) **Recycling of water :** The water obtained after removal of suspended particulate matter and dissolved organic matter (as described in (i) and (ii) above) is recycled. This procedure helps to reduce the discharge of unwanted noxious chemicals in the water bodies.

(iv) **Use of alternative bleaching agents:** Though chlorine is a very effective bleaching agent, it reacts with other industrial effluents to produce toxic chemicals. In view of this, attempts have been made to replace chlorine with other bleaching agents like hydrogen peroxide and ozone. However, it adds to the costs.

(v) **Recycling of waste paper:** It goes without saying that the waste used paper should be recycled. It is very convenient to convert used paper into pulp by treatment with alkali and steam. The pulp, thus, obtained is bleached. This procedure saves a host of other chemicals which are used to convert wood into pulp.

6.4.2.3 Leather Tanning Industry

The raw material in leather industry (hides of cattle and the skins of certain other animals) is mostly the by-product of edible meats, which is produced in slaughter houses. The leather industry is one of the oldest industries. With the advancement of scientific knowledge, leather industry has now developed into a technical industry. Basically, there are only three types of tanning processes.

(i) **Vegetable tanning process:** In this process, sole, belting and saddlery leathers are tanned with infusions of barks, wood, nuts and leaves.

(ii) **Chrome tanning process:** In this process, shoe upper leathers are made from calf, goat, horse or cattle hides tanned with mineral salts. Some leathers are tanned with formaldehyde and some others are tanned with zirconium salts in a process similar to chrome tanning.

(iii) Some leathers are tanned by a combination of vegetable and chrome or in combination with newer synthetic tanning materials.

Whatever process is used, the skin and hides are made free of hair and extraneous tissues before being tanned. This is achieved by steeping the hides in saturated lime solution fortified with sodium sulphide or other materials. The de-haired skin and hides are then subjected to enzyme treatment to remove the final hair debris. In the above process, water is used which should not be hard but it should be free from micro-organisms; for micro-organisms disinfectants are used.

After tanning, the leather is subjected to finishing processes. A number of such processes are in the use. The choice of the process depends on the use for which the leather is put to.

The leather and tanning industry produce effluents which contain small bits of skin tissue, soluble proteins, hair, blood, dirt, salts of many kinds and lime mixed with sulphides, amines, chrome and vegetable tanning agents, sugars, starch, oils and fats, acids, alkalies, dyes and pigments.

The tannery waste contains organic pollutants which are responsible for using the dissolved oxygen in water bodies and so are responsible for destruction of aquatic life. The tannery waste must be evaluated for their biochemical oxygen demand, suspended solids, pH, basicity, colour, turbidity, mineral content etc. The pollution caused by leather tanning industry can be minimised by the following procedures.

(i) **Removal of particulate pollution:** This is carried out as described under paper and pulp industry (see Sec. 6.4.2.2)

(ii) **Removal of dissolved organic matter:** This is carried by the procedure described under paper and pulp industry (see Sec. 6.4.2.2).

The pollution can be minimized by adopting some new procedures in the tanning process. Some of these are as follows:

- In unhairing bath, dimethylamine (DMA) sulphate is used instead of amines. The use of lime is also avoided.
- Use of proteolytic enzymes for loosening the hair.
- Synthetic tanning material is used in combination with natural and mineral tanning materials. Most of the synthetic tanning materials are absorbed by the hide and remain as part of the leather.

6.4.2.4 Agrochemical Industry

Most of the agrochemicals, *viz.* fertilizers, insecticides, herbicides and fungicides are used in large quantities for better and quality crops. These are made by various industrial processes. The discharges from these industries contain various toxic chemicals and when discharged into water bodies are responsible for a number of adverse effects.

Fertilizers are essential plant nutrients. When fertilizers containing sulphates, nitrates and phosphates are added to the soil, some of these are washed off through rain fall, and these agricultural discharges find their way into various water bodies. The excess use of fertilizers distrupt the nitrogen and phosphate balance in water bodies, resulting in increase of BOD of water sources and this may also lead to *eutrophication* of the lakes etc.

Some insecticides, pesticides and herbicides are not biodegradable. When sprayed on crops, they remain adhered to the soil for long periods. During rainfall, these agrochemicals are carried as suspended particles into water bodies and create problems. For a detailed discussion on agrochemicals, see sec. 4.4.6.

6.4.2.5 Petrochemical Industry

The basic raw material for petrochemical industry is petroleum, which is thick and usually dark coloured liquid, found below the earth's crust trapped within rock structure or below the water in the oceans. The origin of the word *petroleum* is from the Latin words *'petra'* (meaning rock) and *'oleum'* (meaning oil).

Following are the various operations performed in the petroleum industry.

(i) **Isolation of Crude Oil :** It involves isolation of crude oil (petroleum) from underground sources (below the surface of earth or oceans). Wastes from production operation by drilling include mud. Even after separation of crude oil, some amount of crude oil remains in the mud and accompanying water (water is used for drilling purposes for cooling the pipes). Though water does not come in contact with oil directly, but some oil may enter water by leakage etc.

(ii) **Transportation :** The crude oil is transported to the refineries through pipelines or oil tankers. Transportation via pipe lines is a safer way, though sometimes leaks in pipes may occur. Leakage of pipe lines under the surface of water in the oceans may lead to ocean water pollution (see Marine Pollution, chapter 7). However, transportation by oil tankers is not very safe. The oil tankers may meet with accidents resulting in spilling of oil which causes pollution. Wastes from cleaning of transportation equipment also require special precautions and treatments.

(iii) **Refining :** Petroleum is a mixture of aliphatic and aromatic hydrocarbons. These are separated by fractional distillation into various fractions (as given in Table 6.5).

Table 6.5 Some important fractions of petroleum

Fraction	Building range	Approximate composition
Gaseous	Up to 30°C	C_1 - C_4
Petroleum ether	30 - 70°C	C_3 - C_7
Gasoline (Petrol)	40 - 200°C	C_5 - C_{10}
Kerosene oil	170 - 290°C	C_{10} - C_{16}
Heavy oil	250 - 320°C	C_{15} - C_{19}
Lubricating oil	320 - 400°C	C_{17} - C_{50}

Water is used in refining operations for cooling etc. The waste water is usually contaminated with various fractions of the fractional distillation of crude oil.

The most important fraction obtained by refining of petroleum is gasoline (petrol) which is used as an automobile fuel. However, the yield of this fraction is quite low. Due to increased demand of this product, the yield of this fraction is increased by cracking. By this procedure, higher boiling fractions are converted into lower boiling fractions. There are two types of cracking. The first is *liquid phase cracking*, in which heavy oil is cracked by heating (470-525°C) under pressure (10-70 atmospheric pressure); the yield is 60-65%. The second is the *vapour phase cracking*. In this process, kerosene oil or any other material is heated at 600°C under 3-10 atmospheric pressure. Following are the products of cracking of some of the hydrocarbons:

$$\underset{\text{Butane}}{C_4H_{10}} \rightarrow CH_3CH=CHCH_3 + CH_3CH_2CH=CH_2 + CH_3CH=CH_2 + CH_2=CH_2 + CH_4 + H_2$$

$$\underset{\text{Octane}}{C_8H_{18}} \rightarrow C_8H_6 + C_6H_{14} + C_4H_{10} + C_4H_8 + C_2H_4 + CH_4 + H_2$$

$$\underset{\text{Dodecane}}{C_{12}H_{26}} \rightarrow C_7H_{16} + C_7H_{14} + C_6H_{14} + C_6H_{12} + C_5H_{12} + C_5H_8$$

The mechanism of cracking is complex. Besides molecular disintegration, it also involves polymerisation, alkylation, cyclisation, aromatisation etc. Some of these are discussed below:

(i) **Polymerisation :** The hydrocarbons - butane, butene isobutene etc. produced during cracking undergo polymerisation. For example, isobutene on dimerisation gives *iso*-octene, which on hydrogenation gives isooctane having high octane number.

Isobutene

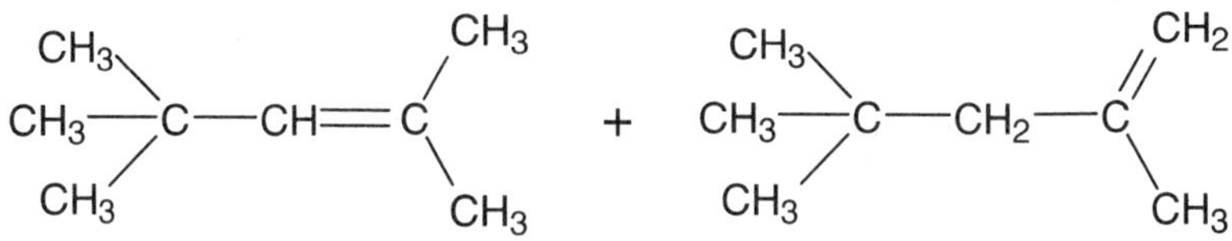

Isooctene

(ii) Alkylation : Highly branched hydrocarbons, which constitute high quality gasoline are formed during alkylation involving addition of saturated hydrocarbons to alkenes. Thus, addition of isobutane to isobutene gives isooctane.

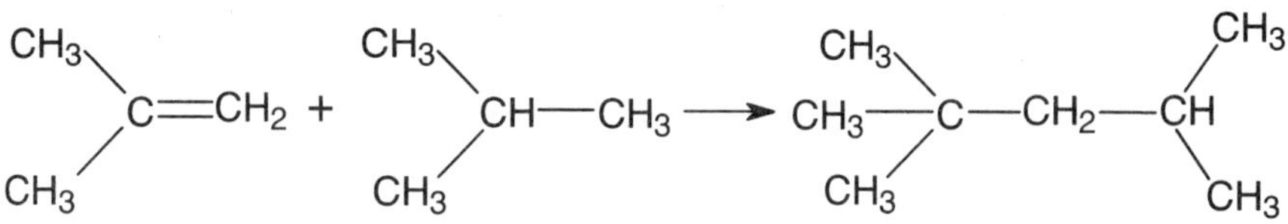

Isobutene Isobutane Isooctane

(iii) Reforming : This process is used to increase the octane number and consists of heating the gasoline at 600^0 under pressure (28-50 atm.). In this process, no cracking takes place, but the molecule is reformed. In reforming, there may be isomerization, dehydration, cylisation, and aromatisation. Some examples are as follows:

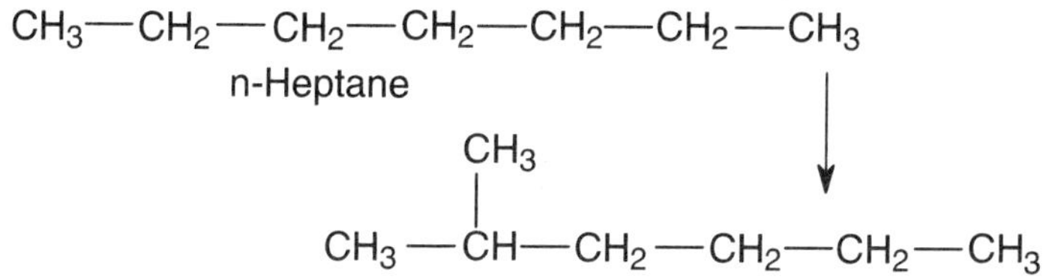

2-Methylhexane

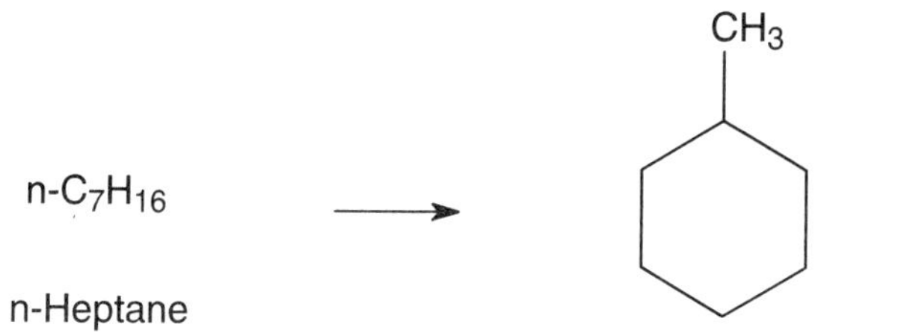

Methyl cyclohexane

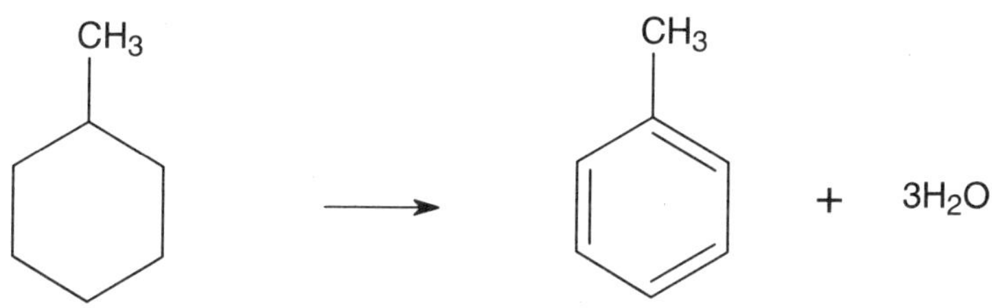

Methyl cyclohexane Toluene

Reforming can also be carried out in presence of catalysts like oxides of chromium, vanadium, molybdenum or aluminium at 150-300^0.

(iv) Production of petrochemicals : A large number of organic chemicals are produced from petroleum sources. These are obtained either from natural gas or any other chemical obtained during the refining of crude oil. All such chemicals that are derived from petroleum sources are called *petrochemicals*. Petroleum is a source of approximately one million organic compounds. The main processes that convert petroleum hydrocarbons into raw materials for chemical industries are (a) removal of hydrogen (Dehydrogenation) to give unsaturated hydrocarbons (alkenes or aromatics), (b) oxidation and (c) chlorination. Wastewaters from petrochemical industries comprise surface run-off, process water and cooling water. Water immiscible oils and solvents with high BOD along with solid matter are common to both oil refineries and petrochemical industries.

The wastewaters from petroleum and petrochemical industries usually contain:

- Chemicals including acids, alkalies, sulphides, mercaptans, ammonia, phenols and suspended solids.
- Separable and emulsified oils are found in wastewater sources. Sulphides are obtained from cracking units; these also contribute to phenolic wastes.
- Many oxidisable solutes present in refinery wastewaters contribute to overall oxygen demand. Bad taste and odours in waste waters are caused by phenolic, naphthenic, nitrogenous and organic sulphur compounds; all these result from treating operations for removal of sulphur, nitrogen and oxygen compounds from crude and distilled fractions.

It is best to treat the wastes discharged from the petrochemical industries before they are allowed to find their way into water bodies. Some of the procedures for waste disposal or treatment are given below

(i) In-House Procedures - Specific precautions and procedures should be adopted to prevent entry of the wastes into the water bodies. These procedures involve recovery of raw materials, intermediates and final end products from the discharges. Reuse of water in cooling towers is helpful. The discharges usually contain oil, which can be separated.

(ii) Water soluble volatile components from wastewater streams are removed commonly by steam distillation in specially designed equipments. The wastewater contains varying amounts of sulphides, ammonia and phenolics depending on the type of the crude oil; these are obtained by the distillation of crude oil and in catalytic cracking. Oxidation of sulphur compounds and ammonia is carried by air.

(iii) Some selected wastes are combusted in specially designed incinerators.

(iv) Chlorinated hydrocarbons obtained in wastes of petrochemical manufacturing processes are removed by absorption in towers.

(v) Addition of certain reagents like aluminium sulphate, ferric chloride and aluminium chloride to waste waters help in chemical flocculation. In this procedure, the fine particles coagulate forming bigger molecules which settle down and can be removed.

(vi) Biological oxidations are extensively used for treating refinery wastewaters.

6.4.2.6 Food Processing Industries

In food processing industries, certain additives are added to improve its appearance, flavour, texture and shelf life. These food additives

- improve or maintain the nutritional quality of food. An example is the addition of vitamin D and A to edible oils including butter.
- improve the shelf life of food. An example is addition of calcium propionate to make bread.

Citric acid is added to preserve fruit juices. Also, artificial colours and flavours are added to make the food more attractive.

Table 6.6 gives some of the additives along with their functions.

Table 6.6 Some common additives used in food industries

Type of additive	Examples	Functions
Preservatives	C_6H_5COONa (sodium benzoate) used in soft drinks, $(CH_3CH_2COO)_2Ca$ (calcium propionate) used in meat products; sugar used in jellies. Sorbic acid ($CH_3CH = CH{\cdot}CH = CH{\cdot}COOH$) used in cheese. Ethylene oxide, to control microorganisms in dried fruits and in spices, Sulphites used to prevent enzymatic browning of fruits and vegetables.	Prevent or retard the undesirable changes caused primarily by bacteria, yeast, and molds.
Antioxidants	Butylated hdyroxy anisole (BHA) 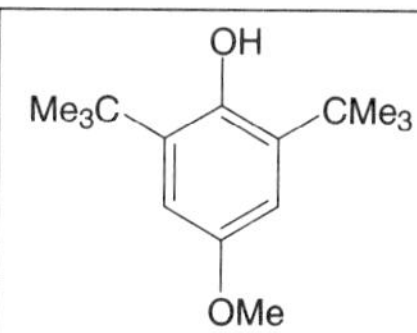	Prevents oxidation of fats, Prevents development of rancidity.
	Butylated hydorxy toluene (BHT) 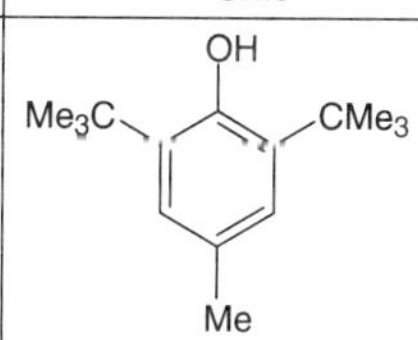	Increases shelf life.
	Propyl gallate 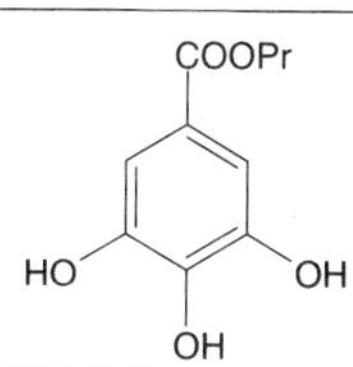	
	Sulphites-besides being preservatives also act as antioxidants	
Sequestrants	Adipic acid $HOOC(CH_2)_2CO_2H$ 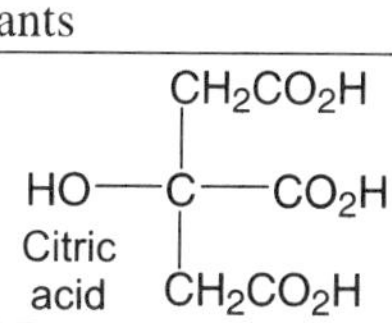	These are metals (e.g. iron, copper etc. present in processed food) scavengers and are generally chelating

	Ethylenediamine tetra acetic acid (EDTA) 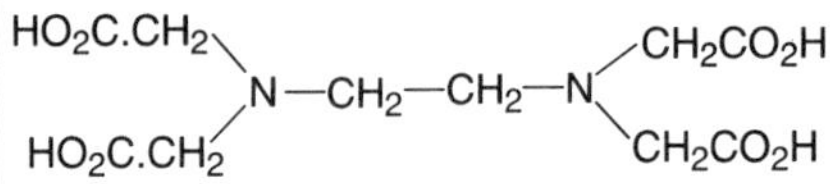	agents. The metal ions if not removed cause deterioration of foods.
Acidulants	Lactic acid [$CH_3CHOHCO_2H$] used in fruit juices, squashes. Phosphoric acid, H_3PO_4 lactic acid, citric acid acts as acidulant as well as sequestrant.	These are organic acids and enhance flavour and control pH, and assist antioxidants
Surface Active agents	Lecithin, mono and diglycerides R = alkyl group 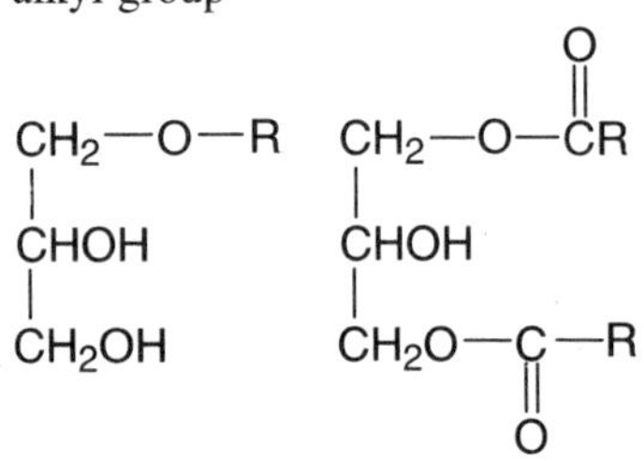 Monoglyceride Diglyceride	These are wetting agents, emulsifiers and lubricants. These make bakery goods to stay longer and keeping fresh.
Bleaching agents	Benzoyl peroxide $C_6H_5-\overset{O}{\overset{\parallel}{C}}-O-O-\overset{O}{\overset{\parallel}{C}}-C_6H_5$ for making milled flour colourless or white. Bromates and iodates are used to bleach bread.	These remove undesirable colours from food stuffs
Natural supplements	Vitamin A added to edible oils. Vitamin B, iron and calcium added to cereals. Vitamin C added to fruit Juices. Vitamin D added to milk Iodine added to salt Lysine (an amino acid) added to wheat flour	These enhance food value of products.
Colorants	Natural colours like turmeric, saffron, paprika and caramel Artificial colourants like indigo, carmine, erythrosine, tartrazine etc.	These make food attractive
Artificial sweeteners	Saccharin, cyclamic acid	These are added when low caloric diet is preferred. Useful for diabetes and weight control measures.
Flavours	Vanillin-imparts vanilla flavour Ethyl butyrate-pineapple flavour Methyl anthranilate-grape flavour	These are added to improve the acceptability of foods.

Though additives are used for improving the quality, appearance, flavour, texture and shelf life of food products, these should be used in right amounts, since some of them may have harmful effects. The harmful effects of some of the food additives are given below:

Sodium Nitrite

It is used as an additive for cured meat products. It prevents botulism in cured or canned meat products. It also acts to stabilize the colour and flavour of the meat. Sodium nitrite is known to inhibit growth of bacterium.

Clostridium Botulinum

This bacterium if not checked, produces poisonous toxins. The nitrite in food products is partially converted into nitrous acid which may react with secondary amines in the meat to yield *N-nitrosamines*.

$$R_2NH + HNO_2 \xrightarrow{H^+} R_2N\text{-}NO + H_2O$$

These nitrosamines are believed to be carcinogenic. The nitrites may find their way into human stomach when cured meat is consumed. The best option is to use minimum amount of nitrite (about 200 ppm per kg of meat product).

Radiation

Radiation is not an additive, but it is a non-chemical method of food preservation. Beneficial effects are extension of storage life of crops, disinfection of cereals, fruits and destruction of pathogens. In fact, irradiated foods retain their freshness, flavour and nutritional value. However, the consumers are concerned by the fear that radiation of food material may generate toxic radiolytic products and may result in the formation of radicals which are very harmful.

BHA and BHT

The antioxidants BHA and BHT exhibit anti-microbial activity against several common food pathogens and viruses. These inhibit the oxidation of food fats, especially unsaturated fats in the body. In experimental animals, it has been found that the liver gets enlarged and the rate of synthesis of DNA and RNA gets slowed down, which in turn reduces the rate of cell division and growth. However, these effects have not been observed in humans. One thing which has been established is that if the oxidation of fats in the body is not prevented, peroxides are produced which are definitely carcinogenic. The recommended daily intake of BHA and BHT is 0.3 mg and 0.125 mg, respectively per kg of body weight.

Diethyl Pyrocarbonate

It is used as an anti-microbial food preservative in fruit juices, wine and beer. Normally, it gets hydrolysed to ethanol and carbon dioxide. However, the beverages usually contain small amount of amino acids and proteins, which slowly hydrolyse to give ammonia. The ammonia so formed reacts with diethyl pyrocarbonate in the pH range 4-9 to form urethane (ethyl carbonate, H_2N-COOEt) which is carcinogenic. WHO has recommended that diethyl pyrocarbonate should be used in those beverages which have a pH lower than 4 and a low concentration of amino acids and proteins. The recommended level of diethyl pyrocarbonate is about 300 mg per kg of food item.

Phosphoric Acid

It is used in soft drinks (cola types). Excessive intake of phosphoric acid reduces the level of calcium and leads to osteoporosis - a condition in which bones becomes brittle and more susceptible to fractures.

Potassium Bromate

It is used as a bleaching agent for flour. In experimental animals, it has been found that in bromate fed rats, there is increased incidence of kidney tumours. Due to this reason, the use of bromide in flour is significantly curtailed. Some countries have banned the use of bromate as an additive.

Artificial Sweeteners

Saccharin and cyclamates are commonly used artificial sweeteners in a variety of foods. It has been found that in experimental animals, both saccharin and cyclamates produced bladder cancer. In case of cyclamate, the metabolite cyclohexamine is a cancer promoter. WHO has recommended a daily intake of 0.43-4.3 mg and 4-10 mg per kg body weight of saccharin and cyclamate, respectively. In some countries pregnant women are advised to avoid saccharin and cyclamates.

6.4.2.7 Dairy Industry

Dairy industry includes processing of milk to produce homogenised pasteurised milk, milk powder, cheese and other products. The wastes from dairy industry resemble other food wastes but do not contain suspended solid materials. These wastes arise during handling of milk at different stages, eg. receiving section, bottling, condensing, dry milk powder manufacture and cheese and butter making along with ice-cream manufacturing. The wastes from dairy plants consist of various dilutions of milk that enter the drainage system. The wastes also include by products such as buttermilk and skim milk that cannot be utilized. Most of the dairy wastes (expect the by-products) can be reduced by proper handling and controls so that there is no spilling or leakage. The by-products such as whey, buttermilk should not be allowed to enter the plant sewer system.

The dairy wastes are composed of organic matter and are high in oxygen demand. If allowed to enter the water bodies, the oxygen in these is considerably depleted causing serious pollution. Milk wastes also contain nitrogen and phosphorous, which are harmful if allowed to enter the water bodies.

The dairy wastes are allowed to settle in settling tank. Here, lactose is converted into lactic acid which precipitates the casein. Air is bubbled into these tanks which is helpful for biological oxidation of organic matter. The solid material is filtered before the wastes are discharged into water bodies. However, if suitable land is available, disposal of milk wastes by irrigation is possible; this is the most convenient and economical procedure of dairy waste disposal.

In units dealing with the manufacture of soymilk, the waste consists of particles of soybeans. These are removed by filtration and then used for cattle feed.

6.4.2.8 Soaps and Detergents

Soaps are sodium salts of fatty acids. They are obtained by the saponification of fats and oils. Synthetic detergents (or *syndets*) are soap substitutes and comprise of surfactant, builder and other ingredients. Soaps are not effective in hard water (there is considerable loss of soap before it is available for cleaning purpose) compared to detergents which can be used in hard water. A common class of detergents is alkylbenzene sulfonates. Both soaps and detergents contain non-polar or hydrophobic (water repelling) group (the fatty acid chain in soaps and the alkylbenzene group in detergents) as well as a polar or hydrophilic (water attracting) group (the carboxylate group in soap and the sulfonate group in detergents). Due to the presence of hydrophobic as well as hydrophilic groups, both soaps and detergents lower the surface tension of water and aid in the wetting of clothes, dishes etc. and in emulsifying and dispersing of soil or dirt particles; such substances are called **surfactants**.

The soaps cause virtually no environmental problems, since they are biodegradable. In aquatic systems, they are degraded by micro-organisms in presence of oxygen into harmless products (CO_2, H_2O, carbonates and bicarbonates). The disadvantage of using soaps is that they do not function efficiently in hard water. The calcium and magnesium ions present in hard water react with soap to produce precipitate of calcium and magnesium salts of fatty acids.

$$2\,RCOO^-Na^+ + Ca^{2+} \longrightarrow (RCOO)_2Ca + 2\,Na^+$$

$$2\,RCOO^-Na^+ + Mg^{2+} \longrightarrow (RCOO)_2Mg + 2\,Na^+$$

This leads to wastage of soaps. Also, the precipitates of calcium and magnesium salts adhere to the fabrics and stain them. Further, in acidic medium, the soaps are precipitated as fatty acids.

The detergents, on the other hand, are not completely biodegradable. When discharged into water bodies, a part of detergents always persists and causes environmental problems. It has been found that detergents having branched chain are less biodegradable than those having a linear carbon chain. Some examples are as follows:

Linear chain detergents (more biodegradable)

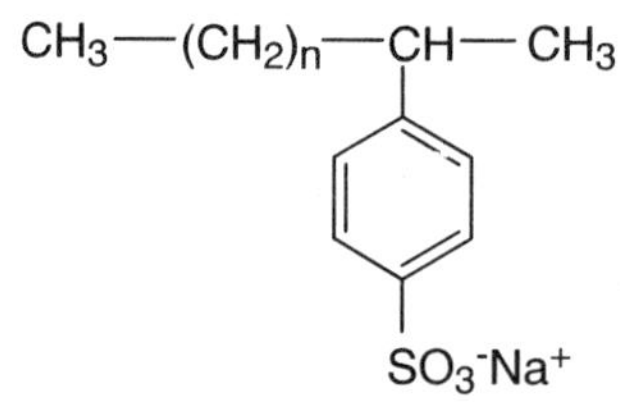

Alkyl benzene sulphonate

$$CH_3—(CH_2)_n—CH_2—OSO_3^-Na^+$$

Teepol

Branched chain detergents (less biodegradable)

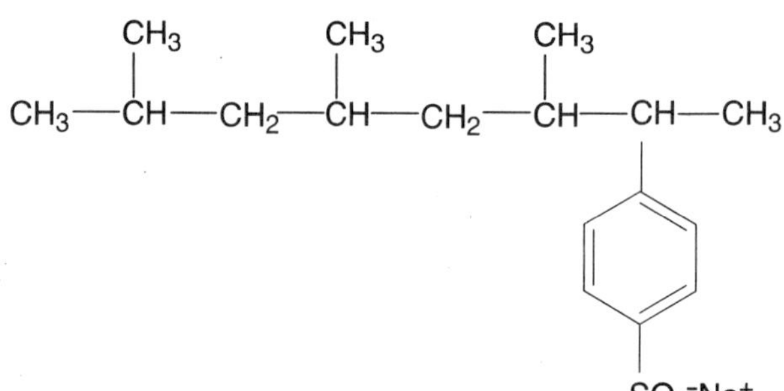

alkylbenzene sulfonate

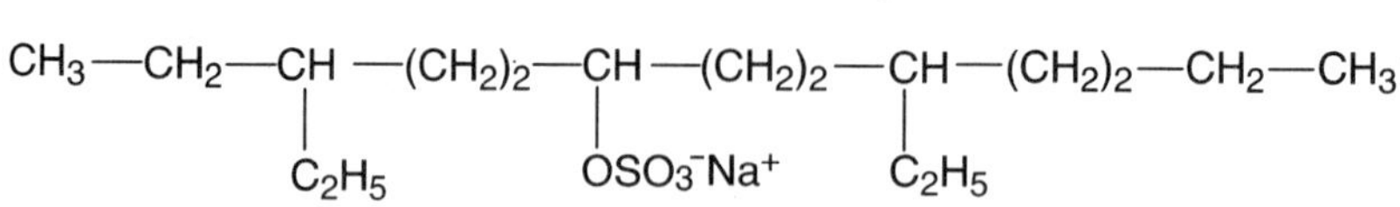

Tangitol 7

Types of Surfactants

The surfactants can be anionic, cationic, non-ionic or amphoteric in nature. Table 6.7 gives example of various types of surfactants along with their applications.

Table 6.7 Types of Surfactants

Types	Examples	Uses
Anionic	Soaps 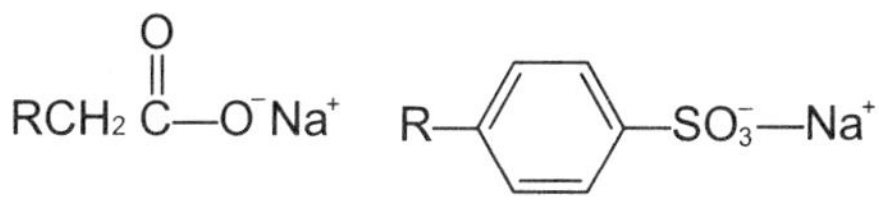 $R{=}C_{10}{-}C_{15}$ Alkyl benzene sulphonate	Most commonly used for toilet soap, laundry detergents and shampoos.
Cationic	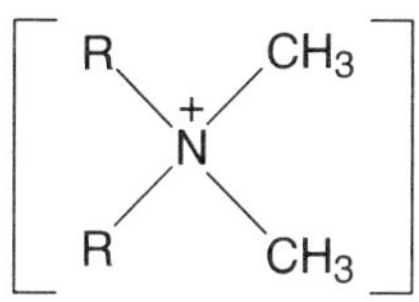 Cl^- $(R{=}C_{10\text{-}18})$ Quaternary ammonium salts 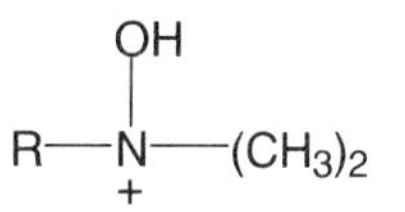 $(R{=}C_{14\text{-}18})$	Used mostly as fabric softeners
Non-ionic	$R(CH_2CH_2O)_n$ H Polyoxyethylenes $(R{=}C_{6-16};\ n{=}3\text{-}15)$	Used for heavy duty laundry detergents
Amphoteric	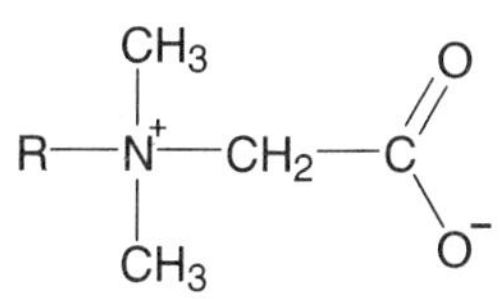	Used for shampoos

Ingredients of Detergents

As already stated, the detergents are of synthetic origin and contain surfactants (mostly anionic like alkyl benzene sulphonates) and builder along with other additives. Some of the additives are given below in Table 6.8.

Table 6.8 Additives of Detergents

Additive	Example	Role of additive
Builders*	Polyphosphates*	Better cleaning action even in hard water
Brighteners	Ultramarine	Improve the appearance of the fabric
Bleachers	Perborates	Removing stains etc.
Enzymes	Amylase	Removing starch , protein based stains
Softeners	Quaternary ammonium salts	Making fabric feel soft

(Note: Builders, the most important additive will be discussed in detail subsequently)

Builders are the most important constituents of detergents and are present to the extent of about 40% of the total weight. Following are the most useful and powerful builders.

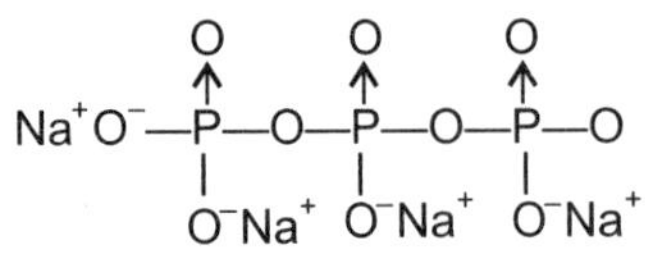

Sodium tripalyphosphate

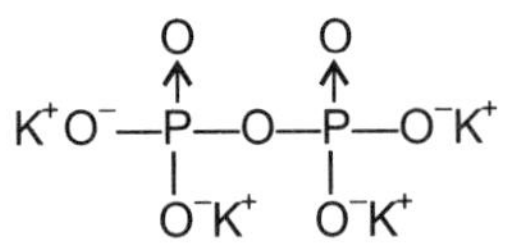

Potassium polyphosphate

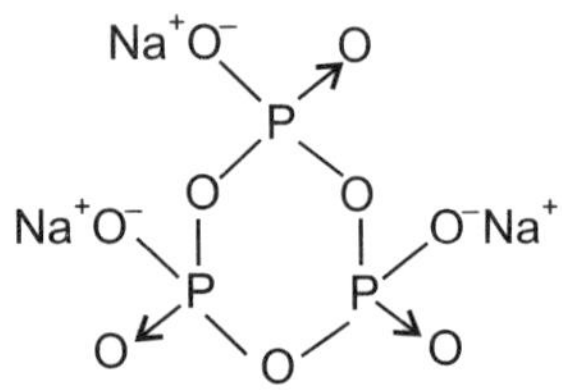

Sodium tripalyphosphate

Following are given some of the common and useful functions of builders

(i) The builders act as sequestering agents. These block the Ca^{2+} and Mg^{2+} ions present in hard water.

$$K_4P_2O_7 + 2\,Mg^{2+} \longrightarrow Mg_2(P_2O_7) + 4\,K^+$$

It is due to this reason that detergents are effective even in hard water.

(ii) Builders along with surfactants exhibit marked synergistic effect; this makes their combination a much better cleaning agent.

(iii) Builders make the wash solutions alkaline -a requisite necessary for effective removal of soil and dirt particles.

(iv) Builders act as buffers– this implies that change of pH during washing is taken care of.

(v) Builders are responsible to make aggregates of surfactant molecules in a concentrated wash solution commonly called **micelles**. These micelles remove oil and grease particles (from the fabric to be cleaned) by dissolving them.

The main problem with polyphosphate additives (builders) is environmental. These cause eutrophication of water bodies. This aspect is discussed in a subsequent section.

Effects of Soaps and Detergents on Environment

Soaps, as already stated, are biodegradable. This implies that when discharged into water bodies, these are acted upon by micro-organisms in presence of dissolved oxygen and converted into harmless products (e.g. carbon dioxide, water, carbonates and bicarbonates). On the other hand, detergents affect the environment. Following are some of the harmful effects of detergents on the environment.

(i) Detergents containing phosphate as builders contribute to the growth of algae in water bodies. This results in the consumption of the dissolved oxygen and the aquatic animals are deprived of oxygen resulting in their death. Thus, phosphate detergents are the main cause of eutrophication of water bodies, particularly lakes.

(ii) Aquatic animals and water plants are severely affected by synthetic detergents. A concentration of 2.5-3.5 percent of common household detergents causes mortality to some types of fish and retards the growth of (or even kill) the water plants.

(iii) Health hazards- The workers in detergent industry are exposed to enzyme additive in the formulation section. The enzymes attack the proteins in the soft lung tissue resulting in allergic conditions and respiratory problems.

(iv) The potable water acquires objectionable taste.

(v) The detergents make the water purification process difficult. This is because, the micro-organisms which oxidise organic pollutants to harmless materials are killed. Also, with detergents, the oily and greasy impurities are emulsified and their removal becomes difficult by skimming process.

The best way to check the environmental problems is to use non-phosphate builders. Some builders which have been tried are EDTA, sodium citrate, zeolite-A [$Na_2O.Al_2O_3.2SiO_2.4.5H_2O$] and nitrilotric acetic acid [$N(CH_2COOH)_3$]. However, none of these is as efficient as the phosphate additives.

6.4.2.9 Hazardous Wastes

Hazardous wastes are mostly the by-products of various industrial set ups. Some of these are known carcinogens and some are suspected carcinogens. Table 6.9 gives various types of hazardous wastes and their sources.

Table 6.9 Hazardous wastes

Type of hazardous waste	Source
Cyanide wastes	Electroplating industries and some metallurgical operations
Metal finishing wastes	Metal polishing industries
Wastes containing water soluble compounds of lead, copper, zinc, chromium, nickel, selenium, barium, antimony (heavy metals)	Extraction of various metals from their ores
Mercury, arsenic, thalium and cadmium wastes (heavy metals)	Metallurgical Processes
Waste from paints, pigments, glue and printing inks	Paint and pigments industries
Dyes	Dye industries and dyeing of fabrics
Oil and oil emulsions Tary wastes and tar residues	Petroleum extraction and refining
Asbestos	Mining and construction industries
Agrochemicals	Agro industries

In addition to the hazardous wastes mentioned above (see table), some more examples of hazardous wastes include:

- plastic materials generate harmful combustion products when disposed off by burning; the toxic materials generated include carbon monoxide, sulphur dioxide, nitrogen oxides, ammonia, formaldehyde, hydrogen cyanide, hydrogen chloride etc.
- the insulating fluid used in the electrical transformer on burning produces toxic materials containing polychlorobiphenyls (PCBs) which take very long time for biodegradation.
- the nuclear waste generated in nuclear plants is highly dangerous and may cause various complications.
- the sludges arising from treatment of waste waters containing heavy metals, toxic organic compounds, oils, emulsions etc. are extremely harmful.
- the discarded containers containing hazardous and toxic wastes are also hazardous waste.
- traces of heavy metals (see table above) such as Hg, Cd, Pb, As, Co, Mn, Fe and Cr are deleterious to aquatic ecosystem. Wastes containing high concentration of toxic metals are

extremely toxic to all organisms. These metals are cumulative poisons and if accumulated in the food chain can have very severe consequences. A detailed discussion on the heavy metal is given in sec. 5.4.12 under toxic elements in water.

Management of Hazardous Wastes

The methods of hazardous waste management include

- Reduction of waste at source
- Recycling, recovery and reuse
- Treatment
- Disposal

The first method, *viz.* reduction of waste at source is the most preferred one. The procedure to be adopted for any process should be the one which generates no or the least amount of wastes. These include

- Recycling of a waste at site
- Improving process, technology and equipment which is the main source of waste generation
- Improvement in plant operations

The common types of treatment are chemical, physical, biological or thermal treatment. Thermal treatment using incineration technique is the preferred procedure of hazardous waste management. Treated hazardous wastes are disposed off on land which is not a permanent solution since it is only a transfer of wastes from one place to another.

6.5 Impact of Water Pollution on Human Health

We now know that water is easily polluted by physical and chemical agents present in the atmosphere, microbial disease causing germs, nuisance organisms like slime, molluscs, sewage and industrial effluents and agricultural based effluents. All these pollutants affect our health by causing various diseases. They also effect our other resources such as animals, forests etc. on which we directly and indirectly depend for food and survival.

6.5.1 Water as a Carrier for the Transmission of Diseases

A number of micro-organisms (eg. Bacteria, viruses, protozoans, insects and helminthes etc) present in polluted water can cause infectious diseases.

These diseases are called water-borne diseases. These water-borne diseases can be present in the form of outbreaks. This usually happens if the dose of contamination is heavy. Cholera and jaundice (hepatitis) usually occur in the form of outbreaks, which can be very explosive. However, when contaminated water is treated or not used, the outbreak subsides (eg. In the case of cholera). The following table gives diseases caused by micro-organisms commonly in the polluted water. These diseases are the important causes of ill health particularly among the people in the developing countries.

Table 6.10 Diseases caused by micro-organisms commonly in pollutant water

Organism	Diseases	Remarks
Ascaris sp.	Nematode worms	Danger to man from polluted water and dried sludge used as a fertilizer
Bacillus anthracis	Anthrax	Found in waste water spores are resistant to treatment.
Brucella sp.	Brucellosis, Malta fever in man, contagious abortion in sheep, goat and cattle.	Transmitted by infected milk by contact, by waste water
Entameoba histolytica	Amoebic Dysentery	Spread by contaminated water
Leptospira sp.	Leptospirosis (Weil's Disease)	Caused by sewer rats
Mycobact.tuberculi	Tuberculosis	Isolated from waste water
Salmonella paratyphi	Paratyphoid fever	Common in waste water.
Salmonella typhi	Typhoid fever	Common in waste water.
Salmonella sp.	Food poisoning	Common in waste water.
Shigella sp.	Bacillary dysentery	Polluted water
Vibrio cholarae	Cholera	Polluted water
Virus	Poliomyelitus hepatitis	Waste water treatment plants
E. coli	Diarrhoea	Polluted water

6.5.2 Chemical Toxicants in Water Affecting Health

We are aware that there are a number of chemicals in drinking water. These may be naturally occurring or reach the water due to human activities. The presence of these chemicals in water do not pose health problems if their levels are below the threshold level. However, if these toxicants are present in excessive amounts, their presence is a matter of great concern since excessive amounts are detrimental to the health. Various chemicals which may pose a threat are: Fluoride, Arsenic, Lead, Nitrates, Pesticides, Cadmium, Chromium and Cobalt. Let us see these chemicals one by one.

Fluoride

The major source of fluoride to man is drinking water. In most parts of India the fluoride content is about 0.5 mg/L^{-1}. There are, however some areas where fluoride content in drinking water may be as high as 3-12 mg/L^{-1}. The recommended level of fluoride in drinking water in India is 0.5 to 0.8 mg/L^{-1}.

High intake of fluoride causes humped back. Continuous consumption causes stiffness of bone joints, particularly spinal cord. It accumulates in bones causing outward bending of legs from knee-called **Knock Knee Syndrome**. Water rich in fluoride cause endemic fluorosis. It was first detected among cattles by the farmers of Andhra Pradesh in early 1930's.

Excess fluoride is removed by Nalgond techniques. It involves the addition of lime and alum in sequence followed by flocculation, sedimentation and filtration.

Arsenic

Arsenic is introduced into water via the dissolution of minerals and ores, from industrial effluent and from atmospheric deposition. A provisional guideline value for arsenic in drinking water of 0.01 mg/l. World's worst affected areas due to arsenic contamination is the Indo-Bangladesh region. WHO has alarmed govt. of India to keep Arsenic related disease under immediate national surveillance, giving it a status similar to that of cancer. WHO prescribed limit to 0.01 mg/l.

At higher levels, As(III) compounds coagulate proteins, possibly by attacking S-bond retaining the secondary and tertiary structure of proteins. In the biological systems, Arsenic compounds complexes will coenzymes resulting in coagulation of biological proteins and uncoupling at phosphorylated sites. The effect on human depends on the arsenic dosage.

Mild Dose – nausea, fainting, salivation, vomiting, burning pain in Stomach.

Higher Dose – diarrhoea, peripheral neuritis, hyperketosis, conjunctivitis.

Chronic dose – severe gastroenteritis, loss of weight, skin lesions, black food disease.

Lethal dose – death due to shock and vascular failure.

Arsenic is a cumulative poison. In animals, it affects the organs like liver lungs causing toxicity to biological systems. Arsenic in water may be carcinogenic in nature.

Lead

The important sources of lead for biological systems are mining, smelting, winning of metals, automobiles etc. In urban areas, citizens are exposed to it through emission of vehicles and food. About 200-300 mg/day of lead intake through food and about 10-15 mg/day through air and water is quite common in urban settlements. Lead is present in tap water to some extent as a result of dissolution from natural sources but mainly from household plumbing systems containing lead pipes. It is satisfying to note that emission of lead from vehicular emissions has ceased since leaded petrol is no more used in cities like Delhi.

When present in <40 mg/100 mL of blood, lead is considered as normal. However >100 mg/100 mL of blood is unacceptable.

Lead gets deposited in bones and is considered as a cumulative poison. Lead is a general toxicant that accumulates in the skeleton. Infants, children up to six years of age and pregnant women are most susceptible to its adverse health effects. Lead interferes with calcium metabolism, both directly and by interfering with vitamin D metabolism. Lead is toxic to both central and peripheral nervous system, inducing subencepalopathic neurological and behavioral effects.

Since most of the lead in drinking water arises from plumbing in buildings and so the remedy consists in removing plumbing and fittings containing lead.

Nitrates

Both nitrate and nitrites are naturally occurring ions that are part of the nitrogen cycle. Naturally occurring nitrate levels in surface and ground water are generally a few mg/L^{-1}. In many ground waters, an increase of nitrate level has been observed due to intensification of farming practice.

The danger of drinking water containing high nitrate content has long been recognized. Live stocks have been effected by well water containing 75-10 ppm of nitrate nitrogen. Human infants have been susceptible, especially bottle–fed infants, of less than 3 months of age are most susceptible.

Nitrate poisoning brings about a condition called **methemoglobinamia**, resulting from change in the hemoglobin of the red blood cells that reduce their capacity to carry oxygen. This is referred to as blue baby syndrome.

The Table 6.11 given below gives the toxic and pathological effects of some heavy metals.

Table 6.11 Hazardous effects of some metals on Human beings

Metals	Impact
Arsenic	– Adversely affect peripheral circulatory systems and CNS, hyperkeratosis, ulcer in gastrointestinal tract, liver cirrhosis.
Barium	– Abdominal pain, diarrhoea, vomiting, excess salivation
Cadmium	– Diarrhoea, growth retardation, malfunctioning of urinogenital system, CNS, Kidney, Liver
Copper	– Sporadic fever, uremia, coma, hypertension
Cobalt	– Diarrhoea, lowering of blood pressure, paralysis, respiratory problems, deforming of skeleton
Chromium (Generally Hexavalent form)	– Malfunctioning of gastrointestinal, CNS and urinogenital systems
Mercury	– Haemolysis, diarrhoea, severe headache, paralysis, desiccation of intestinal epithelium, ultimately death, loss of appetite, convulsions, brain damage, malfunctioning of liver, kidney, anaemia, vomiting
Zinc	– Abdominal cramps, vomiting, diarrhoea, renal damage

6.5.3 Water as a Vector Habitat in Disease Transmission in Humans

Following are the most important water-related vector borne diseases

- Malaria
- Schistosomiasis (found mostly in sub Sahara Africa)
- Dengue fever
- Filariasis
- Japanese encephalitis

6.5.4 Treatment of Waste Water

The waste water is the spent water supply of the community. It comprises of sanitary sewage, domestic sewage (arising from kitchens, bathrooms, laundries, and laboratories etc.) and wastes (discharged into water bodies) arising from industrial units.

The waste waters arising from industrial set ups must be treated appropriately to remove all possible impurities before discharging into water bodies. It is however, best course that the industrial units should use the spent water (after purification) in their own units.

The waste from sewage and domestic discharges must be treated before being discharged into the water bodies.

Methods of Sewage Treatment

The treatment procedures and generally divided into the following three groups,

- Primary Treatment
- Secondary Treatment
- Tertiary Treatment

Removal of Solids, which settle down is called primary treatment. Secondary treatment involves removal of solids with the aid of living organisms (biological treatment). The combination of primary and secondary treatment constitutes complete treatment. The purification of waste water by advanced biological, chemical processes and also the recycling of waste water constitute the tertiary treatment. For details see sec. 5.2.3, sub-sec. iv.

6.6.5 Eutrophication

When sewage and agricultural run-off containing phosphates (phosphates are major ingredients of most of the detergents) or other nutrients enter natural water bodies, they cause over nutrition leading to eutrophication. In such a situation the water bodies became highly productive or eutrophic and the phenomenon is called eutrophication. The ponds or lakes etc. during their early stages of formation are comparatively barren or nutrients-deficient-thus supporting no or poor aquatic life. This stage of these water bodies is known as Oligotrophic. With the introduction of nutrients, there is a stimulated luxuriant growth of algae in water. This growth depletes the dissolve oxygen and causes the death of the other aquatic flora and fauna and ultimately the lake dies.

Marine Pollution

CHAPTER 7

7.1 Introduction

The discharge of waste substances into the sea, which results in harming aquatic organisms, human health and impairment of the quality of seawater is called *marine pollution*. It is normally associated with physical, chemical and biological conditions of seawater.

7.2 Sources of Marine Pollution

The water bodies, *viz*., rivers, seas and oceans are used by mankind as a dustbin for dumping all types of waste materials. These include industrial wastes, heavy metals, agro chemicals, mine tailings, urban and rural sewage, farm and forest run-offs; these wastes include non-biodegradable components, which are the major cause of problems. Even radioactive wastes are dumped into the seas. Infact human activities are basically responsible for marine pollution. The marine pollution is caused by the following:

(i) Dumping of hazardous materials

(ii) Oil pollution

(iii) Pollution by radioactive materials

7.2.1 Hazardous Materials

These have been discussed in secs. 5.4.12 and 6.4.2.9.

7.2.2 Oil Pollution

This is a major cause of marine pollution. At present, the annual consumption of refined petroleum products all over the world is more than 50 billion barrels (about 1600 billion gallons). Such consumption is associated with some losses which can be accidental or intentional and are responsible for marine pollution. It is estimated that the total annual influx of petroleum oil into the oceans is more than 250 million gallons.

The discharge of petroleum oil or its products into the oceans occurs at various stages in the petroleum industry. These include the following:

- Extraction or isolation of crude oil (petroleum) from underground sources which can be below the surface of earth or oceans.
- Transportation – The crude oil is transported through pipelines or oil tankers. Leakage may occur in both the modes of transportation. Major marine pollution occurs during transportation in huge oil tankers, by sea, which may spill oil in the oceans or in case these tankers meet with an accident there may be disastrous oil pollution.
- Refining of petroleum oil in various refineries.
- Manufacture of petrochemicals which are made from various fractions of petroleum refining.

  The oil tankers are cleaned periodically with water and the washings discharged into the seas. Approximately, 3 million tonnes of oil is discharged annually by this process.

For more details see sec. 6.4.2.5.

It is estimated that out of all marine pollution by petroleum oil, about 200 million gallons of oil come into seas annually from natural seepage of the annual discharge of petroleum into the seas, oil spills account for about 8 – 9 percent while oil drilling and extraction contribute about 3 – 4 percent.

7.2.2.1 Composition of Crude Petroleum

The composition of crude petroleum varies with occurrence. However, it is a complex mixture containing hundreds of compounds. These compounds include alkanes (paraffins) (about 25%), cycloalkanes (cycloparaffins) (about 20%), aromatics (about 5%) and naphthalene. Besides these, the petroleum also contains compounds containing oxygen, nitrogen and sulphur. Some typical examples of various types of compounds that occur in crude petroleum oil are given below:

Alkanes

$$CH_3-CH_2-CH_2-CH_2-CH_2-CH_3$$

n – Hexane

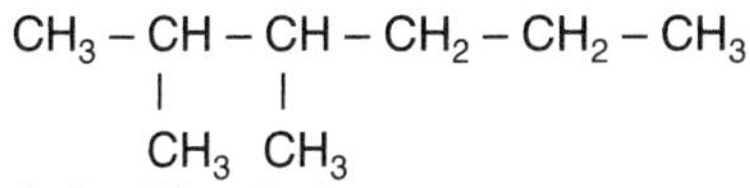

2, 3 – Dimethylhexane

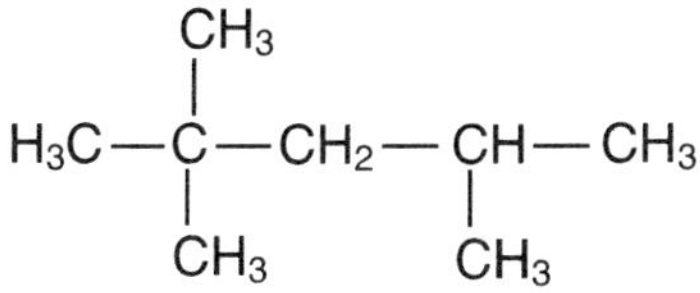

2, 2, 4 – Trimethylpentane

Cycloalkanes

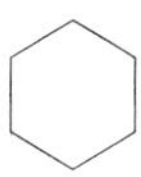

Cyclohexane

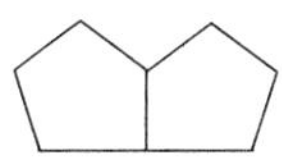

[3.3.2] – Bicyclo-octane

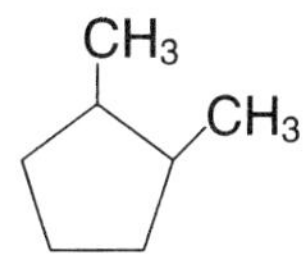

1, 2 – Dimethylcyclopentane

Aromatics

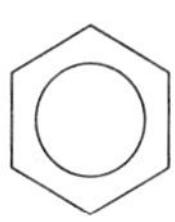

Benzene

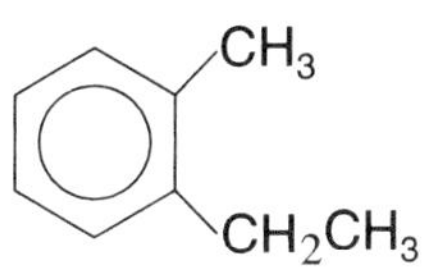

2 – Methylethylbenzene

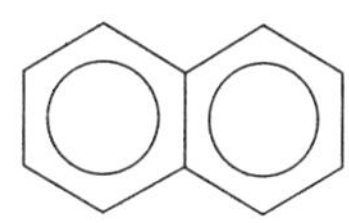

Naphthalene

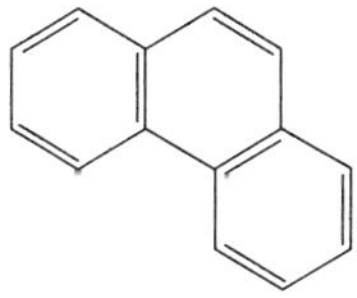

Phenanthrene

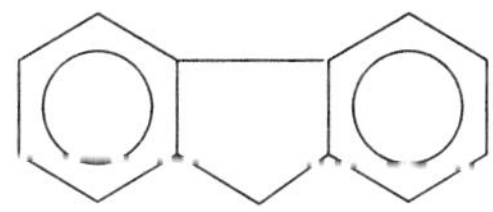

Fluorene

7.2.2.2 Some Major Oil Tanker Accidents

Massive marine pollution in the seas is caused by oil tanker accidents. Following are some of the major oil tanker accidents in the seas.

- In 1967, a Liberian tanker, Torry Canyon got wrecked on the Ollard Rock near Land end, England resulting in the discharge of about 100,000 tonnes of oil into the sea. The spill of oil contaminated more than 200 km of Cornich coast line in England. This oil spill affected even France.

- In 1969, in the United States, offshore oil blew out the coast of Santa Barbara resulting in the discharge of oil at the rate of 1,000 gallons per hour spreading about 10, 000 metric tonnes of oil over the surface of water.
- In 1973, the oil tanker, M.V. Cosmos Pioneer broke into two near Porbunder in Gujarat, India, due to rough weather, releasing about 18,000 tonnes of light diesel oil in the Gujarat coast.
- In 1974, an American oil tanker, 'Transhuron' collided the atolls of the Kilton Island in the Laccadives sea discharging about 5000 tonnes of furnace oil on the beaches.
- In 1976, the oil tanker 'Urquiola' met with an accident near the Spanish coasts La Covuna and discharged about 100,000 tonnes of crude oil.
- In 1989, an oil spill was reported near the Alaska coast. The oil tanker, Exxon Vaaldez crashed on Prince William sound in Alaska discharging 11 million gallons of oil into Alaska's waters.
- In 1992, a super tanker met with an accident near the Northwestern Coast of Spain.
- In 1991, during Gulf War about 300 million gallons of oil in the Persian Gulf were discharged in the water. The oil spill hit the Saudi shores.

The National Academy of Sciences, Washington has estimated that about 8 – 9 million metric tonnes of oil enters the ocean annually.

7.2.2.3 Effects of Oil Pollution

Given below are some of the adverse effects of oil pollution.

(i) The oily surface markedly reduce the rate of oxygen uptake by water.

(ii) Oily slit decreases markedly the penetration of light; this retards photosynthesis in aquatic plants.

(iii) Lichens and algae along the shore lines are destroyed.

(iv) Fishery is hampered.

(v) Soluble aromatics present in oil affect the aquatic organisms. Some of the aromatic compounds are carcinogenic and affect plant metabolism.

(vi) The aromatic compounds in oil are very harmful to mankind. These produce irritation, anesthesia and damage liver and kidneys.

(vii) Birds are particularly vulnerable to damage from oil spills in the oceans. Hundreds and thousands of birds die due to the effects of oil.

7.2.2.4 Control of Oil Pollution in Seas

A number of procedures are available to clean up the mess created by oil spills in the oceans. However, the best course is to avoid oil spills by taking adequate measures. Some of the methods to control the spreading of the oil are given below:

(i) **Skimming** : The heavy oil fraction floating on the surface of the sea are best removed by skimming operation. This is the simplest and the best method. This method should be used in the shortest possible time after the spilling has taken place, otherwise the wind will spread the oil spills over a very large surface and then the skimming becomes very difficult.

(ii) **Use of Absorbents**: Some suitable absorbing materials like polyurethane, saw dust, chopped straw are spread on the surface of oil and then skimmed off.

(iii) **Burning of Oil Slick**: This method though useful causes extensive air pollution and may also cause thermal pollution.

(iv) **Chemical Additives**: These can be used to solidify oil from water surface and then the solid material is easily skimmed.

(v) **Micro - organisms**: The most effective method to clean up oil pollution is by using certain oil – eating bacteria. The oil – eating microbes were used in cleaning up oil spills in the waters of the state of Texas, U.S.A.

7.2.3 Pollution by Radioactive Materials

Radioactive pollutants enter the water bodies (especially seas and oceans) from various sources, some of which are given below:

(i) Mining and processing of ores to produce unstable radioactive substances.

(ii) Nuclear Power Plants

(iii) Nuclear Reactors

(iv) Nuclear Weapons

(v) Use of Radioactive isotopes in medical, industrial and research applications.

(vi) Testing of nuclear weapons in air, under water give radioactive fallout.

7.2.3.1 Types of Radionuclides

Nuclear weapon testing produces, Sr^{90}, a long-lived component of radioactive fallout. It is chemically similar to calcium and accompanies calcium in soil, plants and animals and finally finds its way via food chain into humans in bones and teeth. The presence of Sr^{90} in bones is responsible for disorders in blood – cell formation and causes anemia and more serious disorders.

Nuclear power plants generate liquid wastes (low level radioactivity), liquid and gaseous wastes from fuel elements, fission products and heat. The discharge of radioactive wastes in air and water bodies causes problems in the long run. Table 7.1 lists some of the radioactive materials discharged from various activities.

Table 7.1 Radioactive Materials Discharged

Activities	Radioactive Materials Obtained
Leaching of minerals	^{226}Ra, ^{40}K
Atomic Reactors	^{90}Sr, ^{131}I, ^{137}Cs, ^{140}Ba, ^{141}Cs, ^{85}Kr, ^{60}CO, ^{54}Mn, ^{55}Fe, ^{239}Pu
Neutron bombardment of atomic fuel	Radio-active Pu, Np, Am, Cm, Bk, Cs, Zn, Ru, etc.

7.2.3.2 Effects of Radionuclides

The level of radionuclides found in water is measured in units of pico curies ($=10^{-12}$ curie $= 3.7 \times 10^{-3}$ disintegrations per second). The permissible limit is 3 pico curies. Following are given some of the bad effects due to radioactive pollutants in water.

(i) They cause serious skin cancer, carcinoma, melanoma, breast cancer, leukemia, breakage of DNA and cataracts.

(ii) They cause the so-called radiation sickness, which is characterized by nausea, vomiting, diarrhoea, anorexia, epilation, lethargy and weakness.

(iii) They cause genetic disorders, gene mutations and blood abnormalities.

(iv) The radionuclides present in water get accumulated in soil sediments, water and aquatic ecosystem.

7.3 Thermal Pollution

When the temperature of water - body or air in the atmosphere is increased or lowered, we call it thermal pollution. The temperature of water bodies regulates the aquatic life. If the temperature of oceans is lowered by even one degree, the environment becomes harmful to some corals and reef species. Similar effect is felt if the water temperature increases. Thermal pollution occurs when the heat generated in any process is released into water bodies. Some of the causes of thermal pollution include:

- Forest fires and volcanic eruptions
- Metal smelters, processing melts, petroleum refineries, paper mills, food processing units
- Chemical manufacturing plants, power plants and refrigeration units.

In all the above processes, water is used as a coolant. Water gets heated and the hot water is released into water bodies.

Thermal pollution has profound effect in aquatic system. Some of these effects are

- Reduction in dissolved oxygen thereby affecting aquatic life.
- Increase in BOD.
- Multiplication of bacteria, which in turn became food of protozoa.

The best course to remove chronic thermal pollution is to retain the heated water and effluents discharged by power plants and other industrial units in a holding section and cool them prior to their discharge in water bodies. Alternatively, the heat generated from industrial units can be used for purposes like warming the buildings in colder seasons.

Soil

CHAPTER 8

8.1 Introduction

Soil plays an extremely important role since it produces food for all. The word soil is derived from a Latin word *Solum* meaning earthy materials in which plants grow. It is defined as a weathered layer of the earth's crust with living organisms and their products of decay. Good quality of soil and an appropriate congenial climate besides water are a must for the production of food materials on which the survival of mankind depends. However, human activities, e.g., deforestation, mining, destruction of grazing lands, industrial discharges etc., are responsible for severe pollution of soil. Many pollutants like insecticides, fertilizers, etc. add to the problem of soil pollution. Soil also plays a vital role in environmental chemical cycles.

8.2 Formation of Soil

Soil is described as loose, powdery earth in which plants grow. It is made up of small pieces of rock and decayed plant and animal materials. The pieces of rock or particles of rock were earlier parts of bigger rocks and the animal or plant materials originate from dead animal bodies and plants.

The formation of soil from rocks has taken centuries. The rocks get broken up into pieces in several ways. Some of these are given below:

The crumbling and wearing away of rock is known as **weathering**. The agents responsible for weathering of rocks are running water, moving ice or strong waves. The soil formation started with weathering of rocks by physical, chemical or biological agents. The weathered rocks under the influence of various other pedogenic processes get converted into the so-called matured soil. The physical weathering agents originate from climate. These include temperature, water, ice, gravity and wind. A rock becomes very warm with sun's heat which makes the rock expand. During night, the rock cools more quickly from outside than from inside. This results in the contraction of rocks from outside and cracks develop on its outer layer. The cracked layers of rock break into small fragments as pieces. Water may run into cracks which freezes with a change in temperature. The freezing of water causes the cracks to become deeper. At times, due to falling of seeds into the cracks, sprouting of seeds takes place and the roots of the plant grow; this makes the cracks bigger and larger. Also, lichens, a type of plants which do not have any roots but have acid present in them, cause chemical changes which make the rocks crumble.

The glaciers push bigger pieces of rocks ahead of them as they move along. This pushing helps the rocks to crumble. Water and chemicals present in it dissolve and wear away some kind of rocks. Once the process starts, the remaining rock crumbles. The landslides caused by earthquakes are also responsible for fragmentation of rocks by forces of impact and abrasion. The wind helps in carrying small fragmented particles of rocks to the surface of the earth.

Chemical weathering of rocks occurs along with physical weathering. The chemical weathering agents are water, oxygen and carbon dioxide. Water is an important weathering agent due to its solvent action. Water soluble minerals like limestone, gypsum, etc. from the weathered rock increase the presence of carbon dioxide and organic acids, formed as a result of decay of remains of animals and plants. Water is also responsible for hydrolysis along with carbon dioxide. Thus, the silicates in the weathered rock are hydrolysed.

$$K_2Al_2Si_6O_{16} + 2\,H_2O + CO_2 \longrightarrow Al_2O_3{\cdot}2SiO_2{\cdot}2H_2O + 4\,SiO_2 + K_2CO_3$$

This process releases Ca, Mg, K, Na and silicates into the soil. These materials increase the fertility of resulting soil.

The oxygen reacts with minerals in rocks to produce oxides, which dissolve in water and the resulting hydroxides bring about further weathering. The CO_2 and H_2O combine to form carbonic acid, which is also responsible for the weathering of rocks.

The weathering of rocks involving cracking and crumbling of the rock is the beginning of the soil making. To make the real soil, humus must be added to the sand or fine particles of rock. *Humus* is an organic matter and is obtained from dead animals and plants. The bodies of almost all animals and plants become a part of soil through the action of bacteria. Bacteria cause the plants and animals to decay and make the soil fertile. Many kinds of insects and earthworms help to make the soil rich.

8.3 Characteristics of Soil

Soil is a complex mixture of different types of materials and is derived from the rock, organic matter and other materials carried by geologic agents and biotic community. The soil characteristics depend upon its texture and structure.

Soil texture is dependent on the size of the original rock particles in the soil. Soil texture may be of following types:

(i) **Sandy Soils :** These contain loose and dry sand particles (particle size 1.00 to 0.05 mm) along with clay (10%) and silt (size 0.05 to 0.002 mm). Such soils are poorly aerated, have low water holding capacity and have low nutrients.

(ii) **Clay Soils :** These contain clay particles (size less than 0.002 mm) and inhibit free circulation of air and water. These are unsuitable for cultivation, since they become water logged easily. Clay soils contain sand, silt and clay in the ratio 2: 2: 1.

(iii) **Loam Soils :** These soils are most suitable for vegetation, since these have good aeration properties of larger particles. They contain nutrients and have moisture retaining ability. These contain sand, gravel, silt and clay in the ratio 2: 2: 1: 1.

The manner in which different constituent particles aggregate with each other to form clumps determines the **soil structure.** Sandy soils have bigger particles, do not clump and so lack structure. Large clumps are formed if small clay particles stick together in closed packed manner. Loam soils are good soils and contain a balanced composition of large and small particles. They form clumps of reasonable size. Such soils crumble easily creating passage for air, water and nutrients.

8.4 Soil Profile

Soil is comprised of distinct layers which are called soil horizons. All soil horizons taken together form soil profile. Fig. 8.1 below shows different horizons of soil with increasing depth.

Fig 8.1: Soil Profile

The top layer or 'A' horizon is several inches in thickness. This layer is most important for vegetation and agricultural crops as it is rich in humus. Deforestation over the years is responsible for the loss of topsoil leading to considerable decrease in the agricultural crops.

The next layer is 'B' horizon and is also known as **sub soil**. It receives organic matter, salts and clay particles by leaching from the topsoil. It contains little humus also. Below 'B' Horizon is the 'C' Horizon which is composed of weathered parent rocks from which the soil originated. The lowest layer is hard zone of rocks called **bedrocks**.

8.5 Composition of Soil

The chief components of soil are inorganic matter (90 – 95%) and organic matter (5 – 10%). Besides, soil contains water and air. The composition of soil varies considerably depending on the type of soil.

8.5.1 Inorganic Components of Soil

Most of the inorganic components of soil exist as silicate minerals. The common elements in soil are oxygen (46.6%), silicon (27.7%), aluminium (8.1%), iron (5.0%), calcium (3.6%), sodium (2.8%), potassium (2.6%) and magnesium (2.1%). The silicates which are commonly found in soil are quartz (SiO_2), orthoclase ($KAlSi_3O_8$), albite ($NaAlSi_3O_8$) and epidote ($4CaO\,(AlF)_2O_3{\cdot}6SiO_2{\cdot}H_2O$), Besides iron oxides, FeO(OH), magnetite (Fe_3O_4) are also present in many soils. Certain soils contain titanium and manganese oxides. Another common component of soil is calcium carbonate.

Soil also contains mixtures of minerals as carbonates, sulphates, chlorides and nitrates. Some soils contain clay minerals, which are hydrated aluminium and iron silicates. These clay minerals bind cations such as Ca^{2+}, Mg^{2+}, Na^+, K^+, NH_4^+, so that these cations are not leached by water and are available in soil as plant nutrients.

8.5.2 Organic Components of Soil

The organic components are responsible for the fertility of the soil. The major organic compounds in soil are organic matter, polysaccharides, amino sugars, nucleotides, amino acids, proteins, organo - sulphur and phosphorus compounds. The humic materials constitute the most important compounds which form complexes with a number of cations, so that these are retained in the soil. The major organic compounds in soil are given below:

(i) **Humus :** The humic substances, in general, are high molecular weight macromolecules. These are obtained in soil as degradation resistant residue from plant decay.

(ii) **Lipids :** These are extractable by organic compounds and constitute fats, resins and waxes. The physical properties of soil depend on these components.

(iii) **Saccharides :** These are the major food materials for soil microorganisms. Their examples being cellulose, starches, hemicelluloses, etc.

(iv) **Nitrogen containing Organics** : These provide nitrogen fertility in soil. Their examples include amino acids, amino sugars, etc.

(v) **Phosphorus containing Compounds** : These are source of phosphorus in soil. Their examples are phosphate esters, phospholipids, etc.

8.5.3 Water and Air in Soil

Water is essential for the sustenance and growth. It is an important medium for transporting essential plant nutrients from soil into the roots of plants and other part of the plant body. The excess of water from plant leaves finds its way into the atmosphere by the process of *transpiration*. In the soil, water stands at a certain level commonly called **water table**. If the water table is too low, the plants are unable to get the required amount of water. On the other hand, if the water table is too high, the plant drowns. Most of the crops except rice cannot grow on water logged soils.

Water comes in the soil mainly through dew, rain, snow, hail, irrigation, etc. Water may also be present in soil as water of crystallization of mineral grains and water of hydration of clay mineral particles. Free water is held in the soil by adhesion and cohesion.

The air in the soil contains oxygen, nitrogen and carbon dioxide. Air in the soil contains more moisture and carbon dioxide but less of oxygen as compared to that present in the atmosphere. The composition of air in soil is affected by temperature, pressure, rainfall, wind, etc. The decay of organic matter in the soil consumes O_2 and produces CO_2. As a result, soil contains less oxygen (about 15% compared to 21% in the atmosphere) and more CO_2. The presence of CO_2 in soil lowers the pH of ground water.

8.6 Macronutrients and Micronutrients in Soil

It is well known that plants require nutrients for their growth and they are also beneficial to mankind. The nutrients that are needed by organisms in large amounts are called **macronutrients** and those which are needed in trace amounts are called **micronutrients**.

The macronutrients that are present in soil are carbon, hydrogen, oxygen, nitrogen, phosphorus, calcium, copper, iron, magnesium, sulphur, sodium and potassium. The atmosphere and water are the sources of carbon, hydrogen and oxygen. Nitrogen may be obtained by some plants directly from the air with the help of nitrogen fixing bacteria. The other macronutrients are present in the soil. Nitrogen, phosphorus and potassium (NPK) are generally added to the soil as fertilizers.

Soil may become deficient in calcium due to uptake by plant, leaching by carbonic acid, etc. Calcium deficient soils are treated with lime to provide the necessary supply of calcium to plants. The latter process is called **liming**.

For plants, the essential micronutrients are aluminium, boron, bromine, zinc, cobalt, iodine, copper, iron, manganese, zinc, vanadium and molybdenum. These elements are required at trace levels. If present at higher levels, they have toxic effects.

From soil, most of the nutrients find their way into plants and when the plants / crops are consumed by humans; these nutrients are used in various life processes. When plants, animals and human beings die, the nutrients return to the soil. In this way, most of the elements are recycled.

8.7 Sources of Essential Elements in Soil

The essential elements on which the fertility of soil depends are *nitrogen*, *phosphorus* and *potassium*.

Nitrogen is made available to the soil in the form of organic material resulting from the decay of dead plants and animals and human wastes. The organic material is hydrolysed to NH_4^+, which in turn is oxidized to NO_3^- by various natural processes in soil. The plants use NO_3^- from soil. When nitrogen is added to soil in the form of fertilizers (like ammonium sulphate, urea, etc.) nitrifying bacteria convert NH_4^+ into NO_3^-, which is used by plants. Some leguminous plants have the inherent ability to fix atmospheric nitrogen through nitrogen fixing bacteria in their root nodules. Legumes are capable of adding considerable quantities of nitrogen to the soil.

Phosphorus is essential for the growth of plants. The main sources of phosphorus are rocks from which it finds its way into soil by weathering or mining. The decay of dead organisms also releases phosphorus in the soil. In soils, the plants utilize phosphate as $H_2PO_4^-$ and HPO_4^{2-}.

Potassium is essential for plants. It activates some enzymes and plays a vital role in the water balance in plants. It is, of course, available in the earth's crust but is not easily available to plants. The fertilizers mostly used contain nitrogen, phosphorus and potassium (NPK) as the major components. These provide the essential elements in the soil.

Soil Pollution

CHAPTER 9

9.1 Introduction

Soil is the dumping ground of most of the waste products – domestic, human, animal, industrial and agricultural. Every year the solid wastes dumped into the soil are increasing at an alarming rate all over the world. Large number of hazardous chemicals and several thousand tons of wastes are ultimately dumped on the land. These are leached by municipal and industrial wastes and are responsible for pollution of ground water (also see sec. 6.3.1). The problem of soil pollution is compounded by the use of agrochemicals, e.g., pesticides, fungicides, bactericides, insecticides, biocides, fertilizers and manure. Besides these, the soil is polluted by deadly pathogenic organisms. The soil pollution differs from air and water pollution in the sense that the pollutants in soil remain in direct contact with the soil for relatively longer periods. In fact, the soil has been heavily polluted as a result of industrial revolution and green revolution. Basically humans are responsible for the pollution of land.

9.2 Sources of Soil Pollution

The main sources of soil pollution are as follows:

- Industrial Wastes
- Urban and Domestic Wastes
- Radioactive Pollutants
- Agrochemicals
- Chemical and metallic pollutants
- Biological agents
- Acid rain

9.2.1 Industrial Wastes

Industrial wastes are the major cause of soil or land pollution. These wastes are discharged from textile industry, paper and pulp industry, leather tanning industry, petrochemical industry, food processing industry, soap and detergents industry, dairy industry, metal processing industries, drugs, glass, cement and engineering industries. It is estimated that about 50% of the raw materials become waste products in industry. These pollutants pollute not only land or soil but also water bodies (For details see sec. 6.4).

With the advancement of new technology, new types of industrial wastes are produced and discharged on to the land. These waste products are tipped on soil, increasing the extent of soil pollution. The furnaces of industries like thermal and electric power plants produce 'fly ash' i.e., unburnt blackish substance which severely pollute air, water and soil.

The pollution of soil affects and alters the chemical and biological properties of soil. These pollutants enter the food chain from the soil, disturb the biochemical processes and seriously affect the living organisms.

Industrial sludges are more dangerous than industrial solid wastes. The industrial sludges vary enormously in composition and include boiler scales which contain calcium carbonate and flue gas sludge. This flue gas desulphurization sludge results when limestone or calcium hydroxide slurries are used to trap sulphur dioxide from gases discharged in coal fired power plants. These wastes also contain toxic volatile elements like arsenic, selenium, mercury, lead and cadmium.

9.2.2 Urban and Domestic Wastes

The urban and domestic wastes combined are referred to as **refuse**. This comprises of garbage and waste materials like plastics, glass, metallic containers, fibres, papers and a number of discarded products. It is estimated that in UK and India each, about 15 million tonnes of domestic wastes are disposed off into the land and in USA about 4,00,000 tonnes of solid waste is produced daily. The situation is so alarming that the land fills in use for more than 25 years cannot take any more refuse.

The urban and domestic wastes are more harmful since most of these cannot be easily degraded. Certain items like paints, varnishes and oils etc. create more soil pollution problems. The leaches from the dumping sites and sewage along with industrial effluent are extremely harmful and dangerous. These leaches contain poisonous gases along with partially decomposed materials from food remains, vegetables, toxic chemicals and pathogenic bacteria which can be very harmful and may cause various diseases. These are responsible not only for soil pollution but also for pollution of water bodies.

A practice that is followed in some countries for disposal of urban and domestic wastes is its incineration but this procedure is very expensive.

9.2.3 Radioactive Pollutants

The main sources of radioactive pollutants in soil are as given below:

(i) **Radioactive fall - out** resulting from testing of nuclear weapons. Considerable amounts of fission products, particularly C-14, Sr -90, I-131, Cs-137, etc. are dispersed in the atmosphere as a result of nuclear weapon testing. These nuclides spread all over the globe and ultimately settle on earth's surface as radioactive fall – out, from where these find their way into the food chain. Most of the nations of the world are in favour of complete Nuclear Test Ban Treaty. However, so far only few countries have signed this treaty.

(ii) **Radio-isotopes** – These are used in research, industry and medicine and are also generated by the operation of nuclear reactors for generation of power and the nuclear fuel processing plants. (For more details see sec. 1.5.2). Nuclear reactors produce waste containing Ruthenium – 106, Iodine-131, Barium-140, Lanthenium-140, Caesium – 144, Promethium – 444, along with Strontium – 90 and Caesium – 137. All the nucleotides get ultimately deposited in the soil.

(iii) **Natural Radioactive Nuclides** – The earth's crust contains some radioactive nuclides like U – 238, Th – 234, Ra – 226 which continually emit radiation. The food crops grown in the soil and the drinking water percolating through the soil also contain some radioactive nuclides like Potassium – 40, Carbon – 14 and Rn – 222. These enter the body through food chain.

(iv) **Accidents in Nuclear Reactors** – These are, of course, rare but if take place, they become the major cause of emission of radio isotopes or nucleolides into the air and the land. Typical disastrous nuclear accidents are the Three Mile Island nuclear plant accident (USA, 1979) and Chernobyl disaster (USSR, 1986).

9.2.4 Agrochemicals

The agrochemicals like fertilizers, pesticides (insecticides, fungicides and herbicides) etc. are commonly used for agricultural crops. These agrochemicals particularly pesticides (which are very harmful and are not biodegradable) find their way into food chain to be consumed by humans and animals and cause a number of health hazards. (For more details, see sec. 4.4.6).

9.2.5 Chemical and Metallic Pollutants

The chemical and metallic pollutants result from a large number of industries including textiles, pesticides, paints, dyes, soap, synthetic detergents, tanneries, drugs, batteries, cement, asbestos, rubber, petroleum, paper and pulp, steel, glass, electroplating and metal extraction and processing industries (For details see sec. 4.4.1, 4.4.2 and 4.4.3).

A number of trace elements such as Fe, Co, Ni, Cu, Zn, Ba, Pb, V, Mn, Ni, As, Hg, Mo and Si are added to the soil in one form or the other. Synthetic chemicals and fertilizers are a source of trace metals which are added to the soil. For example, As, Pb, Cd are the common trace metals in rock phosphate and also occur in super phosphate fertilizers. Sources of some metallic pollutants and their effects are discussed in sec. 4.4.3).

9.2.6 Biological Agents

The biological agents are responsible for the contamination of soil and crops. These include pathogenic organisms which occur naturally in contaminated soil or are discharged into the soil by excreta of birds, animals and humans. In addition to this, faulty sanitation, municipal garbage, waste water are also responsible for the discharge of biological agents into the soil. Intestinal parasites cause the most serious soil pollution.

9.2.7 Acid Rain

We know that air contains SO_2 and NO_2 as pollutants (see sec. 4.3.2 and 4.3.3). Sulphur dioxide present in the air undergoes catalytic and photolytic oxidation to form SO_3 which reacts with moisture present in the atmosphere to produce sulphuric acid (H_2SO_4).

$$SO_2 + O \longrightarrow SO_3$$
$$SO_3 + H_2O \longrightarrow H_2SO_4$$

The NO_2 present in the atmosphere reacts with rain water or moisture in the presence of O_2 or O_3 and produces nitric acid (HNO_3).

$$4\,NO_2 + 2H_2O + O_2 \longrightarrow 4\,HNO_3$$

$$2\,NO_2 + H_2O + O_3 \longrightarrow 2\,HNO_3 + O_2$$

H_2SO_4 and HNO_3 formed as above come down the atmosphere in the form of acid rain. This damages the crops and plants and makes the soil acidic which is harmful for subsequent crops. For more details see sec. 4.5.1.

9.3 Harmful Effects of Soil Pollutants

It has already been stated that soil pollution is the result of the dumping of waste products – domestic, human, animal, industrial and agricultural. The harmful effects of various pollutants are described below:

9.3.1 Effects of Industrial Pollutants

(1) Industrial pollutants comprise of a variety of chemicals which are very toxic to living beings. Industries like paper and pulp, textile, steel, fertilizers, pesticides etc. release metallic wastes, oils, greases, plastics, plasticizers, heavy metals most of which are non – biodegradable. So these pollutants or toxic materials are transferred to different organisms via food chain and are responsible for undesirable effects.

(2) Industrial effluents if discharged in the sewage system, will hinder the biological purification mechanism of sewage treatment resulting in a number of soil and water borne diseases.

(3) The organic compounds particularly amino acids, albumins and gelatins if discharged in soil, undergo putrefaction by bacterial action releasing sulphur and phosphorus compounds. These compounds produce gases like hydrogen sulphide and sulphur dioxide which are responsible for the foul smell of soil.

(4) Metallic contaminants (e.g., Hg, Pb, Zn, As, Cd, Cr, etc.) destroy bacteria and beneficial micro organisms in the soil. These metals accumulate in the soil and are fatal to living organisms.

(5) Soluble salts in the soil make soil unfit for agricultural cropping. It is believed that about 45% of the agricultural land is damaged by soluble salts.

(6) The chemicals from industrial wastes turn the soil acidic or alkaline which becomes unfit for agriculture. It is estimated that about 30% of agricultural land is rendered unfit for further crops.

(7) Some of the industrial wastes contain pathogenic bacteria. For example, the wastes, from tanning industries contain pathogen *Anthrax bacilli*.

9.3.2 Effects of Urban and Domestic Waste Products

(1) Sewage which is an excellent medium for the growth of pathogenic bacteria, virus and protozoa, can cause cholera, bacillary dysentery and even typhoid. The chances of mixing with ground water even after the treatment of sewage arc quite high. The ova and larva of many worms are parasitic to man and may be transmitted to soil through urine and faeces.

(2) Solid wastes are responsible for offensive odour and cause clogging of ground water filters.

9.3.3 Effects of Radioactive Pollutants

(1) These pollutants are the worst pollutants. The volatile materials are first released into the atmosphere and are then washed down by rain on to the soil and water bodies. This results in the soil getting radioactivity, which in turn is passed to plants and finally to the humans.

(2) In water bodies, the aquatic flora and fauna absorb radiations which are also ultimately passed on to the humans.

(3) Radioactive pollutants produce great human misery.

9.3.4 Effects of Agrochemicals

Agrochemicals like fertilizers, pesticides and biocides etc. are used to increase the soil fertility and crop production. However, excessive use of agrochemicals is disadvantageous as will be clear from the following effects:

(1) Fertilizer enriched soils do not support microbial flora.

(2) With the use of NPK fertilizers, there is about 25% decline in protein content in corn, maize, gram and wheat crops. In fact, this practice leads to malnutrition.

(3) Use of potassium fertilizers decreases the content of carotene and vitamin C in vegetables and fruits.

(4) Though fertilized soil produces big size vegetables and fruits but such produces are more prone to pests, insects and diseases.

(5) Excessive use of nitrogenous fertilizers leads to accumulation of nitrate in the soil. These nitrates are passed on to humans. Being soluble, the nitrates enter in the ground water. Drinking nitrate rich water can cause diarrhoea and cyanosis (blue jaundice) in children. In the human body, the nitrates are converted into nitrosamines and nitroso compounds which are suspected carcinogens and may be responsible for stomach cancer.

In case the nitrate fertilizers are used beyond safe amounts, they affect children less than 2 months old who become susceptible to a fatal disease '*methemoglobinemia*' or '*blue baby syndrome*'. In this disease, the nitrates interfere with the oxygen carrying capacity of the blood. Children who drink water containing excessive amounts of nitrate harbour the wrong type of bacteria which reduce nitrate to nitrite ions which damage the respiratory and vascular systems ultimately causing the death of the infants.

(6) Phosphate fertilizer like diammonium phosphate (DAP) is considered detrimental to the crop production.

(7) Excessive fertilizers are drained off into water bodies causing *eutrophication*.

(8) Farmers use pesticides to increase yield. However, they are hazardous to humans, animals, fish and livestock and they also reduce the yield of the crops.

(9) The level of pesticide residues like DDT, dieldrin, etc. in fruits, milk, eggs, vegetables beyond the permissible levels are responsible for causing diseases like cancer, sterility and even death.

(10) Organo chlorine pesticides, e.g., polychlorinated biphenyls (PCBs) are very dangerous. These accumulate in the soil and plants which ultimately enter human bodies. PCBs cause deformities in foetus, nervous disorders, liver and stomach cancer in animals.

(11) Organo phosphate pesticides cause muscular weakness, tremors and dizziness in animals.

(12) Pesticide poisoning from soil is a global problem. Excess absorption leads to greater accumulation of acetylchloline in the body. Chronic absorption damages liver and kidney causing malfunctioning. It also causes cancer and deformation of brain tissues.

9.4 Soil Erosion

It has already been stated that the formation of soil, as it exists today, has taken centuries (see sec. 8.2). The top soil is the most important as all agricultural activities depend on it. The upper most layer of the soil is a vital component since all the nutrients required by plant are present in this layer. At times, if it so happens that the top soil is dissipated or taken to other places by water or wind, then this situation is referred to as 'soil erosion'. The soil which has been eroded becomes unfit or unsuitable for vegetation or agricultural production. It has been found that on global scale, the fertile land is lost by erosion at the rate of about ten million hectares per year.

9.4.1 Causes of Soil Erosion

Soil erosion is a natural process and is caused by the flow of water over the fields and by wind. Following factors influence the rate of soil erosion.

(1) Deforestation and loss of vegetation cover increase the rate of erosion.

(2) Cultivation on slopes of mountains increase the rate of erosion.

(3) Soils with fine texture and low organic content are more susceptible to erosion.

9.4.2 Effects of Soil Erosion

Soil erosion depletes the inorganic nutrients like nitrogen, phosphorus, potassium and calcium present in soil. It also uproots the organisms living in the soil – this results in the loss of organic nutrients. The soil erosion also leads to the reduction in soil productivity.

9.4.3 Control of Soil Erosion

Formation of soil is an extremely slow process and it takes 200 – 1000 years to form one inch of topsoil. In view of this, it is very necessary to conserve the soil and prevent its erosion. Following are some important methods of controlling soil erosion.

(i) **Rotation of crops:** This procedure ensures that some part of the land is continually covered by vegetation.

(ii) **Tillage at right angle to the slope of land:** This is known as **contour farming**. This procedure creates a series of ridges that slow down the flow of water and thus prevent the soil erosion. This method is useful on gentle slopes.

(iii) **Strip Farming:** This is used on steep slopes. The procedure involves alternating strips of closely sown crops. Alternative bands of wheat and soya beans are commonly sown. The closeness of the strips retards the flow of water and so prevents soil erosion.

(iv) **Using wind brakes:** It is the practice of planting trees or other plants that protect the bare soil from the full force of winds. Wind brakes decrease the speed of wind and hence reduce the quantity of soil that the wind can carry away. Thus, soil erosion by wind is reduced.

9.5 Reclamation of Soil

It has already been stated (sec. 4.5.1 and 9.2.7) that acid rain makes the soil acidic. Also, use of excess fertilizers pollutes the soil. For example, when ammonium sulphate is used as fertilizer again and again into the soil, NH_4^+ ions are used up by successive crops but SO_4^{2-} ions get accumulated into the soil. Being acidic, SO_4^{2-} ions make the soil acidic which renders it unfit for plant growth.

In case $NaNO_3$ or KNO_3 is used as fertilizer again and again, the NO_3^- ions are used up by the successive crops, but Na^+ and K^+ ions get accumulated into the soil. These cations render the soil alkaline and so it becomes unfit for subsequent crop cultivation.

We also know that fertilizers are used to increase crop production and a lot of water is used for irrigation. In case, there is not good drainage system for the outlet of unused water, the soil becomes highly **saline** (salt containing). This is because the salts formed get accumulated in the soil. The saline soil becomes unfit for growing crops.

For agricultural crops, the pH of the soil should be near neutral. So acidic or basic soils are not good for cropping. In such a situation, the acidity or alkalinity must be removed from the soil. This is called **reclamation** of soil.

If the soil becomes too acidic, it may be reclaimed by addition of lime ($CaCO_3$). This process is called **liming**.

$$\text{Soil]}\,H^+ + CaCO_3 \longrightarrow \text{Soil]}\,Ca^{2+} + CO_2 + H_2O$$

In drought – prone areas with low rainfall, the soil may turn too alkaline due to the presence of basic salts like Na_2CO_3. Alkaline soils can be reclaimed by treating with $Al_2(SO_4)_3$ or $Fe_2(SO_4)_3$ which on hydrolysis sets free the acids.

$$2Fe^{3+} + 3SO_4^{2-} + 6H_2O \longrightarrow 2Fe(OH)_3 + 6H^+ + 3SO_4^{2-}$$

Alkaline soils can also be reclaimed by addition of sulphur which is oxidized in soil by bacterically mediated reactions to H_2SO_4.

$$2S + 3O_2 + 2H_2O \longrightarrow 4H^+ + 2SO_4^{2-}$$

Saline soils can be reclaimed to a desired level by the removal of salts which are responsible for salinity. The procedure consist in leaching the soil with good quantity of water filled by drainage of water. The water dissolves the salts which are thus drained out.

9.6 Treatment of Soil Pollutants

The nature of pollutants depends on the specific industries, e.g., polymers and plastic, asbestos, tobacco, petroleum, mining and agrochemical industries, (for details of pollutants, see secs. 4.3.1; 4.3.2; 4.3.3; 4.3.6). Besides these, the pollutants are also discharged into the land by industries such as textile, pulp and paper, leather, food processing, dairy, soap and detergents etc. (for details, see secs. 6.4.2; 6.4.2.1; 6.4.2.2; 6.4.2.3; 6.4.2.4; 6.4.2.5; 6.4.2.6; 6.4.2.7; 6.4.2.8). The harmful effects of various pollutants and their prevention have also been discussed in the relevant sections mentioned above.

The industrial wastes are the source of soil pollution due to their disposal on the land. The industrial units adopt one or more of the following procedures to take care of their wastes and its disposal.

(i) Treat their wastes to remove suspended materials and then discharge the effluents in the water bodies.

(ii) Discharge untreated effluents into the sewage system which has to be processed with domestic sewage by municipal sewage treatment plants. This, of course, has to be done with the concurrence of the municipalities.

(iii) Dumping of solid wastes into land dumping sites and discharging liquid wastes into sewers.

(iv) The chemical treatment of industrial wastes e.g., coagulation and precipitation to remove suspended matter.

(v) Screening, oxidation, flocculation and sedimentation of effluents followed by biological treatments.

(vi) Removal of highly toxic compounds by ion – exchange, electro – dialysis or absorption by active carbon before their disposal into the land.

The sewage and domestic waste consists of dirty water and contains human and animal excretions (e.g., urine and faeces). The discharge of large quantities of sewage into land and water bodies causes pollution. It is, thus, necessary to treat the sewage and domestic waste. The procedure is based on biological decomposition of non – toxic organic wastes using bacteria under aerobic conditions i.e., in the presence of plenty of oxygen.

$$\text{Organic load} + O_2 + \text{bacteria} \xrightarrow[\text{in water}]{} CO_2 + \text{other products}$$

As a first step, the solid wastes are removed from sewage and domestic waste by screening followed by settling of the sludge. Subsequently, the residual liquid is subjected to biological oxidation of soluble organic material through a bed of microbes in activated sludge. Any sludge formed is again removed. The sludge may be used as fertilizer and the remaining water is chlorinated before discharging into water bodies.

9.7 Control of Soil Pollution

(1) It should be mandatory for industrial units not to dump their wastes on to the land. In fact, the industries must ensure that there is minimum of waste products (see sec. on green chemistry as an alternative for reducing pollution in Chapter 16). As far as possible, the waste products could be recycled or used to make useful products.

(2) The domestic and urban garbage wastes should be properly managed by the municipalities.

(3) Animal refuge and agricultural wastes can be utilized as manure and for the production of biogas – that in turn may help to generate electricity.

(4) Materials like paper, glass, metal scraps and some type of plastics can be recycled.

(5) Biological methods of pest control can also reduce the use of pesticides and so minimize soil pollution.

9.8 Indicators of Soil Pollution

Plants can be used to indicate the nature of the soil (whether acidic, basic or saline) and other pollutions. Besides plants micro organisms are the rapid detectors of environmental pollution. This aspect form the subject matter of subsequent sections.

9.8.1 Plants as Indicators of Soil Pollution

Plants require a number of macro-nutrients and micro- nutrients (see sec. 8.6). Besides these, there are other elements which are essential for plant growth like sodium for *Chenopodiaceae*, cobalt for *Fabaceae*, aluminium for *ferns*, silicon for *diatoms* and gold for *Equisetum*. Certain plants can withstand the presence of heavy metals in higher concentration, e.g., *silene* plants to zinc, liver worts and mosses to copper, serpentine to chromium and nickel, Fe, Stucca species to lead and Acidophilic grasses to aluminium.

Some plants are called **soil indicator plants** since they reveal the quality of substrate (on which they grow) in the soil. (F.E. Clements, 1920, Ecologist Division of Plant Biology, Carnegie, Institute of Washington, USA).

Plants belonging to the species of *sphagnum* can grow in acidic soil (*p*H 3.5 – 8.5) e.g., bog soils. Such plants are considered as strongly **acidophilic** and are sensitive to alkaline soils. Certain grasses like *Deschampsia flexuosa*, *Calluna vulgaris*, *Sarothamus scoparis* are also indicators of acidic soil; however, these can grow in neutral soil and can tolerate weakly alkaline soils.

Alkaline soils have high concentration of Na and K ions. Strongly alkaline soils are injurious to **glycophytes.** Bacteria of the *Basiphilic* species do not survive if the *p*H of soil falls below 6.0. *Tussilage ferfare* is not affected in the *p*H range 7.0 – 8.5; it can also grow in neutral soil and can tolerate *p*H 4. It is known as **Basiphilic acido – tolerant species.**

Saline soils (*p*H below 8.5) have accumulated salts and interfere with the growth of non – halophytes. Salinity of soil is also indicated by the distribution of different species starting from *Chenopodiaceae* (weakly alkaline) to highly alkaline mangrove vegetation.

Liverworts and some species of mosses are called **copper mosses** since they are associated with copper bearing substrates. These have been used for locating copper deposits. Copper mosses give information about metallic content in soil to geologists.

Plants provide an excellent means for indicating soil pollution, since they develop toxicity symptoms for a particular pollutant. Certain plant species like wolffia, chara, utricularia grow well in polluted waters, while *Anthoxanthum*, *Agrotis*, *Festuca* and *Impatiens* are used as metallic tolerana of zinc, copper, lead and cadmium, respectively. Lichens and mosses are also used to indicate pollution. Leaf cabbage indicates accumulation of polycyclic hydrocarbons in the soil. Tolerant plant species are good indicators to monitor the incidence of particular soil pollution. Certain plant species besides acting as bio – indicators are also used as pollutant scavengers.

9.8.2 Organisms as Bio-Indicators

Some species of organisms are used as bio–indicators. Such species are referred to as **bio–sensors** or **bio–accumulators**. These can concentrate toxic metals, pesticides and organic industrial components in their tissues like gills, flesh, muscles, liver, bones, etc. (Also see Section 6.4.2.9).

9.8.3 Micro-Organisms as Bio-Indicators

Environmental pollution in soil as well as in water can be conveniently detected by some micro – organisms. Some organisms are sensitive to foreign materials and some others help in the bio decomposition of pollutants. The presence of toxic agents is indicated by the elimination of sensitive species. In petroleum belts and sediments, detection of polar lipids of archeobacteria is a convenient monitoring system. For soil pesticides, cyanobacteria is effectively used as bio – indicator. As an example, *Nostoc microscopicum*, *Haplosiphon welwitschii* and *H. confervaceus* indicate pesticide pollution of Dithane, BHC, Aldrex, Deltan, Roger and Phorate, etc. in the soil. Oil spillage can be monitored by *Actinomycetes*, *bacteria*, *yeast* and *some filamentous fungi*. In fact, waste oil can be absorbed by some species like *Humicola*, *Verticillium*, *Scolecobasidium* and *Mortierella*.

Hazardous Waste and Its Management

CHAPTER 10

10.1 Introduction

The word *hazardous* indicates something dangerous. Many of the wastes generated by human activities fall into this category. Our modern lifestyles need a large number of household items, electronic gadgets, chemicals, paints, etc. Their manufacture generates a large variety of hazardous waste. According to the Environment Protection Agency, USA, the four characteristics of hazardous waste are *corrosivity, ignitibility, reactivity* and *toxicity*.

A ***corrosive*** substance can corrode or destroy a material. Thus, most acids fall into the category as they can corrode metals, materials and even burn the skin.

An ***ignitable*** material bursts into the flames easily. The ignitable materials pose a fire hazard, irritate skin and eyes and give off harmful vapours. Some examples of ignitable substances are paints, polishes, gasoline, etc.

A ***reactive*** substance has a tendency to undergo violent chemical change. Such substances include explosives, pyrophoric materials, cyanide– or sulphide–bearing wastes and materials which can react with water.

A ***toxic*** material is the one which is poisonous in nature. It can cause illness and even death if swallowed or absorbed through the skin. Some examples of this category of compounds are pesticides, weedicides, household cleaners etc.

The hazardous wastes have detrimental effects on the environment. They affect humans and other organisms as well. Their intentional dumping or unintentional release by accidents has been a matter of global concern. Another related global issue is the management of *radioactive waste*.

Incidents like Chernobyl, Bhopal gas tragedy and Love canal have created the awareness about the potential dangers of hazardous and nuclear wastes and have raised many safety issues associated with them.

The use of nuclear power as an alternative source of energy is being reconsidered as the damage caused by such accidents, although rare, is very huge and has lasting harmful effects on the generations to come.

In this chapter, we will first discuss radioactive waste – its generation, effects and management followed by hazardous waste including biochemical waste. Lastly, certain transboundry issues related to hazardous and radioactive wastes would be explained and international agreements for controlling such wastes would be covered.

10.2 Radioactive Waste

Radioactive waste is unique in nature because radioactive substances exhibit radioactivity i.e., they emit particles and energy in the form of radiation. The adverse effects of radiation on the health of human beings, other organisms and environment are a cause of concern for all.

10.2.1 Types of Radiation and their Characteristics

The emission of radiation is going on since the creation of the universe. The natural sources of radiation are *cosmic rays* which are high energy protons and electrons coming from the Sun and the space. The amount of cosmic radiation reaching the Earth varies with the altitude. The amount of radiation which can be tolerated by human beings is 0.01 roentgen / day.

Besides cosmic radiation, various radionuclides such as actinium, radium, thorium, uranium and radioactive isotopes of various elements such as carbon and potassium which are widely spread in nature, also emit radiation called **terrestrial radiation**. Cosmic radiation and terrestrial radiation are together known as **background radiation**.

In terms of nature of emission, the radiation is of three major types: *alpha particles*, *beta particles* and *gamma rays*. An alpha (α) particle consists of two protons and two neutrons. It is a helium nucleus in other words. The alpha particles have least penetrating power as compared to beta and gamma rays. In human tissues, they penetrate about 0.005 – 0.008 cm. Beta particles are electrons and have mass of 1/1840 of a proton. They are more penetrating than alpha particles. Gamma radiation is the most penetrating of all the three types. It is a form of electromagnetic radiation. Gamma rays travel long distances and are very energetic.

Each radioisotope has its characteristics emissions. Some emit only one type of radiation whereas others emit more than one type. **Radioisotopes** are those isotopes of an element which spontaneously undergo radioactive decay by emission of radiation.

An important concept associated with radioisotopes is their of **half – life**. It is the time required for one - half of a given amount of the isotope to decay. Some radioisotopes have very long half - life. For example, U^{235} has a half - life of 700 million years. Carbon - 14 has a half - life of 5570 years which is an intermediate value. Some other radioisotopes have very small half - life. An example of this category is Po^{218} which has a half-life of just about 3 minutes.

Various units are used for describing radioactivity. A commonly used unit for radioactive decay is *Curie* (Ci). It is named after Marie Curie who along with her husband Pierre Curie, discovered radium. They also discovered *Polonium* which was named after Poland, the homeland of Marie Curie. A curie is the amount of radioactivity from 1 gram of radium – 226 that undergoes about 37 billion nuclear transformations per second.

In the International System (SI) of measurement, radioactive decay is measured in *Becqueral* (Bq). One Becquerel is activity of a quantity of a radioactive material in which one nucleus decays per second.

Thus, one curie is 3.7×10^{10} decays per second.

When considering the environmental effects, the actual dose of radiation is important which is measured in *rads* (rd) and *rems*. The corresponding SI units of absorbed dose of radiation are *grays* (Gy) and *sieverts* (Sv); 1 gray is equivalent to 100 rads and 1 sievert is 100 rem. Rem or sievert denote effective equivalent dose whereas the energy retained by exposed living tissue is called *radiation absorbed dose* (or rad). While dealing with small doses of radiation, *millirem* (mrem) or *millisievert* (mSv) are used. For gamma rays, the commonly used unit is roentgen (R) and the SI unit is Coulomb per kilogram (C/kg).

10.2.2 Sources of Radioactive Waste and Radiation

Here, the focus is on emission of radiation and generation of radioactive waste by human activities. These are as follows:

(a) Many radioisotopes are used for medical purposes. Also, X – rays are used very commonly for diagnostic purposes. Thus, the patients and radiologists face the risk of exposure to radiation. Such an exposure ranges from 0.7 to 0.8 mSv / year.

(b) The nuclear explosions carried out both for testing nuclear weapons and their actual use as in Hiroshima and Nagasaki release a large amount of radiation, heat and dust. The radioactive dust falling on the Earth after such explosions is known as **radioactive fallout**. ^{137}Cs, ^{131}I and ^{90}Sr are some of the radioisotopes produced during nuclear explosions. These, after the nuclear fallout, reach the earth and enter into the food chain. They also undergo biomagnification and have different half-life periods in different organisms.

(c) Nuclear power plants also use radioactive materials for the generation of power. The **nuclear fuel cycle** starts with the mining and processing of uranium and it has various stages such as enrichment and processing of uranium to controlled fission, reprocessing of spent fuel and disposal of nuclear waste shown in Fig. 10.1. All of these generate some amount of radiation.

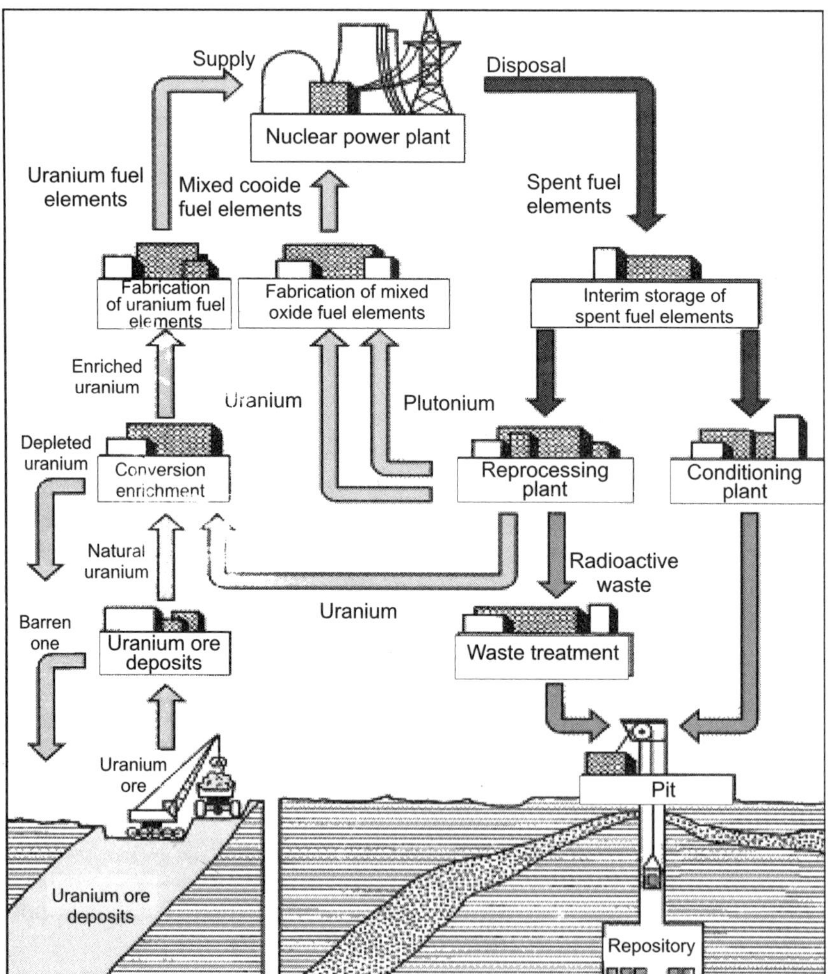

Fig. 10.1: Various Stages of Nuclear Fuel Cycle

The mining and milling operations leave a large quantity of tailings which contain radioactive materials. Similarly, the enrichment and fabrication processes also generate radioactive waste which must be carefully handled. Since the nuclear reactor is the most visible part of the facility, people are more concerned about the location of the plant and the risks or hazards associated with its proximity to their towns.

Certain countries such as India reprocess the spent fuel to recover uranium and plutonium whereas others like USA does not reprocess the spent fuel. The disposal of radioactive spent fuel also poses a lot of problems. At last, the nuclear plants may need modernization or decommissioning after they have been in operation for 20 to 30 years. Then, the contaminated equipment poses storage and disposal problems.

Further, the transport of radioactive materials needs special care. In addition, there is always a possibility of smuggling of weapon grade materials to other countries or to terrorists which must be taken care of.

Inspite of all care and precautions, there may be accidental leak of radioactivity from nuclear power plants. Various such incidents have occurred in the past. They have both immediate and long term effects on human health and environment.

Before knowing the details about these accidents, it would be beneficial to understand the functioning of a nuclear reactor. The generation of power using a nuclear reactor is shown in Fig. 10.2.

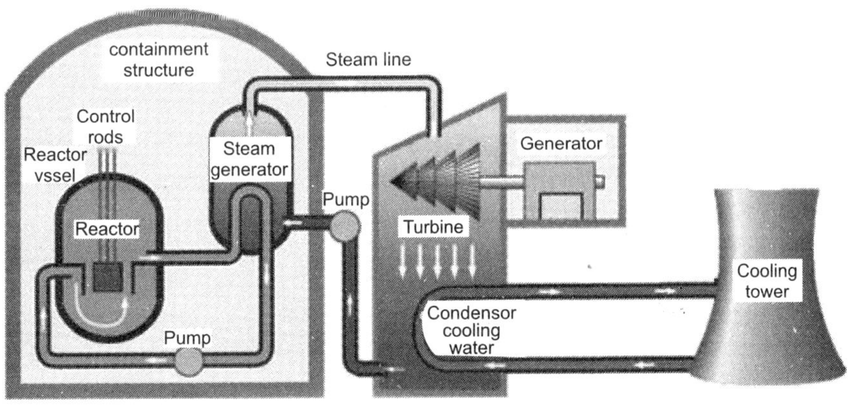

Fig. 10.2: Power generation using Nuclear Reactor

The fission reaction is carried out in the reactor. Its main components being the *core*, *control rods*, *coolant* and *reactor vessel*. The core comprises fuel and moderator. It is enclosed in a heavy stainless steel reactor vessel. The entire reactor is contained in a reinforced concrete building for extra safety and security.

The enriched uranium pallets are placed in fuel rods. These are packed together into fuel subassemblies in the core. A minimum concentration of the fuel is required to make the reactor *critical* which means that the chain reaction is self – sustaining. The control rods regulate the chain reaction by capturing the neutrons. When control rods are pulled out, the rate of reaction increases whereas when they are inserted in, the rate of reaction becomes slow. The heat produced by fission reaction is taken away by the coolant. Water can also be used as a coolant. It also functions as a moderator and slows down the fast moving neutrons. The other moderators used are heavy water and graphite.

The rest of the assembly contains pump, condenser, turbine and generator which use the heat produced during nuclear fission to generate steam and then electricity.

10.2.2.1 Accidents at Nuclear Power Plants

Although the chances of nuclear power plant accidents are very low, but accidents do happen and cause an irreversible damage. Some such accidents are described below:

(a) Three Mile Island, Harrisburg, USA

This accident occurred on 28th March 1979. During the night, the reactor 2 (Fig. 10.3) was working on full power when the cooling water supplies were cut off and begin to drain out. The operators, who were unaware about this, shut off the emergency pumps. This caused the overheating of the reactor which came close to meltdown. However, the situation was controlled and 100,000 people had to evacuate the surrounding area. The accident released radioisotopes and radiation

Fig. 10.3: The Reactors at Three Mile Island.
Source: Wikipedia

of 1mSv into the environment. However, the radiation was intense near the site. Even on third day after the incident, 12mSv / hour was recorded near the site. The authorities were, however, not prepared to deal with the situation. Also, the long – term effects of the low doses of radiation were not known. However, this incident created an awareness about the importance of tackling emergency situations like this and better management of the nuclear power plants.

(b) Wind Scale / Sellafield, UK, 1957

A major fire broke out at Windscale plutonium production plant which caused contamination in vegetation and milk supplies of adjoining area due to radiation leak. However, most of the radioactivity was trapped in the filter of the chimney in the plant (Fig. 10.4). The accident was kept confidential for 30 years. In 1973 also, there was contamination problem and the plant was shut down. A decade later, accidental discharge of radioactive 'crude' took place into the Irish Sea which was discovered by Green peace.

Fig. 10.4: Windscale reactor

(c) Chernobyl, Ukraine, 1986

On April 26, 1986, the control rods of the fourth reactor were pulled out during the night shift as a part of deliberate but unauthorized experiment. The temperature of the reactant core went extremely high. The graphite in the core caught fire and this melted the part ot the fuel. The pressure of hydrogen built up, led to the cracks in the ceilings and explosions removed the top of the building above the reactor. The fire produced a cloud of radioactive particles which went about 5000 ft in the atmosphere (Fig 10.5(a) and (b)).

Fig 10.5: (a) Chernobyl Nuclear Plant **(b)** The Accident

The radioactive cloud spread north – westwards and affected a large area of Ukraine, Byelorussia, Russia and Northern Europe. When high levels of radioactivity were recorded by workers of the nuclear power plant in Sweden, the accident became known to the public.

It was reported that 237 people suffered from acute radiation sickness and 31 people were dead. In the 30 km zone, approximately 115,000 people were evacuated and about 24,000 people received an average radiation of 0.43Sv. However, a total of about 3 billion people in the Northern Hemisphere received varying amounts of radiation. The radiation exposure would cause about 16,000 more deaths in 50 years time due to long term effects of radiation. Also, it is being observed that later, people were suffering from diseases like leukemia, thyroid, cancer, etc. due to radiation exposure.

The vegetation of the area within 7 km of the power plant was severely damaged. The pine trees showed the presence of radioactivity even after four years of the accident. The contamination of soil and water made the area inhabitable. It was estimated that the total cost of the accident would cross $ 200 billion. It was the biggest nuclear power plant accident happened so far. The studies on long-term effects on exposure to radiation are still going on.

The above accidents have raised many questions about the safety and proper handling of the nuclear power plants. Even with best of the solutions, the risk of accidents would always be there and in case the accident occurs, the damage to the mankind would be enormous.

10.2.3 Effects of Radiation

Since radiation has high energy associated with it, it can break the chemical bonds of any medium through which it passes. When radiation passes through living cells, it breaks the molecules present in them. The biological effects of radiation can be classified as *somatic* or *genetic effects*.

(a) Somatic Effects

The direct effects of radiation on body cells and tissues are called *somatic effects*. They may be immediate or delayed. They often include the conversion of normal cells to cancer cells and other changes that lead to a decline in life expectancy. A dose of 400 – 500 roentgen* on whole body is fatal in about 50% of the cases. Large doses can lead to radiation sickness which shows vomiting, diarrhoea, nausea and often death. The delayed effects take some time to develop and include leukemia, cardiovascular disorders, eye cataracts, premature aging and reduced longevity.

* : Roentgen, a unit for measuring the dose of radiation is the amount of X-rays or γ-rays which will produce ions charging electrostatic unit of electricity in 1 cubic centimetre of air.

(b) Genetic Effects

Genetic effects change the genetic make up of sex cells and they are transferred to the offspring. Radiation can damage chromosomes, genes and DNA which may induce cancer.

10.2.4 Management of Radioactive Waste

The radioactive waste is generated as a result of various activities associated with power generation using nuclear reactors, fuel processing plants, hospitals and research institutions. In addition to this, tailings which are materials removed by mining activity but not processed and remain at site, also constitute hazardous waste and need management.

Radioactive waste may be classified as *high – level waste* or *low – level waste*.

A ***high – level waste*** is extremely toxic and demands urgent attention. The spent uranium fuel comes under this category of waste. When the uranium – 235 concentration decreases from 3% to 1%, it can no longer be used for power generation and is called *spent fuel*. The splitting of U^{235} gives radioactive isotopes of smaller elements. However, some uranium atoms also capture neutrons to form heavier elements like plutonium. These heavier – than – uranium, i.e. *transuranic* elements take longer time to decay. They remain in high – level waste for thousands of years to come. For example, Plutonium – 239 has a half – life of 24,000 years whereas the smaller fission fragments such as strontium – 90 and cesium – 137 have half – lives of about 30 years.

Reprocessing separates residual uranium and unfissioned plutonium from fission products. These can again be used as fuel.

High-level wastes are hazardous to humans and other life forms. They have high radiation levels which are fatal even if the direct exposure is for short time. For example, even after ten years, the surface dose rate of a spent fuel assembly exceeds 10,000 rem / hour whereas the fatal whole body dose for humans is about 500 rem.

Also, if the high level waste components enter into ground water or rivers, they can enter into food chains.

Low-level wastes have sufficiently low concentrations of radioactivity. If handled properly, they do not pose much environmental hazard. Low-level waste includes residuals or solutions from chemical processing, solid or liquid plant waste, sludges and slightly contaminated equipment, tools, plastic, glass, wood and other materials. They are commonly disposed of in near-surface facilities and are monitored regularly from safety point of view.

The high-level radioactive waste is generally stored in geologic formations. The associated points which need to be considered for storage are as follows:

(a) Identification of proper storage sites which meet the criteria of ground stability and slow movement of ground water.
(b) Intensive survey for determination of geologic and hydrologic characteristic of the site.
(c) Prediction of future changes of variation in climate, ground water flow, erosion and earth movements for such sites.
(d) Correctness of the above predictions over thousands of years since the high level wastes would be stored there for such long periods of time.

Many geologic formations have remained undisturbed for million of years. It seems feasible to isolate radioactive wastes from the environment by burying in these repositories. Such a storage facility must be built at least 600 metres below the surface in a stable geologic formation. The location should be checked with respect to various factors such as rock type, rock strength, earthquake activity, ground water movement, surface characteristics, etc. The radioactive waste is solidified into a durable, leach resistant material such as vitrified glass. The solidified waste is sealed in corrosion resistant containers which are placed in the repository having a corrosion resistant lining. The repository is filled with additional material which have high sorptive characteristics.

The burial under deep-sea trenches is also one alternative for radioactive wastes. But again objections were raised for sea dumping because it would lead to concentration of radioactivity in marine organisms and hence eventually the entire food chain would be affected.

In India, the survey and exploration of uranium, thorium, rare metals and rare earths etc. is done by Atomic Minerals Directorate for Exploration and Research (AMD). The uranium mines are located at Jadugada, Bhatin and Naruapahar – all in Jharkhand. The mining and processing of uranium ores and mineral sands are done by the Uranium Corporation of India Ltd. (UCIL) and Indian Rare Earths Ltd. (IREL), respectively.

There are seven heavy water plants under the Heavy Water Board. The fabrication of fuel bundles for power reactors is done by the Nuclear Fuel Complex (NFC), Hyderabad. The fuel reprocessing plants are located at Trombay, Tarapur and Kalapakkam.

The low medium level wastes are retreated in eco-friendly ways. The high level wastes which are generated in very small quantities are immobilized in glass matrix by vitrification. A Waste Immobilisation Plant (WIP) is in operation at Tarapur and two more plants are coming up at Trombay and Kalapakkam. The vitrification process involves the use of sodium borosilicate glass matrix with some modifiers. The process has been developed by BARC. This storage facility can store the radioactive waste generated by the operation of two nuclear reactors of 220 MWe capacities each for 40 years. The waste management facilities also exist at Rajasthan and Kalapakkam. The safety aspects such as radiological safety, industrial safety, occupational health, fire safety and environmental protection are all taken care of. The safety standards are at par with those recommended by various international bodies such as International Atomic Energy Agency (IAEA) and the International Commission on Radiological Protection (ICRP).

The Atomic Power Stations at Tarapur, Narora, Kakrapara and Kalapakkam have obtained the Environmental Management System (EMS) certification as per ISO – 140001.

10.3 Hazardous Waste

Hazardous wastes arise as a result of various industrial operations, energy production, metallurgical operations and health services. The hazardous waste contain organo-chlorine compounds, organophosphate compounds, heavy metals, oils, phenols, acids, fluorides, abrasives, solvents, caustics, etc. In addition to chemicals, hazardous wastes also include biochemical waste arising from hospitals, nursing homes and laboratories. The hazardous waste poses several environmental problems and health risks. Therefore, its disposal demands special attention. If not managed properly, it can cause several problems.

10.3.1 Incidents of Improper Disposal and Accidents

Several incidents of improper disposal have been reported in the past. Some of them caused major problems and are known worldwide. These are explained below:

1. Love Canal

In 1976, in a residential area near Niagara Falls, New York, the trees and gardens began to die, rubber articles disintegrated and toxic substances began to ooze out from the soil.

The story of Love Canal actually began about 100 years back. In 1880, William T. Love started digging a canal which was never completed. A small remnant of the canal remained in 1900s. When it used to get filled with rainwater, it served as a favourite swimming spot. In 1942, the Hooker Chemical Co. signed an agreement with the canal's owner, i.e., Niagara Power and Development Corporation, for dumping wastes in the canal. Hooker bought the sight in 1946 and for six years to comes, it dumped about 21,000 tonnes of highly toxic and carcinogenic wastes in the canal which included lindane and dioxin (Fig. 10.6).

Fig. 10.6 : Love Canal

Later in April 1953, Hooker sold the land and got free from the liability for possible damage caused by dumping. Eventually several homes and an elementary school were built on and near the site after filling and sealing the dump. In 1954, the construction workers removed the top filling of the soil, and rusting and leaking barrels became visible. Children playing near the site suffered chemical burns and died also. The problem continued for years. Later in 1976 – 77, heavy rains created more problems. Tests near the site detected 82 chemicals and many of them were suspected carcinogens. Officials admitted that not much was known about the impact of these chemicals. In 1978, it was found that 29% of women suffered from miscarriages and children were born with birth defects. The cases of asthma, urinary, ear and sinus infections increased. People also reported respiratory diseases, rashes and headaches.

The school was closed down and the families were evacuated. The clean up operations and relocation costed $275 million to the government.

2. Minamata

In the middle of twentieth century, a strange illness began to occur in the Japanese costal town, *Minamata*. Birds lost their coordination and fell to the ground, cats started running in circles and foaming at the mouth, fishermen suffered from fatigue, irritability, headache, numbness and difficulty in swallowing. People reported blurred vision, lost muscular coordination and became hard of hearing. Some experienced problems in gums and had diarrhoea. Later, 43 people died and 111 were severely disabled. 19 babies were born with congenital defects.

The cause was later found to be mercury released by a vinyl chloride factory into the Minamata bay. The microbial action converted the released inorganic mercury into the organic methyl mercury. The methyl mercury could pass readily through cell membranes in contrast to the inorganic mercury. Methyl mercury on transport by RBCs could enter into the entire body and also damage brain cells. The organic mercury was absorbed 100 times faster than the inorganic one by fish and its retention was also longer. These fish when eaten by the people of Minamata bay resulted in the observed effects.

3. Bhopal Gas Tragedy

The leakage of methyl isocyanate (MIC) gas from the Union Carbide Factory in Bhopal, India killed thousands of people and caused immense sufferings. The detailed effects of this incident are described in Chapter 15.

The above incidents have drawn international attention to the problems associated with hazardous substances whether arising from dumping of wastes or from accidents during production. Various bilateral, regional and global agreements came into existence for control and management of hazardous waste. One such convention is the *Basel Convention* which is discussed in detail in sec. 10.6.

10.3.2 Biomedical Waste

The biomedical waste comprises used needles, syringes, blood bags, urobags, glucose bottles, surgical gloves, bandages, chemicals, body parts and radioactive substances. Such a kind of waste is generated in hospitals, nursing homes, pathological labs and research institutions.

The biomedical waste poses special problem of spreading diseases such as AIDS, hepatitis, fevers, septicemia and other infections. It has a potential danger for the workers, public and the environment.

The Biomedical Waste (Handling and Management) Rules, 1998 prescribe the segregation of waste at source, colour coding of waste bags, incineration of infectious waste, careful disposal of waste sharps and shredding of plastics to prevent their reuse.

The schedule I categorises biomedical waste into 10 categories and recommends the methods of their treatment and disposal. The treatment of biomedical waste involves processes such as *incineration*, *autoclaving*, *microclaving*, *hydroclaving* and *shredding*.

- **Incineration** is a controlled combustion process where waste is completely oxidised and harmful microorganisms present in it are destroyed or denatured by high temperature.

- **Autoclaving** is a low – heat thermal process where steam is brought into direct contact with waste in a controlled manner for sufficient duration to disinfect the waste.
- In **microclaving**, microbial inactivation is done by thermal effects of electro – magnetic radiation having frequency between 300 and 3,00,000 MHz.
- **Hydroclaving** is similar to autoclaving except that the waste is subjected to indirect heating by application of steam in the outer jacket.
- **Shredding** cuts the waste into smaller pieces to make the waste unrecognizable. It prevents its reuse and signifies that the waste has been disinfected and can be disposed off.

10.4 Management of Hazardous Waste

Hazardous waste is produced in all countries. The amounts of wastes to be managed are enormous. There has been several practices for managing such wastes. The three main challenges in the waste management are as follows:

(i) Produce less waste.
(ii) Convert hazardous substances to less hazardous ones.
(iii) Proper and safe disposal of wastes

These are discussed below in detail:

(i) Reduction in the Amount of Waste

It is the safest and the least expensive option. The amount of hazardous waste generated by manufacturing and other processes can be reduced by changes in the chemical processes involved, equipment used, raw materials used, etc.

Recycling and *reuse* also eliminates hazardous waste. For example, certain solvents and acids can be recovered and recycled. This practice yields economic benefits both in terms of reducing cost of starting materials and cost involved in the handling of hazardous waste.

Also certain industries can mutually benefit by taking wastes from one industry and using it by another industry as raw materials.

(ii) Conversion to less Hazardous Substances

The toxicity or hazardous characteristics can be reduced by *physical, chemical* or *biological methods.*

Physical methods such as absorption on charcoal or resin filters for absorbing toxins can be used. Distillation is also useful at times. Precipitation and immobilization are also helpful in making wastes suitable for long term storage.

Chemical methods such as neutralization, oxidation, reduction, ion – exchange, etc. detoxify the waste.

Biological methods involve the use of bacteria, soil microorganisms and other species in reducing the hazardous characteristics of the waste.

Biodegradation can be accomplished by aerobic and anaerobic methods.

Incineration is a physical method of treatment. It is very important and has many advantages. It involves heating of the waste to 1200°C for quite sometime to complete destruction. The volume of the waste is reduced by at least 90% and the resulting waste is safer to store in a landfill or other storage site. Incineration removes volatile, combustible and mobile organic matter from the wastes by converting them to CO_2, water and other gases. It is a quick, easy and permanent solution for many hazardous wastes and can be used for even mixture of wastes.

(iii) Disposal Methods

The final and ultimate step in the management of hazardous wastes is its permanent storage. But after storing it, we should be able to regularly monitor its safe placement by drawing the samples. The two important methods of disposal are *secure landfills* and *deep – well disposal.*

(a) **Secure Landfills :** Land filling has been a very common and popular option for disposal of hazardous waste. But, the landfills should be *secure.* The modern *secure landfills* are specially designed for safe disposal of the waste. Such a landfill is shown in Fig. 10.7.

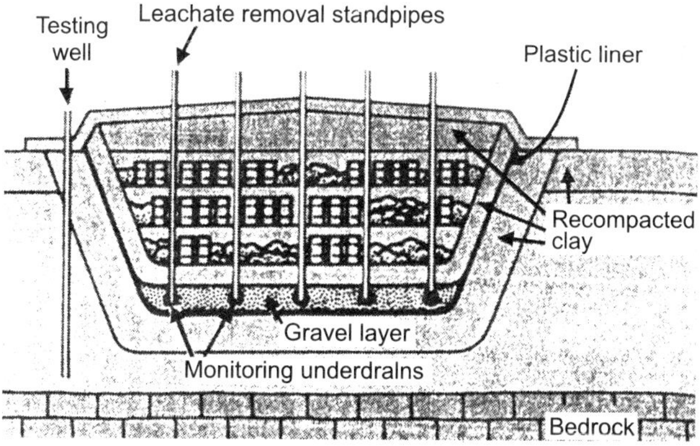

Fig. 10.7 : A Schematic Diagram of a Secure Landfill

The bottom of the 'secure landfill' is lined with compacted clay followed by a layer of gravel. The clay is flexible and resists cracking in case the ground shifts. A thick polythene liner is then placed over the gravel layer. A layer of soil or absorbent cushions the inner layers. The wastes are packed in drums and put into the landfill. The filled landfill is then finally covered with clay, plastic and soil. Vegetation is then placed for stabilizing the surface and improving the appearance. Sump well is there to collect the liquids present by leakage of drums, if any, or rainwater filtering into the landfill.

(b) **Deep well Disposal :** It involves injection of waste into deep wells. The waste is injected into a permeable rock layer several thousand metres below the surface, in geologic basins topped by relatively impervious, fracture resistant rock. This method requires careful monitoring after disposal.

Although, ocean dumping was used for disposal of wastes but due to marine pollution, this practice has been stopped.

10.5 Transboundary Issues

Although millions of tonnes of hazardous waste is managed according to the proper disposal methods, the major fraction of waste is dumped illegally or is exported to other countries. The illegal transfer of hazardous waste from advanced industrized countries to developing countries has become a matter of conflict and is a serious *transboundary issue*. Various such incidents have come to light in last few decades. They are discussed below:

(a) **The Khian Sea :** On 31st August 1985, the cargo ship *Khian Sea* loaded with 15,000 tones of municipal incinerator ash left Philadelphia. Its various stoppage points were Fort Louderdale (Florida), Puerto Rico, the Antilles, the Dominican Republic, Jamiaca, Panama before it dumped 2000 tonnes on shore in Haiti in January 1988. The representatives of the environmental group Greenpeace alerted the local residents to the potentially dangerous levels of arsenic, mercury, dioxin and other toxins in the waste. The authorities then ordered the ship to take its objectionable goods elsewhere.

The ship wandered from place to place for 24 months to get rid of its toxic materials. It travelled in the Caribbean, across to West Africa, around the Mediterranean, through the Suez Canal, past India and over to Singapore.

During its journey to eleven countries in four continents, the description of the cargo changed from *non – hazardous ash* to *general cargo* to *bulk construction material* to *topsoil ash fertilizer*. The name of the ship also changed from *Khian Sea* to *Felicia* to *Pelacano*.

Somewhere in the ocean, it dumped the remaining material and returned unladen in November 1988.

(b) **Toxic Wastes to Koko, Nigeria :** During 1987 and 1988, 4000 tonnes of mixed chemical waste containing about 150 tonnes of highly toxic PCBs was exported to the river port of Koko, Nigeria from Italy. It happened on the pursuance of an Italian businessman to local landowner who agreed to store the contents for US $ 100 per month. The landowner was able to get the necessary documents with the help of a local firm, health inspectors and custom officers. Some of the drums started leaking and smell spread over. The rice crop got contaminated and the reports of premature births and deaths consuming it, started coming. The evacuation of Koko was announced. 40 people were arrested for the conspiracy and death penalty was imposed for importers of toxic wastes.

The toxic waste was loaded onto two ships for sending it back to Italy. But, one of the ships, the *Karin B*, was refused entry to Ravenna and was banned from British ports and was prevented entry in French territorial waters. After negotiations, the wastes from the two ships were finally incinerated in the UK.

There are many more such cases and the list does not end with the above two incidents.

(c) In 1980, shipments upto 5 million tonnes of toxic and mixed wastes were proposed to Benin for $ 12.5 million to the benefit of the economy.

(d) Guinea received 15,000 tonnes of toxic incinerator ash from the USA in 1988 but returned it after protests.

(e) In 1988, Italy targeted its 2400 tonnes of toxic waste to Djibouti which was refused. Later, the waste was transported to Venezuela and Syria and then back to Italy. Finally, it was incinerated in UK.

(f) Gabon received nuclear waste from Canada and the USA.

(g) In 1988, over 4000 tonnes of hazardous waste from Italy, Netherlands and West Germany were illegally dumped in Rumania.

(h) In 1988 – 89, Austria incorrectly labelled barrels containing cyanamide, chlorinated solvents and PCBs and dumped them in Poland.

(i) In 1983, 27,000 tonnes of toxic waste were illegally dumped by a West German firm in Belgium for saving disposal costs.

The above examples of *falsification*, *illegal dumping* and *corruption* raise many questions about environmental aspects and health risks for humans.

The developing countries were treated as dumping grounds by the developed nations by providing money for storing their hazardous wastes.

Such issues related to proper description as well as export / import of hazardous waste raised serious concerns and led to the formulation of certain conventions which are discussed in next section.

10.6 International Conventions

Various agreements in the form of conventions have evolved for solving the problems related to the hazardous waste. These have been accepted by many countries and are being implemented worldwide.

1. **Basel Convention** : In the late 1980, a tightening of the environmental regulations in industrialized countries led to a dramatic rise in the cost of hazardous waste disposal. The 'Traders' of toxic waste searched the cheaper ways of getting rid of hazardous waste by shipping it to developing countries and Eastern Europe.

In 1982, UNEP decided to deal with the growing problem of transboundary transportation and disposal of toxic wastes after an ad-hoc – working group of environment experts met in Paraguay. UNEP issued *Cairo Guidelines and Principles* for the Environmentally Sound Management (ESM) of Hazardous Waste. A Draft Convention was established in June, 1987. The convention was adapted in 1989 at a conference held in Basel, Switzerland and was ratified by 135 members countries and the European Union (as on April 2000). The convention came into force in 1992. It covers toxic, poisonous, explosive, corrosive, flammable, ecotoxic and infectious wastes. The main objectives of the convention are as follows:

- Minimization of the generation of hazardous waste in terms of quantity and hazardousness.
- Disposal of the waste as close to the source of generation as possible.
- Reduction in the movement of hazardous waste.

Under this convention, the transboundary movement of hazardous or other wastes can take place only upon prior written notification of the State of export to the competent authorities of the States of import and transit (if appropriate). Each shipment of the hazardous or other waste must be accompanied by a movement document from the point at which a transboundary movement begins to the point of disposal. The shipments made without such documents are illegal. Also, there is an outright ban on the export of these wastes to certain countries. The transboundary movements can take place, if the State of export does not have the capability of managing or disposing of hazardous waste in an environmentally sound manner.

Each member country which is a Party to the Convention, is required to report information on the generation and movement of hazardous waste. The information is compiled by the Secretariat and presented in the Annual Report.

Regarding the training and capacity building, the Convention has established Regional Centres for Training and Technology Transfer in many countries including India. These centres help these countries in the implementation of the Convention. They also encourage the introduction of cleaner production technologies and the use of environmentally sound waste management practices.

The 1995 Ban Amendment calls for prohibiting exports of hazardous wastes from countries listed in a proposed new annex to the Convention (Annex VIII Parties that are members of the EU, OECD, Liechtenstein) to all other Parties to the Convention.

In 1998, the Technical Working Group of the Convention agreed on the lists of specific wastes characterized as hazardous or non – hazardous. These lists were later adopted by Parties to the Convention.

The Protocol on liability and compensation was adopted in December, 1999. It established rules on liability and compensation for damages caused by spills of hazardous waste during export, import or disposal. Besides Basel Convention, two more international conventions are there which are as follows:

2. **Rotterdam Convention** on the Prior Informed Consent Procedure for certain Hazardous Chemicals and Pesticides in International Trade.
3. **Stockholm Convention on Persistent Organic Pollutants (POPs)**

India signed this Convention in May 2002. It seeks to eliminate production, use, import and export of 12 POPs wherever techno – economically feasible and in the interim period restrict the production and use of these chemicals.

10.7 Hazardous Waste in India

In India, the Hazardous Substances Management Division (HSMD) under the Ministry of Environment and Forests is the nodal point for implementation of policies and programmes on management of chemical emergencies and hazardous substances. This division is also the nodal point for above three international conventions.

There are 1580 major Accident Hazard Units in 234 districts of 19 States / Union Territories in India as far as the chemicals are concerned. The chemical safety is governed by the following rules:

- The Manufacture, Storage and Import of Hazardous Chemical (MSIHC) Rules, 1989.
- The Chemical Accidents (Emergency Planning, Preparedness and Response) Rules, 1996.
- Under the Public Liability Insurance Act, 1991 as amended in 1992, all MAH units handling chemicals in excess of the threshold quantities referred to in the Schedule, are mandated to take an insurance policy and deposit an equal amount in the Environment Relief Fund to ensure immediate payment to chemical accident victims (MoEF, 2003 – 04).

There are 13011 hazardous waste generating units which are spread over 373 districts in India. They generate about 4.4 million tonnes of hazardous waste. Maharahtra, Gujarat and Tamil Nadu account for more than 63% of total hazardous waste.

Recently, the Supreme Court of India has given detailed directions for immediate closure of industries operating in violation of Hazardous Waste Rules, preparation of a National Inventory on Hazardous Wastes by Central Pollution Control Board, inventorization of illegal dump sites of hazardous wastes and preparation of rehabilitation plans for the same, disposal of hazardous wastes lying at various ports / ICDs / Docks of India, amendment of Hazardous Wastes Rules, 2003.

Subsequently, the MoEF has constituted a monitoring committee on Hazardous Waste Management in November 2003. The Committee has submitted its first quarterly report to the Supreme Court in January, 2004.

Various rules which are in operation in India are as follows:

- Hazardous Wastes (Management and Handling) Rules, 1989, as amended in 2000 and 2003.
- Biomedical Wastes (Management and Handling) Rules, 1998 / 2000 / 2003.
- Batteries (Management and Handling) Rules, 2001

These regulate the collection, channelization and recycling and import of used lead acid batteries. These rules make the return of used batteries mandatory for consumers.

The manufactures / assemblers / re-conditioners / importers of lead acid batteries are responsible for collecting used batteries against the one sold.

Noise Pollution

CHAPTER 11

11.1 Introduction

Everyday we hear different kinds of sounds. Some of them are pleasant to us but others cause irritation. Even the music which is enjoyed by one person can cause disturbance to others. When does a sound become noise? What is the difference between sound and noise? We will deal with such questions in this chapter. Various sources of noise would be described and the effects of noise pollution on human health would be highlighted. Finally, various steps and measures intiated to control noise pollution would be explained.

11.2 Sound and Noise

Imagine a world without sound. It would be very difficult to live in such a world. Sound is a medium for communication. We share our thoughts, feelings and information with others using sound. Light music for entertainment, sounds of birds, falling of water in streams are very soothing and give us happiness.

However, a loud music, sound from a loud speaker, horns of vehicles on a busy road, barking of dogs, crying of a child etc. rather irritate a person. All these sounds are unwanted. Hence, we can say that **noise** is an unwanted sound. Even the slightest unwanted sounds are very annoying if they continue for a long time.

Noise is inseparable part of our everyday life. People do not easily become accustomed to noise. Everyone has a different tolerance to noise. The tolerance may also vary according to time, mood, physical and mental health of a person or the type of activity one is involved in. Also at times, one enjoys loud music when he himself has put it on whereas it becomes irritating when somebody else is responsible for it.

11.3 What is Sound?

Sound is a form of energy which travels in the form of waves. The waves are created when molecules in air undergo compression and rarefaction alternately. Sound waves move outward from the vibrating source. Hence, they weaken with the distance from the source.

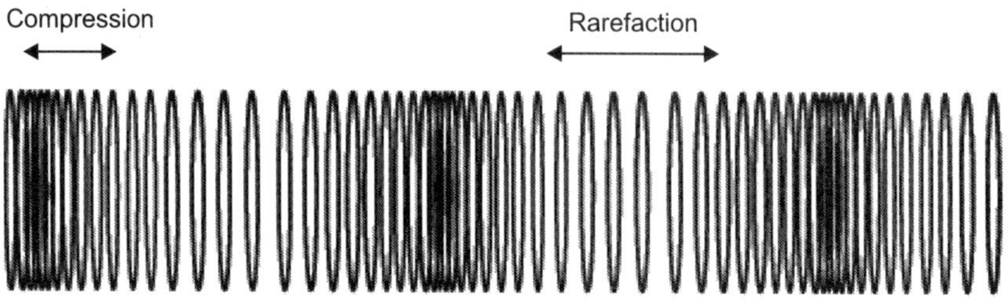

Fig 11.1: Compression and Rarefaction

There are several physical properties associated with sound of which 'frequency' and 'intensity' are very important. **Frequency** (or pitch) is a measure of how low or how high a sound is. It is measured in terms of waves per seconds or cycles per second (cps). It also represents the number of compressions which pass a given point per second. Cycles per second (cps) are commonly known as hertz (Hz), after the name of the physicist Heinrich Rudolf Hertz.

The human ear is sensitive to the frequency range of 20 Hz to 20,000 Hz. Sounds with frequency below 20 Hz are not detected by the human ear and are called **infrasonic** whereas sounds having frequency above 20,000 Hz is called **ultrasound**. Frequencies between 2000 Hz and 8000 Hz are considered most annoying whereas those below 500 Hz or above 10000 Hz are less annoying. The '**intensity**' of sound or its loudness is the energy delivered by sound per unit area. It is expressed in *decibel* (dB) units.

11.4 The Decibel Scale

Decibel is literally one-tenth of a *bel*. The bel is named after Alexander Graham Bell. A **bel** is the 'logarithm of a ratio' of any two sound (or electrical) intensities. Thus,

$$1 \text{ decibel} = 10 \text{ bel} = 10 \log_{10} (I/I_0)$$

where I is the intensity of sound which we want to measure, and

I_0 is the intensity of reference sound.

The bel had its origin in the Bell Telephone Labs where it was used as a convenient way to express power losses in telephone lines as power ratios. Thus, decibel is log of the ratio of two powers. It was originally called the 'transmission unit' or TU by the bell Labs people.

Although decibel is used to measure sound level, it is also widely used in electronics, signals and communication.

If the ratio between the powers (P_1 and P_2) of two sounds is two, then the difference in dB is 10 log (P_2/P_1) =10 log 2 = 3 dB.

Similarly, if the power of second sound were 10 times the first, then the difference in dB would be 10 log (P_2/P_1) = 10 log 10 =10 dB.

Note that the decibel scale is logarithmic and each 10 decibel increase means 10 fold increase in noise intensity. A 10 decibel increase is perceived as the doubling of loudness. Thus, if we compare sounds of 20 and 30 decibels, then the 30 decibel sound is 10 times more intense and has double loudness as compared to 20 decibel sound. Similarly, if we compare 20 decibel sound with 40 decibel sound, then we find that 40 decibel sound is 100 times more intense and has 4 times more loudness as compared to 20 decibel sound.

The lowest sound that the human ear can detect is 0 dB. It is also the **threshold of hearing**. But what does 0 decibel mean? When the intensity of sound to be measured is equal to that of the reference sound, then

$$10 \log I_2/I_1 = 10 \log 1 \text{ (when } I_1{=}I_2)$$

$$= 0 \text{ dB (since log 1 is 0)}$$

Thus, 0 dB does not mean that there is no sound.

A sound of 0 dB corresponds to 0.02 m Pa pressure. This is a very low pressure and is 2 ten billionth of an atmosphere.

The human ear does not respond to all frequencies equally. It is more sensitive towards the frequency range of 1000 to 4000 Hz as compared to very low or higher frequencies. Thus, sound meters are fitted with filters having responses similar to human ear. If the 'A weighting filter' is used, then the sound pressure level is given in dB (A) or dBA units. These units are widely used because sound pressure level is easy to measure in dBA scale. The other filters such as B or C type are also used which yield dBB or dBC scales of measurement.

The decibel levels of some familiar sounds are given in Table 11.1.

Table 11.1 Some Common Sounds and their Characteristics

Source	dB	Intensity Factor	Loudness	Effect
Threshold of hearing	0	1	Barely audible	
Rustling of leaves	10	10		
Breathing				
Whisper	20	100	Very Quiet	
Library	30	1000		
Bedroom	40	10,000	Quiet	
Average Room in home	50	100,000		
Average conversation	60	1,000,000	Moderately	Intrusive
Vacuum cleaner	70	10,000,000	loud	
Average city traffic				Annoying
Alarm clock,	80	100,000,000	Very Loud	Possible
Average Factory				hearing damage
Diesel Truck,	90	1,000,000,000		8 hrs hearing
Food Blender				damage
Jet aircraft at 300 m	100	10,000,000,000	Uncomfortably	
Steel mill	110	100,000,000,000	Loud	
Automobile horn at 1 m	120	1,000,000,000,000		
Thunder clap,	130	10,000,000,000,000		Pain threshold
Jet take off at 100 m	140	100,000,000,000,000	Painful	
Jet take off	150	1,000,000,000,000,000	Eardrum	
Rocket engine	180	1,000,000,000,000,000,000	rupture	

11.5 Structure and Functioning of Ear

Ear consist of three parts- *outer*, *middle* and *inner* out of which only the outer part is visible. The outer ear collects sound waves and these waves then pass through auditory canal to the tympanic membrane (or eardrum) which vibrates, see Fig 11.2 (a).

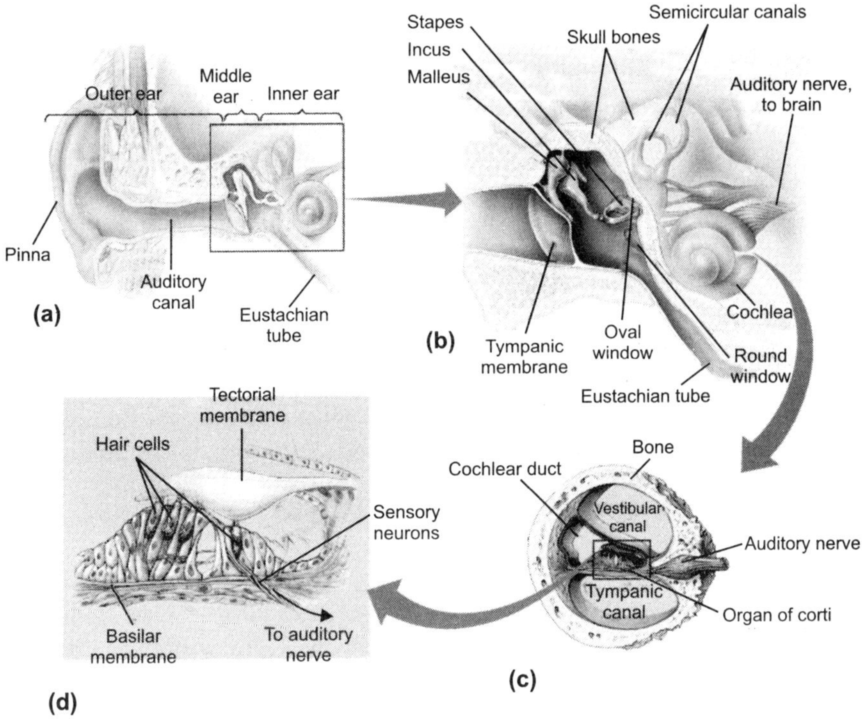

Fig. 11.2 : Human Ear and its Parts

These vibrations are then transferred to cochlea by three small bones called ossicles which are shown in Fig 11.2(b). The vibrations then travel in the fluid of cochlea and cause vibration of tectorial membrane Fig 11.2(c) . In the inner ear, the organ of corti runs along the entire length of the basilar membrane. The vibration of basilar membrane causes distortion of hair cells which generate a characteristic pattern of impulses in the nerve cells associated with each hair cell. The nerve impulses are sent to brain along the auditory nerve which interprets the sound Fig 11.2 (d).

Normal hearing depends upon the proper functioning of three parts of ear. If the outer ear does not function properly, then conductive hearing loss is there. When the auditory nerves are damaged, then neural hearing loss occurs. The sensory hearing loss is caused by the damage of inner ear (or cochlea) and it is mostly the loss associated with noise.

An explosion of 140 dB can cause permanent deafness because it destroys the hair cells which do not regenerate. Loud and persistent sounds can cause premature degeneration of organ of corti. Recent studies suggest that continuous long-term exposure to noise levels as high as 55 dB can permanently damage hearing. Such noise levels are very common in mining and construction industries, factories etc.

11.6 Sources of Noise

Noise can originate from a variety of sources. These sources can be *natural* or *man-made*. The natural sources could be thunder, sea-shores etc. but exposure to them is not very regular and the number of people exposed to it form only a small percentage of the total population.Noise pollution caused by man-made sources is the major contributor and needs urgent attention. These sources of noise are as follows:

(i) Industries

Various industries such as iron and steel, automobiles, power plants, textiles, petroleum, fertilizers etc. involve operations such as drilling, mining, forging, blasting, boiler use etc. The workers are exposed to the noise levels of 80-90 dB in such industries.

(ii) Domestic Sources

Our homes are full of gadgets which generate noise. These household gadgets include TV, radio, music systems, telephones, air-conditioners, coolers, washing machine, vacuum cleaner, food processors, pressure cookers etc .The noise generated by them has become a routine for us.

(iii) Transport and Traffic

Both air traffic and surface transport are responsible for the noise pollution. Taking of jet planes generates a lot of noise. It is a nuisance to those who live in areas near the airports where there is a lot of air traffic.

Surface transport is one of the major sources of noise pollution in big cities. The number of commuters and vehicles is increasing enormously day by day. The horns from cars, buses, trucks, auto rickshaws, bikes, two-wheelers cause a lot of nuisance. The average noise levels on city roads are reported around 90 dB. Similar noise levels exist near railway tracks.

(iv) Construction Activities

The developmental activities and urbanization has led to a great deal of construction work which includes construction of bridges, metro rail services, flyovers, subways, roads, residential and commercial complexes. Large equipments, mixers, cranes operate at construction site and cause a lot of disturbance in the adjoining areas.

(v) Festivals and Religious Activities

India is known for its diverse cultures. The people here celebrate quite a good number of festivals with great pomp and show. These often generate a lot of noise. Deepawali being one such festival.

There are other religious functions and activities which use public address systems such as loud speakers and other aids for attracting public attention. Although, these practices are liked by many but may cause problems to others.

(vi) Workplaces

Typewriters, Telephone rings, hot discussions, mobile phones, general conversations, strikes, union meeting etc. are some routine sources of noise which do not allow concentration at work and hence hamper the efficiency and productivity.

(vii) Market Places, Malls, Restaurants, Fairs, Exhibitions

There is no quiet corner anywhere. Wherever you go, places are full of people. There is problem of honking horns, people chatting, children crying or youngsters making fun. All these noises are annoying. They disrupt pleasure, cause fatigue and affect peace of mind.

11.7 Effects of Noise Pollution

The WHO suggests that noise can affect human health in a number of ways. These include *physical*, *physiological* and *psychological effects*.

(a) Physical Effects

These are high blood pressure, other cardiovascular changes, problems in digestive system and general fatigue. Prolonged exposure to noise levels 80 dB and above can cause deafness.

(b) Physiological Effects

These arise due to interference in the biological functioning of the body due to noise pollution. They include anxiety, insomnia, hypertension, nausea and giddiness. Noise levels in the range 120-150 dB can affect respiratory system and cause dizziness, loss of physical control etc.

A loud noise can also cause increased secretion of hormones of the pituitary gland *viz.* adrenocorticotropic hormone (ACTH), which stimulates adrenal gland for further secretion of several other hormones. This chain of secretions enhances sensitivity of the body to adrenalin, increase in level of blood sugar, suppression of immune system and a decrease in the ability of liver to detoxify blood.

Chronic noise may also lead to abortions and congenital defects. Frequent exposure to noise levels 85 dB or above can lead to increased pulse rate, higher cholesterol levels and violent behaviour. The noise levels of 70 dB can cause constriction of blood vessels in skin.

(c) Psychological Effects

These include loss of working efficiency resulting in reduced output, absentism and higher rates of accidents and injuries . Noise can also interfere with sleep which in turn affects overall mental and physical health of a person.

Noise levels of 45-50 dB can cause changes in sleep patterns and affect deep sleep. Noise levels of 50 to 60 dB can also interfere with communication. At times, it may not be possible to hear warning signals.

Even very low levels of noise can cause annoyance to some people. It can lead to ill temper, quarrel or enmity.

In addition to humans, noise can be equally adverse to wildlife also. Animals in the zoo are quite sensitive to the noise in their surroundings and their behaviour can be affected by noise.

11.8 Control of Noise Pollution

Noise is a controllable pollutant. There is a growing concern about increasing noise pollution amongst people. Since the towns and cities in India are becoming more and more densely packed, the noise pollution is likely to increase both during the day time and in the night.

Noise pollution can be minimized using certain control measures. Changing the vehicle design, changes in tyres or road surfaces and eliminating old and noisier vehicles can reduce noise due to surface transport. Similar changes in design of machinery and equipment used in industries and construction work can help in checking noise pollution.

If possible, the residential areas should be distant from railroads, highways, airports, industries and factories.

Use of sound absorbing materials such as glass wool, felts, porous blocks, perforated files and blocks, carpets etc can also minimize noise.

Noise barriers, mufflers, silencers etc. reduce the transmission of sound. In addition, earplugs, earmuffs and helmets can be used for personal protection from noise.

A buffer zone of vegetation is also helpful to curtail noise to some extent. In addition to this, it is also necessary to educate people about noise and its harmful effects. Such an awareness would encourage them to exercise self-control and reduce noise pollution.

11.9 Legal Provisions and their Implementation

Noise pollution is becoming a global concern today. Every country is making rules and regulations to check noise pollution. In India also, various control measures have been initiated to control the noise pollution.

The Central Pollution Control Board constituted a Committee on Noise Pollution control. This committee recommended noise standards for ambient air, automobiles, domestic appliances and construction equipment. These standards as notified in Environment Protection Rules (1986), are given in Table 11.2.

Table 11.2 Noise Standards

Code		Day Time/ dB	Night time/ dB
A	Industrial Area	75	70
B	Commercial Area	65	55
C	Residential Area	55	45
D	Silence Zone	50	40

The daytime is between 6 a.m. and 9 p.m. whereas night time is 9 p.m. to 6 a.m.

The silence zone is referred to as area upto 100 meters around premises such as hospitals, educational institutions and courts. The silence zones are declared by the Competent Authority. Use of vehicular horns, loudspeakers and bursting of crackers is banned in these zones.

Mixed categories of areas should be declared as one of the above mentioned areas under the categories listed in Table 11.2 by the Competent Authority and for them the corresponding standards shall apply.

Noise standards for automobiles, domestic appliances and construction equipment have been notified in part 'E', schedule VI of Environment (Protection) Rules, 1986, as amended on 19th May 1993. These are given in Tables 11.3 and 11.4.

Table 11.3 Noise Limits for some types of Vehicles

Category of Vehicle	Noise limit/ dB (A)
(a) Motorcycles, scooters and three wheelers	80
(b) Passenger cars	82
(c) Passenger or commercial vehicles upto 4 MT	85
(d) Passenger or commercial vehicles above 4 MT and up to 12 MT	89
(e) Passenger or commercial vehicles exceeding 12 MT	90

Table 11.4 Noise Limits for Domestic Appliances and Construction Equipment

Category of Domestic Appliances/Construction Equipment	Noise limit/ dB (A)
(a) Window air conditions of 1 tonne to 1.5 tonne	68
(b) Air coolers	60
(c) Refrigerators	46
(d) Diesel Generators for domestic purposes	85-90
(e) Compactors (rollers), Front Loaders, Cranes (movable), Concrete mixers, Vibrators and Saws	75

Code of Practice for Controlling Noise from Sources other than Industries and Automobiles:

1. Public Address System

(i) License must be obtained by all parties intending to use loud speakers or public address system for any occasion.

(ii) Public address system and loudspeakers should not be used at night between 9 p.m. to 6 a.m. except in closed premises.

(iii) Loudspeakers should be directed at the audience and not away from audience (i.e. not towards the neighborhoods).

(iv) Loudspeakers should not be allowed for advertisement and commercial activities.

(v) The permitted strength of power amplifier should be just adequate to cover the audience, and noise level beyond the boundary limit of the noise sources in the premises should not be increased by more than 5dB(A) above the ambient noise level.

2. Aircraft Operations

(i) Aerodrome should be located away from the city and growth of the city should not be allowed to extend upto the aerodrome.

(ii) Aeroplanes should takeoff in direction radially away from the city.

(iii) During boarding and unboarding operations, the plane should be sufficiently away from the airport buildings.

(iv) Night time operations should be minimized.

(v) During maintenance and repair of the aeroplanes, workers should use ear muffs.

(vi) Portable silencers should be used in the plane intake as well as exhaust during idling period at the airport.

3. Construction Activities

(i) Acoustic barriers should be placed near construction sites.

(ii) The maximum noise levels near the construction site should be limited to 75 dB (A) Leq (5 min) in industrial areas and to 65 dB (A) Leq(5 min) in other areas.

(iii) There should be fencing around construction site to prevent people coming near the site.

(iv) Materials need not be stockpiled and unused equipment should be placed between noisy operating equipments and other areas.

(v) Constructing temporary earth bund around the site using soil etc. which normally is hauled away from the construction site.

4. Burning of Crackers

(i) Manufacture and sale of crackers having an impulsive noise of more than 90 dB at 5 meters distance from the site of bursting should be banned.

(ii) Manufacture and bursting of joined crackers should be banned.

(iii) Bursting of crackers should be banned between 9 p.m. and 6 a.m.

(iv) Bursting of crackers may be permitted only during public festivals.

In the year 2004, the following two strict rules were introduced for Diwali:

(i) The crackers of more than 125 dB will be banned and

(ii) The crackers packets will contain decibel level printed on them.

Manufacturers of crackers in Sivakasi in Tamilnadu and other places were informed about these rules. The steps for the enforcement of these were undertaken. It was declared that the pollution levels would be monitored at 40 places in Delhi.

Recently, Patna High Court has also directed that all the four zones of Patna be made free from noise pollution.

At present, various notification and rules have been brought to control the noise pollution in the country. They are mentioned below:

(1) Ambient noise standards were notified in 1989 which formed the basis for the State Pollution Control Boards to initiate action against violating sources.

(2) The vehicular noise standards, notified in 1990 are being implemented by the Ministry of Science and Technology to reduce Traffic noise. These standards have been made more stringent *vide* a notification in September 2000 and will be effective from January 2003.

(3) Noise standards for diesel gensets were prescribed in December 1998. The Government has been pursuing the State Pollution Control Boards, generator manufactures and major users for implementation of these standards. These standards are being revised *vide* notification dated 30th December 2002 and are applicable at manufacturing stage from 2003. In this notification, noise limits for vehicles at manufacturing stage applicable from 1st April 2005 have also been stated.

(4) Noise standards for petrol and kerosene generator sets were notified in September 2000 and these are effective since September 2002. The sale of these gensets will be prohibited if not certified by testing agencies, identified for the purpose.

(5) The Noise Rules, 2000, regulate noise due to Public Address system/ Loud speakers and also prescribe procedures for noise complaint handling.

(6) The Central Pollution Control Board is studying aircraft noise at IGI airport Delhi. This will be followed by the development of guidelines/ standards for aircraft noise.

(7) The noise standards for fire crackers were enacted on 5th October 1999 and from time to time various notifications and steps are initiated to implement them.

Climate Change

CHAPTER 12

12.1 Introduction

Often, the difference between the terms *climate* and *weather* is not clearly understood. The term *climate* refers to the long-term average weather conditions over a period of time for a given place whereas *weather* indicates the state of atmosphere for a given place at a given time.

The change in climate is a slower process as compared to changes in weather and the climate change occurs over longer time scales. Space wise also, the climate has wider range whereas weather may change over very small distances. The study of climate – its development, change and factors affecting it, is called **climatology**.

12.2 Role and Importance of Climate

Climate is the result of complex interactions between atmosphere and other environmental systems such as lithosphere, hydrosphere and biosphere. Climate governs the geographical distribution of ecosystems. It is also one major factor for soil formation which is linked to food production. It also dictates the energy needs of the people for heating or cooling purposes. Climate also strongly affects availability of water resources, snow and ice cover, floods and draughts, distribution and types of plant and animal species. It is also related to the phenomena of weathering and desertification.

The relationship between climate and above environmental systems is so interlinked that the impact of any disturbance at one place can be observed at some other place or system. Also, this relationship is very complex and many such interconnections are yet to be fully understood.

The human beings and climate are interdependent and both affect each other in one way or the other. Climate dictates the ways of living of people. Climate governs natural resources and hence is closely linked to occupation, food habits, economic activities and health of the people. A favorable climate attracts the migration of people from other areas whereas the harsh climate is a negative and repelling factor.

Thus, climate has many direct and indirect impacts on human beings. It also equally controls the vegetation, plant life and animal resources of a place. Even the marine life is not spared by the type of climatic conditions prevailing over a particular region.

12.3 Factors Affecting Climate

Climate of a region is never constant. Many factors control the climate and their combined effects result in a particular type of climate at a particular place. Some of these are:

(i) Solar Energy

The Sun is the ultimate source of energy for our planet. At the same time, it is this energy only, which has made the earth habitable. The solar energy received by the earth is responsible for maintaining the temperature suitable for the existence of various life forms. The variation in the amount of solar radiation received is the reason for the existence of different temperature and air pressure zones in different regions of the Earth.

(ii) Oceans and Water Bodies

About 75% of the Earth is covered by water bodies such as oceans and seas. The availability of water vapours and precipitation influence the climate by their effect on air temperature.

In addition, *ocean currents* also redistribute the heat. Ocean currents are the movement of large amounts of warm water from lower to higher latitudes and that of cold water from higher to lower latitude for replacing it. These currents strongly affect the temperature and the humidity of adjacent air mass which affect the climate and weather conditions of adjacent coastal and continental areas along the path of the current.

Water bodies stabilize temperature because they heat and cool more slowly than the land. The variation of air temperature on water bodies is less as compared to that on land. Also, the coastal areas do not have extreme temperatures as faced by the land masses. The land areas have comparatively hotter summers and colder winters.

(iii) Altitude

The atmosphere gets heated through the earth's surface. As we go up, the temperature decreases. Moisture laden air cools on moving up and further condensation of water vapours ultimately leads to precipitation.

(iv) Wind Systems

These are land and sea breezes as well as mountain and valley breezes. Land warms up and cools faster than water. Therefore, there is a low pressure over land during daytime. Thus, air flows from sea to land causing the *sea breeze*. The reverse happens at night causing the *land breeze.*

Similarly, warm air rises up the mountain slopes and valleys during the daytime whereas in the night, cool air descends down the slope.

In addition to the above natural factors affecting the climate patterns, human activities are changing the climate in a much drastic way and at a much faster rate. These include pollution in general and air pollution, in particular. The harmful effects of air pollution have been dealt in Chapter 4. Air Pollution affects human health, plant and animal life and buildings as well. It also causes the occurrence of smog, ozone depletion and global warming.

Industrialization and developmental activities including urbanization has caused deforestation, loss of soil cover etc. and has replaced vegetation by roads and buildings. Such changes are normally accompanied by the increase in temperature, winds, relative humidity and clouds as well as fog. Also, there is an increase in the amount of gaseous oxides and dust particles in the atmosphere. All these have an effect on the climate patterns.

12.4 The Global Climate

Various zones can be identified on earth according to the variation in latitude. These zones have typical climates and associated characteristic vegetation patterns. These climate patterns are mainly *tropical* (hot and wet), *temperate* (mild and variable) and *polar* (cold). The climate near equator is hot and that near poles is cold. Besides temperature, there are various other factors such as pressure, humidity, precipitation, wind, solar radiation and cloudiness which affect the climate. Various classification schemes are available which further sub-divide the climatic zones.

12.5 Climate Change

The climate of earth has been changing since millions of years but the changes during the last two centuries are of very alarming nature. The trends show the following:

(1) Global temperatures are rising. In the last century, the average land temperature has arisen by 0.45-0.6°C.

(2) Precipitation has increased by 1% over the continents in the last century. There was an increase in the rainfall in high altitude areas whereas a decrease was noticed in many tropical regions.

(3) An increase in sea level by 15-20 cm has been observed. The contribution of thermal expansion is 2-7 cm whereas melting of ice has led to 2-5 cm increase.

In Chapter 4, it has been explained that greenhouse gases such as carbon dioxide, methane, ozone, nitrous oxide and water vapour permit the passage of incoming solar radiation but partially absorb and re-emit outgoing infrared terrestrial radiation. This *natural greenhouse effect* raises the average temperature of the earth from – 18° C to +15° C and makes the life sustainable on earth.

Ever since the industrial revolution began about 150 years ago, the human activities have led to an increase of green house gases (GHGs) in the atmosphere. These activities are energy generation from fossil fuels which emits carbon dioxide, industrial activities, change in land use patterns such as deforestation, biomass burning etc. These activities release GHGs in the atmosphere.

Since pre-industrialization, the atmospheric concentration of CO_2 has increased from nearly 280 to 358 ppmv (in1994), methane from 700 to 1720 ppbv (in 1994), N_2O from 275 to 312 ppbv (in 1992) and CFC-11 from zero to 110 pptv (in 1992). The increase is about 31%, 151% and 17% in the first three cases in the period 1750-2000 as per Intergovernmental Panel on Climate Change (IPCC, 2001).

Experiments and measurement records indicate that since 1860's the global average temperature has already increased by 0.6± 0.2°C. The *Third Assessment Report* of IPCC states that there is new and stronger evidence that most of the warming observed over the last 50 years is attributable to human activities.

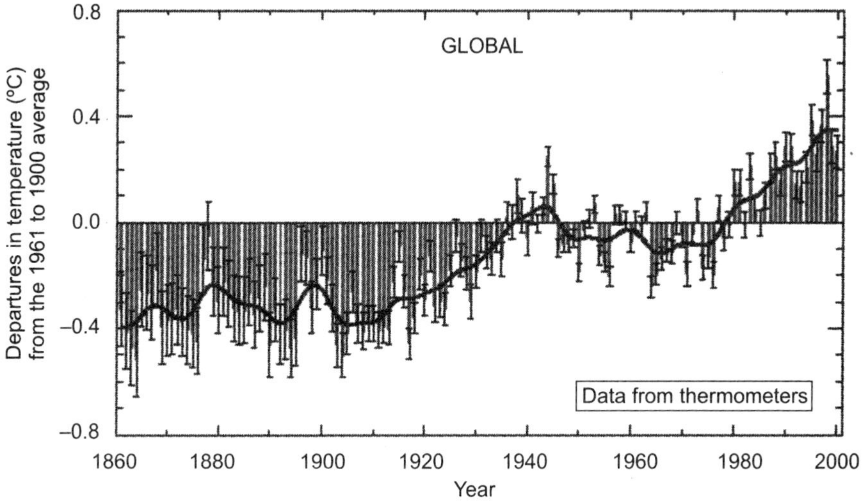

Fig 12.1: Variation of Earth's Surface Temperature Since 1860's

It has been observed that the last two decades of the 20th century were the warmest particularly the year 1998. Various researches and studies show that from 1990 to 2100, the surface temperature would increase by 1.4°C to 5.8°C. This magnitude of warming is greater than that experienced over last 10,000 years. This increase is even higher than that predicted by the *Second Assessment Report* which said it to be 1°C to 3.5°C. Such a fast rate of change of temperature will have strong effects on the climate and the ecosystems may not be able to adapt to such fast changes.

Also, the global mean sea level is projected to rise by 0.09 to 0.88 meters. The reason being thermal expansion of the oceans and melting of glaciers as well as polar ice sheets.

12.6 Effects of Climate Change

It is very difficult to exactly tell the effects of climate change since many uncertainties and variation in variables such as precipitation, evaporation, storms, sea levels etc. are involved. The effects on human health, ecosystems and agriculture are yet to be clearly understood because of complex interrelationship of various factors affecting climate change.

The available studies indicate that the impact would be more severe for Asian and African countries than those situated at higher latitudes. In terms of GDP, these impacts would be of the order of 8.5% for South Asian and African countries whereas for OECD countries, the order would be 1.6%. The effect on the world as a whole would be of the order of 1.5 to 2%.

12.6.1 Climate Change in Asia

Climate change is a matter of special concern for Asian countries because they frequently face the problems of floods, droughts, cyclones and typhoons. These phenomena greatly affect the economy and human settlements in these countries.

Table 12.1 gives some statistical data for the Asian region.

Table 12.1 Some Basic Statistical Data

Parameter	Value	Year
Population (thousands)	3,588,877	1998
Population Density (per 1000 hectares)	1,130	1996
Total Land Area (in thousand hectares)	3,085,414	1995
Total Forest Extent (in thousand of hectares)	474,172	1995
Total Commercial Energy Production (petajoules)	122,438	1995
Traditional fuel Consumption (petajoules)	10,308	1995
Total CO_2 emission (thousand tonnes of CO_2)	8,270,648	1995
Per capita CO_2 emission (tonnes)	2.3	1995

The region extends from 80° longitude (70°E to 150°E) to 40° latitude (from 30° N to 10° S). The region is physically diverse and is rich in natural biodiversity. The present population being about 1.6 billion, which is expected to increase to 2.4 billion by the year 2025.

The climate is dominated by the summer southwest monsoon and the winter northeast monsoon. They affect the precipitation which in turn is linked to water availability for agricultural and domestic use. The tropical cyclones in the northern Indian Ocean and north-western Pacific Ocean also affect the regional weather. The El Nino Southern Oscillation (ENSO) phenomenon also has influence on climate and sea level.

A special report of the IPCC Working Group II (The Regional Impacts of Climate Change: An Assessment of Vulnerability) has described the impacts under the following categories.

(a) Impacts on Temperature and Rainfall

It is predicted that the warming would be faced through out the region but its extent would be lower than the average global warming. The proximity to sea would cause regional variations. Thus, the warming would be least in the islands and coastal areas of Indonesia, Philippines, coastal South Asia and Indo-China while it would be greatest in the continental areas of South Asia and Indo-china except from June to August in South Asia where reduced warming could occur.

The rainfall would be maximum over South Asia and the Indo-China peninsula in April-September but would be minimum over Indonesia and near Australia.

(b) Impacts on Ecological Systems

It has been expected that the ecosystems in the mountains and planes in Asia would shift upwards. At high elevations, weedy species would displace trees. There will be increased soil erosion in Greater Himalayas and hence, the rate of vegetation change would be slow. There will be complex changes in distribution of healthy rainforest and drier monsoon forest. It is expected that in Thailand, the tropical forest area would increase from 45% to 80% of the total forest cover whereas in Sri Lanka, the tropical rainforest would decrease by 2-11% and tropical dry forest would increase by 7-8%.

Coastal ecosystems such as coral reefs could be bleached by increase in sea water temperature. Similar phenomenon was observed during the ENSO event in 1983 when coral reefs in Indonesia faced severe bleaching.

The other coastal ecosystem, which would be affected by sea level rise and increase in temperature, are mangroves. About 40% of the total mangrove area i.e. about 75,000 km^2 lies in tropical Asia. Out of this 40%, 23% is in Indonesia. Bangladesh has the largest continuous mangrove area in the world. The Sunderban region is already facing a threat to many mammals, birds, reptiles and amphibian species due to sea level rise. This area would also be vulnerable to cyclonic storms and surges.

(c) Impacts on Agriculture

Many cereal and tree crops of the region have been studied with respect to changes in temperature, moisture and CO_2 concentration. It is expected that the yields of rice, wheat and sorghum would increase with the increase in CO_2 concentration but there would be region wise variation in growing seasons, crop management, storage and distribution. The agricultural areas of tropical Asia are vulnerable to frequent floods, droughts, cyclones and storms which damage life, property and affect agricultural output. These factors were responsible for loss of about half a million tonnes of rice in Bangladesh during 1962-88 which amounted to about 30% of its average annual food grain exports.

(d) Impacts on Coastal Zones

In coastal areas, rise in sea level and increase in rainfall would lead to the following effects:

- Land loss and population displacement
- Increased flooding of low lying coastal areas
- Loss of agricultural yield and employment
- Decrease in coastal tourism by erosion of sandy beaches

The large delta regions of Bangladesh, Myanmar, Vietnam and Thailand, and low-lying areas of Indonesia, the Philippines and Malaysia are vulnerable to coastal erosion, land loss, sea flooding, inundation etc. The Ganges- Brahmaputra delta is one of the world's most densely populated areas and would face serious drainage and sedimentation problems due to sea level rise besides coastal erosion and land loss.

The economy and food security of the region heavily depends upon aquaculture and marine fisheries. Fisheries account for about 4% of the GNP in the Philippines and it employs about a million people. Fish populations are strongly affected by even slight changes in temperature, salinity, wind speed and direction, ocean currents etc. A rise in sea level would cause saline water to penetrate further inland which would increase the habitat of brackish-water fisheries.

According to the studies, a 1m rise of sea level would lead to land loss in Bangladesh, Indonesia, India and Malaysia. Table 12.2 shows the extent of damage.

Table 12.2 Effects of Sea Level Rise

Country	Land loss (km^2)	People displaced
Bangladesh	30,000	15 million
India	6,000	7 million
Indonesia	34,000	2 million
Malaysia	7,000	—

(e) Impact on Hydrology and Water Resources

The Himalayas are very critical for the availability of water to continental monsoon in Asia. The increase in temperature and variation in precipitation would accelerate recession of glaciers and increase the danger from floods. There would also be an increase in the peak flows in the rivers and increased sediments in turn would have major effects on hydropower generation, agriculture and urban water supply. The water would increase in snow fed rivers for short term but in the long run, there would be a stress on water resources due to lesser availability and increasing demand.

(f) Impacts on Human Health

The occurrence and extent of vector- borne diseases would increase. The climate change would increase the spread of malaria, dengue and schistosomiasis which are reasons of mortality and morbidity in Tropical Asia. The fatalities would be more in the newly affected areas. The epidemic potential would increase for malaria (12-27%), dengue (31-47%) and schistosomiasis (11-17%).

Similarly, water-borne and water related infectious diseases would also show an increase with the higher temperature and higher humidity conditions and a lower water quality due to water crisis.

12.6.2 Impacts of Climate Change on India

India is highly vulnerable to climate change. Our economy heavily depends upon climate sensitive sectors such as agriculture and forestry. India is already facing problems of increasing population, rapid urbanisation, industrialization and economic development. The climate change would further increase these pressures.

It is expected that doubling of CO_2 concentrations would increase the temperature by 2.33°C to 4.78°C. The frequency of heavy rainfalls would increase in South and Southeast Asia. The annual mean maximum and minimum surface air temperatures would increase by 0.7°C and 1°C by 2040s as compared to 1980s. This increase would be higher for adjoining oceans. The specific impacts are as follows:

(a) Impact on Agriculture

Agriculture and related activities constitute the single largest component of Indian economy. It contributed 27% of GDP in 1999-2000. The agricultural export amounted to 13-18% of the total exports (Ministry of Finance, 2002).

A decrease in rice yield by 3-15% has been reported by a rise in temperature by 1.5°C and a 2 mm increase in precipitation. Another study has reported a variation of -22 to 18% under different climatic conditions for soybean fields considering +2 to +4°C rise in temperature and ± 20 to ± 40% change in precipitation.

Many researches and studies are undergoing to assess the impacts of various agencies at different levels.

(b) Impacts on Forestry

Forests contribute 1.7% to the GDP of India (MoEF sources) Non- timber forest products yield 40% of total forest revenues and contribute 55% of forest-based employment. Nearly 55 million people in and around forests depend upon non-timber products for their sustenance (TERI, 1999).

The geographical distribution, composition and productivity is largely determined by the climate. Thus, forest ecosystems would be greatly affected by the climate change.

The increased temperature and rainfall could result in increased productivity, migration of forest types to higher elevations and conversion of drier forests to moister types.

Another study showed that teak productivity would decline due to decrease in the soil moisture content. It also indicated the decrease in the productivity of moist deciduous forests from 1.8 m^3 / ha to 1.5 m^3/ ha.

(c) Impacts on Coastal Zones

India has a low-lying densely populated coastline which extends to about 5700 km. It is vulnerable to rise in sea level. A·1m rise in sea level would inundate 1700 km^2 of prime agricultural land. Another study estimated that a total of 5763 km^2 area and 7.1 million people would be affected if no protection measures are taken. The economic loses would be very-very high. Also coastal infrastructure, oil explorations and tourist activities would be at risk.

The frequency and intensity of storms and cyclones would increase. The typical cyclone of 1999 in Orissa took a toll of about 10,000 people.

(d) Impacts on Water Resources

India has large amount of total water resources and rainfall but they are not evenly distributed. Thus, there is an imbalance in the demand and supply of the quantity and the quality of water.

Climate change would further aggravate the water crisis. The lower rainfall and increased evaporation would lead to less run off, affecting the availability of freshwater. The soil moisture and aridity level would also be affected.

Studies show that melting of snow in the high Himalayas would cause floods in catchments. The impact would be more on the western side. There would be an increased risk of floods in North India during wet season due to rise in snowline because of increased temperature.

12.7 Monitoring Climate Change

Because of the growing concerns about climate change world over, the World Meteorological Organisation (WMO) and United Nations Environmental Programme (UNEP) in 1988 established Intergovernmental Panel on Climate Change (IPCC) for assessing the gravity of the problem. The *First Assessment Report* of IPCC was completed in 1990 which highlighted the threat of climate change.

In December 1990, the UN General Assembly decided to start negotiations which actually commenced in February 1991 and were concluded in 15 months time. These were called *United Nations Framework Convention on Climate Change* (UNFCCC). The Convention was adopted in May 1992. It was opened for signatures in Rio at the UN Conference on Environment and Development. It was ratified by 50 countries and came into force in March 1994.

The ultimate objective of the Convention was to achieve the stabilization of greenhouse gas concentrations in the atmosphere at a level that would prevent dangerous anthropogenic interference with the climate system. Such a level should be achieved within a time frame sufficient to allow ecosystems to adapt naturally to climate change, to ensure that food production is not threatened and to enable economic development to proceed in a sustainable manner.

In 1997, the Parties to the UNFCCC adopted the *Kyoto Protocol*. It required Annex I countries (developed countries and economies in transition) to reduce their overall GHG emissions by 5.2% below their 1990 levels.

Kyoto Protocol created three cooperative mechanisms- *Emission Trading*, *Joint Implementation* and *Clean Development Mechanism*. These mechanisms allow member counties some flexibility in meeting their commitments with respect to reduction in emissions.

The *Clean Development Mechanism* is important for India as it allows developed countries to use certified emission reduction credits from project activities in developing countries, and also assists developing countries in achieving sustainable development objectives.

The *Sixth Conference of Parties* (COP 6) was held in Bonn on 16-27 July 2001. In COP 6, political agreements were made to help countries move towards adoption of the Kyoto Protocol. The Protocol was formalized into the *Marrakesh*

Accords in October-November 2001. The *Seventh Conference of Parties* (COP 7) was held at Marrakesh, Morocco during 29 October-10 November 2001.

The *Marrakesh Ministerial Declaration* reaffirms that economic and social development, and poverty eradication are the first and overriding priorities of the developing countries. It also stresses the importance of capacity building, and developing innovative technologies through public and private sector involvements.

The financing mechanisms for climate change related activities in the developing countries include the following:

- Introduction of an *adaptation fund* to finance adaptation of projects and programmes in developing countries. It derives funding (2%) of the share of proceeds from the clean development mechanism projects.
- Establishment of a *special climate change fund* from Annex I Party Contributions for funding activities complementary to those funded by *Global Environmental Facility* (GEF).
- Introduction of a *Least Developed Country Fund* for supporting a work programme for least developed countries.

Today, 186 Governments are parties to this Convention.

12.8 UNFCCC and India

India signed the UNFCCC on 10th June 1992 and ratified it on 1st November 1993. India does not have binding GHG mitigation commitments because of its small contribution to green house problem and low financial & technical capacities.

The nodal agency for climate change issues is the Ministry of Environment and Forests (MoEF). Working Groups have been constituted for UNFCCC and Kyoto Protocol. Article 12 of the Kyoto Protocol provides for the Clean Development Mechanism (CDM). India acceded to the Kyoto Protocol on 26th August 2002.

India has taken following initiatives for improving the understanding about climate change and to comply with the requirements of the UNFCCC (*Source* - MoEF):

(1) *Preparation of the initial National Communication to UNFCCC by the Government of India*: All parties are required to communicate a national inventory of GHGs and a general description of the steps taken for the implementation of the Convention.

The GHG inventory is being prepared for the base year 1994 and covers the following sectors:

- Energy
- Industrial processes
- Agriculture
- Forestry
- Waste

The detailed estimation of each of the sector's GHG emissions and identification of country specific emission factor would be done.

(2) The Government of India is supporting the study of the Asian -Least cost Greenhouse Gas Abatement Strategy (ALGAS). The study developed a national inventory of GHG sources and sinks, and identified potential mitigation options.

Country specific emission factors have been developed - for methane emission from paddy cultivation, CO_2 emissions from Indian coal etc.

(3) An extensive methane measurement campaign coordinated by National Physical Laboratory was undertaken in 1991. The measurements were done for major paddy growing regions of the country under different rice environments for the whole cropping period.

(4) The Technology Information, Forecasting and Assessment Council has been established under Department of Science and Technology which facilitates the transfer of environmentally sound technology.

(5) The Participatory Forest Management Strategy secures the rehabilitation of degraded areas, conservation of biodiversity along with sharing of benefits with local people.

(6) Establishment of Standing Committee for monitoring development in coastal ecosystems for all coastal States and Union Territories according to the Coastal Zone Regulation Notification of 1991.

(7) Generation of information regarding vulnerability to climate change in collaboration with research organizations and academic institutions.

(8) Replacement of cyclone detection radars with state of the art Doppler Weather Radar in a phased manner.

(9) Information and data collection - the Indian Meteorological Department (IMD) collects information about climatic parameters at surface and upper air observatories throughout the country by 559 surface observatories, more than 8000 rainfall monitoring stations, 100 satellite based data collection platforms in remote areas, 203 voluntary observing ships, 10 cyclone detection radars and 17 storm detection radars.

India has also ratified the Kyoto Protocol on 26 August 2002.The Protocol comes into force only after a minimum of 55 Parties to the Convention included in Annex I accounting for at least 55% of total CO_2 emissions of Annex I parties, ratified the Protocol. 77 countries accounting for 36% of 1990 emissions of Annex I parties have ratified the protocol.

The 8th conference of Parties (COP 8) was held in New Delhi from 23rd October to 1st November 2002.

12.9 Other Aspects of Climate Change

There are various other aspects and issues related to climate change. There are certain other factors also (discussed below) which affect the climate changes. But we have not understood them very clearly so far. Therefore, it will be very difficult to predict the future climate unless we get a clear wholesome picture of all the factors and their interconnections.

1. Role of Aerosols

Since the industrial revolution, the emission of aerosols to the atmosphere has also increased. Aerosols are particles having diameter less than 10 μm. They tend to remain in the atmosphere by collision with the air particles which overcome the gravitational attraction.

It has been observed by researches that aerosols emitted from coal (sulphates) may cause global cooling because they act as seeding for clouds. They provide surface for water to condense on. The resulting cloud formation reflects the incoming solar radiation. Similar cooling was observed after the 1991 eruption of Mount Pinatubo in the Philippines. About 20 million tonnes of sulphur dioxide was released which circles the globe for about 2 years cooling the earth's atmosphere.

Such volcanic eruptions emitting ash and aerosols lower the temperature. The global warming was counterbalanced by the above cooling in 1991-92. By 1994, when the aerosols had settled, the global temperature returned to previous higher levels.

2. Nino and ENSO

Nino refers to the child, the Christ child because these events often begin at the coast of South America near Christmas time. They cause a disruption of the ocean-atmosphere system in the tropical Pacific. During Nino, the eastern equatorial Pacific Ocean becomes unusually warm and the westward moving equatorial current reverses. The rise in temperature of sea surface waters off the South American coast has ecological consequences. Warm surface waters inhibit upwelling of nutrient-rich cold water which supports marine ecosystem and major fisheries. During Nino years, the rainfall follows warm water eastward, there is high precipitation and floods in Peru while droughts and fires are observed in Australia and Indonesia. Warm waters change global atmospheric circulation which alters the weathers in regions far away from the tropical Pacific.

Nino events occur at intervals of two to seven years and last for about 12 to 18 months. Upwelling releases CO_2 to the atmosphere. Nino events reduce the oceanic CO_2 outgassing and thus perturb the global CO_2 cycle.

The Nino of 1982-83 was very strong. These conditions developed in 1991-92 and also in 1993. Nino caused floods, droughts and killed thousands of people. It also damaged crops, buildings and other structures and caused heavy loss.

Nino events alternate with El Nina events in which ocean water get unusually cool and cause effects opposite to Nino events.

3. External Factors Affecting Climate Change

(i) Variation in Solar Output

There are changes in solar energy associated with 11-year sunspot cycles or 22-year solar magnetic cycles. Historical dates suggest that colder conditions in Europe coincide with periods of reduced or no sunspot activity.

(ii) Variation in Earth's Orbit

The earth orbits around the Sun. The Yugoslav astronomer M. Milankovitch described variations in earth's orbit in 1920s. They result by three interacting effects which are as follows:

- Variation in the shape of the elliptical orbit (over 95,000 year cycle)
- Variation in the tilt of the Earth's axis of rotation (over a 42,000 year cycle)
- Variation in the time of year when the Earth is closest to the Sun (over a 21,000 year cycle)

The analysis of ocean cores shows that these three cycles appear in the oxygen isotope and faunal records. It shows that they are associated with the timing of glacial and interglacial phases.

(iii) Continental Drift

The continents have redistributed over the surface of the Earth over millions of years. This has caused the creation of major oceans and mountain belts, which affect the climate.

Around 450 million years ago, the Sahara desert was located near the South Pole and it was glaciated. Similarly, Britain was located much further south in latitudes 20°-30°, where trade winds created desert conditions. Thus, over the time, there have been drastic changes over global climate both for continents and oceans.

Biodiversity

CHAPTER 13

13.1 Introduction

Life exists in many diverse forms. There are plants, animals, microorganisms and other life forms. They all occur in different morphological forms and exhibit a wide spectrum of sizes, shapes, colours etc. The range includes from very tiny plants to large trees and bacteria to elephants. Similar to the diversity in these life forms, there is diversity in their habitats. The living forms exist in deep oceans, dry deserts, cold snow covered regions and on high mountains.

Life originated in very primitive form and such a diversity has evolved over billions of years. Each species existing on the Earth has its importance and a unique place in the ecosystem. These species have diverse roles and uses. In fact, many of these species are yet to be explored. Many of the species which existed earlier do not occur today. Also, various new forms which were not present earlier, have evolved during the course of time. Thus, old species are continually being replaced by new ones. This keeps on happening at all times and would happen in future also. Many factors are responsible for this change. But during last few decades, the rate of disappearance of various species has been very fast. Obviously, the new species do not appear at such a fast rate. Thus, something beyond the natural process of evolution of species is responsible for such loss of biodiversity.

There are many reasons for this loss of diversity. It is a matter of serious concern and demands immediate attention and appropriate corrective measures.

13.2 The Richness of Biodiversity

The exact number of species existing on the Earth is not known to anyone.

There are only estimates and these numbers also keep on changing with time because new species get added to the list of known species. This is because the tropical rain forests which are a home for a large number of species, have not been very much explored so far.

Some estimates put the number of species between 5 to 50 million whereas the others predict it to be somewhere between 1.4 million to 100 million. As per year 2000 data, only 1.7 million species have been described.

About half of the world's total species are present in just 7% of the land area. Thus, the distribution of species is not uniform. It is estimated that the tropics alone have 5 million species.

In fact, insects and plants constitute most of the species. Many of the insects are tropical beetles of rain forests. Mammals, however, form only a small fraction and the number of species of mammals is about 4000.

The diversity of species involves *genetic diversity*, *habitat diversity* or *species diversity*. The **genetic diversity** refers to the total number of genetic characteristics where as the **habitat diversity** is the diversity of habitats per unit area. The **species diversity** further comprises *species richness*, *species evenness* and *species dominance*. The species richness indicates the total number of species, the species evenness means relative abundance of species and the species dominance signifies the most abundant species.

The distribution, more particularly the uneven distribution of biodiversity, is a result of variation in climatic conditions, environmental factors, location of a place or a habitat and human interference and activities. The natural relationships between species, i.e. competition, symbiosis, predation, dominance etc. also govern the diversity.

13.3 The Importance of Biodiversity

Our survival depends on a large number of organisms. Some of the benefits are directly visible to us while some others are indirect. Every organism or species has its unique role and importance. At times, we are not even aware of the significance of a particular species until it is eliminated and we see harmful effects arising from its absence in the complex natural network.

Our needs of food, fuel, shelter, fibres, medicines and other products are met by various biological resources. Of course, plants are primary producers and we depend upon them for meeting our diverse requirements. Since ancient times, about five thousand plant species have been used as food by humans. But at present, less than 20 species feed the majority of the world's population and out of these, only three or four carbohydrate crops are staple diet for majority of the people. Many people also eat seafood, meat and other animal products.

Besides food resources, we derive wood from trees which is used as a fuel by a large number of people. It is also used in making houses, furniture and other household products. In addition to this, we get many useful products such as medicines, dyes, resins, gums, tannins, latex, fruits, seeds, oils, nuts etc. from plants.

India has a long history of using medicinal plants. The therapeutic value of medicinal plants was recognized long back and they have been in use in the traditional systems of medicine since ancient times. India has about 7500 species of plants having medicinal value. The use of herbs in medicines, cosmetics and other products is increasing day by day world over. Many of the synthetic drugs which we use today were, infact, obtained from the plant sources in the earlier times. Some examples being quinine, digitalis, aspirin, taxol and extracts of guggal. In fact, many coral reef species produce toxins for self-defense and they are being explored for such uses. About 40% of the drugs we use today have their origin from plants, animals or microbes.

The biological diversity provides diverse ways of pollution control. Plants, fungi and bacteria remove toxic substances from air, water and soil. Carbon dioxide and sulphur dioxide are removed by plants. Carbon monoxide is oxidized by soil fungi and bacteria. Nitric oxide is incorporated into the nitrogen cycle. The purification of atmosphere by trees and forests is well known. In addition, forests also check floods and prevent soil erosion. They increase the underground water supply and are responsible for the humidity of air. Besides, they provide home for wildlife and a large number of species. Also, forests have immense economic importance.

In addition to the utilitarian and economic importance, biological diversity has many secrets which are yet to be revealed. Many plant species can serve as potential food crops. The way we have domesticated the wild strains of wheat and other grains, we can create new genetic hybrids from other wild strain also. Many wild varieties can be explored for their disease resistance and their genetic characteristics can be introduced into other crops to yield disease-resistant varieties.

Similarly, other diverse uses would become clear once these unknown resources unfold. The presence of a vast gene bank, yet unexplored, can provide many useful varieties of strains which can be developed to meet our diverse needs. In fact, we need to conserve this gene pool so that it is not lost from our hands forever.

Biodiversity has its own ecological importance. Each species forms a special part of the complex natural web and has a definite role to play. Thus, each species present in an ecosystem is important and each ecosystem is important in its own right. Therefore, it is not wise to play with or disturb the biodiversity unless we are fully aware of its consequences.

In addition to its above values, biodiversity attracts a large industry of tourism which has its associated economic benefits for a particular geographical location. The social development and economic benefits of local area people are greatly affected by large number of tourists who visit that land for its scenic beauty, appealing landscape, pleasant climate or cultural importance. The beauty of biodiversity is felt and appreciated by everyone. It gives an immense pleasure to experience the range of biodiversity existing on the Earth. This also leads at times one to ponder about oneself, his or her role and place in such a vast stage of nature. This may affect the way of one's thinking and lead to a change in the attitude making it more positive towards nature and its conservation.

After all, there is a moral justification also for the conservation of existing biodiversity. Whether useful or not, every species has a right to exist and human beings should not temper the nature's creation just because our lust for comforts and list of requirements is unending.

13.4 Biodiversity of India

India has very rich biodiversity. The habitats range from tropical rainforests to alpine vegetation and from temperate forests to coastal wetlands. India has about 2.4% of the world's area and it has 7.3% of global fauna; the number of faunal species being 89,451 (MoEF, 1999). The important features of India's biodiversity are as follows:

(1) There are two major realms, *viz.* Palaearactic and Indo-Malayan. The three biomass areas include the tropical humid forests, the tropical dry/ deciduous forests and the warm desert / semi deserts.

(2) India has the following ten bio-geographical zones:
   - (a) The Trans- Himalayan
   - (b) The Himalayan
   - (c) The Indian Desert
   - (d) The Semi-arid Zones
   - (e) The Western Ghats
   - (f) The Deccan Peninsula
   - (g) The Gangetic Plain
   - (h) The North-East India
   - (i) The Islands
   - (j) The Coasts

These are shown in the map of India in Fig. 13.1.

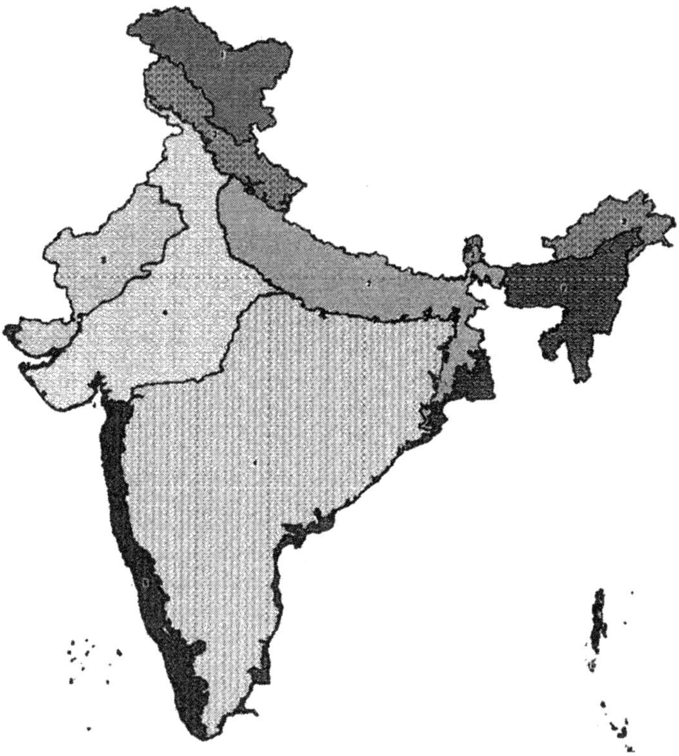

Fig. 13.1: Biogeographical Zones of India

(3) India has five world heritage sites (Table 13.1), twelve biosphere reserves (Table 13.2) and 6 Ramsar wetlands.

(4) India is one of the twelve centres of origin of cultivated plants.

(5) The protected areas amount to 1.53 lakh sq km and comprise 88 national parks and 490 sanctuaries.

Table 13.1 India's Heritage Sites

S.No.	Name	State
1.	Kaziranga National Park	Assam .
2.	Keoladeo Ghana National Park	Rajasthan
3.	Manas Wildlife Sanctuary	Assam
4.	Nanda Devi National Park	Uttar Pradesh
5.	Sundarban National Park	West Bengal

Table 13.2 Biosphere Reserves of India

S. No.	Name	Area (sq km)	Location	Date of Notification
1.	Nilgiri	5,520	Part of Wynad, Nagarhole, Bandipur and Madumalai, Nilambur, Silent Valley and Siruvani Hills (Tamil Nadu, Kerala and Karnataka)	01.08.86
2.	Nanda Devi	5,860.69	Part of Chamoli, Pithoragarh, Almora Districts (Uttaranchal)	18.01.88
3.	Nokrerk	820	Part of Gora Hills (Megalaya)	01.09.88
4.	Manas	2,837	Part of Kokrajhar, Bangaigaon, Barpeta, Nalbari, Kamprup and Dareng district (Assam)	14.03.89
5.	Sunderbans	9,630	Part of delta of Ganga and Brahamputra river system (West Bengal)	29.03.89
6.	Gulf of Mannar	10,500	Indian Part of Gulf of Manar between India and Sri Lanka (Tamil Nadu)	18.02.89
7.	Great Nicobar	885	Southernmost islands of Andaman and Nicobar	06.01.89
8.	Simlipal	4,374	Part of Mayurbhanj district (Orissa)	21.06.94
9.	Dibru Saikhowa	765	Part of Dibrugarh and Tinsukhia district (Assam)	28.07.97
10.	Debang-Debang	5,112	Part of Siang and Debang Valley (Arunachal Pradesh)	02.09.98
11.	Pachmarhi	4,926.28	Part of Betual, Hoshangabad and Chindwara districts (Madhya Pradesh)	03.03.99
12.	Kanchanjanga	2,619.92	Part of Kanchanjanga (Sikkim)	07.02.2000

Source: MoEF 2000

A large variety of species are **endemic** to India. It means they are native to India and are not found at any other place. About 33% of the recorded flora of India is endemic to the country, see Tables 13.3 and 13.4. They occur in the North-east, Western Ghats, North- west Himalaya and the Andaman and Nicobar Islands mainly.

Table 13.3 Endemic Plant Species of India

Group	No. of Species
Pteridophyta	200
Angiosperms	4950

Table 13.4 Endemic Animal Species of India

Group	No. of Species
Mollusca-Land	878
Mollusca-Fresh water	89
Insecta	16,214
Amphibia	110
Reptilia	214
Aves	69
Mammalia	38

There are 49,219 plant species, out of which 5150 are endemic. These species are distributed over 141 genera under 47 families. These endemic species amount to about 30% of the world's recorded flora. According to the Botanical Survey of India and the Zoological Survey of India, there are 47,000 species of plants and 81,000 species of animals. Tables 13.5 and 13.6 give details of plant and animal species recorded in India and the World.

Table 13.5 Number of Plant Species

	Species		% in India
	India	World	
Bacteria	850	4000	21.25%
Viruses	Not known	4000	21%
Algae	6500	40000	16.25%
Fungi	14500	72000	20.14%
Lichens	2000	17000	11.80%
Bryophyta	2850	16000	17.80%
Pteridophyta	1100	13000	8.46%
Gymnosperms	64	750	8.53%
Angiosperms	17500	25000	7%

Source: MoEF, 1999

Table 13.6 Number of Animal Species

	In India	In World	% in India
Protista	2577	31259	8.24
Mollusca	5070	66535	7.62
Arthropoda	68389	987949	6.90
Other Invertebrates	8329	87121	9.56
Protochordata	119	2106	5.65
Pisces	2546	21723	11.72
Amphibia	209	5150	4.06
Reptilia	456	5817	7.84
Aves	1232	9026	13.66
Mammila	390	4629	8.42

The crop diversity of India is also quite rich. It includes 167 cultivated species and 320 wild relatives of crop plants. India ranks seventh in terms of contribution to world agriculture. It is also considered as a centre of origin of 30,000-50,000 varieties of rice, pigeon- pea, mango, ginger, turmeric, sugar cane etc.

Wetlands, Mangroves and Coral Reef

India has a wealth of *wetland ecosystems* which are directly or indirectly connected to the river systems. **Wetlands** are the lands transitional between terrestrial and aquatic systems. In wetlands, the water table is usually near the surface and the land is covered by shallow water. Wetlands are effective in flood control, wastewater treatment, reducing sediments and recharging of aquifers. In winters, a variety of birds take shelter in wetlands and breed there. They are also suitable as a habitat for fish, flora and fauna. They act as a buffer against the effects of hurricanes and cyclones. They also stabilize shoreline and check soil erosion by sea. In addition, they are a source of durable timber, fuel wood, fruits, vegetables, traditional medicines and protein rich fodder for cattle. Thus, wetlands are of great economic, aesthetic and scientific value. In India, there are 27 wetlands which are present in 15 States. There are 19 sites which have been designated as Ramsar sites and 6 more are proposed for the designation.

Mangroves are salt tolerant forest ecosystems. Mangrove plants survive in high salinity, tidal extremes, strong winds, high temperature and muddy anaerobic soil. These conditions are hostile for the other plants. Mangroves occur all along Indian coastline. They protect coastal people from cyclones and storms and prevent soil erosion. They provide medicines and fuel wood. They are a habitat for flora and fauna including crocodiles and tigers.

India has nearly 7% of the world's mangrove area. Sunder bans in West Bengal is the world's largest mangrove. It has also been included in the World list of Biosphere Reserves by UNESCO.

Many mangroves are being cleared and have been unsustainably exploited. The mangrove area is shrinking due to various activities such as surface run off, siltation, encroachment, discharge of wastewater and industrial effluents. The National Committee on Mangroves and Coral Reefs is taking care of their protection. The five coastal States of West Bengal, Orissa, Andhra Pradesh, Tamil Nadu and Gujarat are cyclone prone states and mangrove conservation is being actively done in them. At present, there are 35 mangrove areas which have been identified by MOEF for intensive conservation and management.

The mangroves contain 69 species under 42 genera and 28 families. Out of these, two species are endemic to India.

India has four **coral reef** areas in the Gulf of Mannar, Gulf of Kutch, Lakshwadeep Islands and Andaman & Nicobar Islands. Their conservation and management is being implemented since 1987. There is a National Coral Reef Research Centre at Port Blair.

13.5 The Threatened Biodiversity

A large number of species are under the threat of extinction. According to the International Union for the Conservation of Nature (IUCN), a worldwide organization with headquarters in Switzerland, the number of species under various categories such as *endangered, vulnerable, rare* etc.is increasing day be day world over.

According to the IUCN Red list 2000, the number of Critically Endangered primates has increased to 19 from 13 in 1996. Similarly, there is an increase in the number of Critically Endangered (CR) and Endangered species (EN), see Table 13.7.

Table 13.7 Increase in number of CR and Endangered Species

	CR		EN	
	1996	**2000**	**1996**	**2000**
Mammals	169	180	315	340
Birds	168	182	235	312
Reptiles	10	24	28	47
Turtles (Fresh water)	10	24		

Madagascar having the most Critically Endangered and Endangered primates has lost 90% of its original vegetation. The Philippines also has lost 97% of its original vegetation and has the highest number of Critically Endangered and Endangered birds.

Indonesia has the highest number of species of threatened mammals. The number of species threatened in India is 80 while Brazil and China has 75 and 72 threatened species of mammals, respectively. The number of threatened bird species in various countries is listed in Table 13.8.

Table 13.8 Threatened Bird Species

Country	Number
Indonesia	115
Brazil	113
Colombia	78
China	76
Peru	75
India	74

The number of plant species is also declining in South and Central America, Central and West Africa and South East Asia. The number of threatened plant species is 5,611. Since this number is based on only 4% of the world's total described plants, the actual number would be far greater.

Malaysia has the largest number of threatened plant species (681). For Indonesia, Brazil and Sri Lanka, the corresponding numbers are 384, 338 and 280 species, respectively.

13.5.1 The Indian Scenario

Table 13.9 gives threatened species of plants and animals in India as per the Red list of IUCN, 2000.

Table 13.9 Threatened Species in India

Category	Plants	Animals
Extinct	7	0
Extinct in the wild	2	0
Critically Endangered	44	18
Endangered	113	54
Vulnerable	87	143
Lower Risk- Conservation Dependent	1	10
Lower Risk- near Threatened	72	99
Data Deficient	14	31

According to IUCN, 2000, India is second in terms of number of threatened mammal species and it also ranks second in the number of the most threatened birds.

The number of threatened species by taxonomic group is listed in Table 13.10.

Table 13.10 Threatened Species by Taxonomic Group

Taxonomic Group	Number
Mammals	86
Birds	79
Reptiles	25
Amphibians	3
Fish	3
Molluscs	2
Other Invertebrates	21
Plants	244
Total	459

13.6 Causes of Extinction

There are many reasons for the extinction of species leading to the biodiversity loss. They are biological, environmental and human activities. Out of these, the human activities deserve serious consideration as the extinction caused by the first two factors is less severe. The different causes of extinction are as discussed below:

(1) **Random Variation in Population:** If a species has low abundance, the variations in its (low) birth and death rates can lead to its extinction. This is called *population risk.*

(2) **Environmental Factors:** The variation in physical or biological environments and changes in predator, prey, competitor or symbiotic species may lead to extinction especially if the species are rare and isolated.

(3) **Natural Catastrophe:** A sudden change in environment such as earthquake, flood, fire or storm can greatly affect the survival of biodiversity of the area.

(4) **Genetic Risk:** The small populations of a species have less genetic variability. Such a population is more vulnerable to various factors (environmental as well) causing extinction as compared to large

populations which have enough genetic variability to withstand such risks.

(5) **Human Activities:** Although human beings have affected the extinction of species since ancient times by clearing of forests and land for settlement, the rate of extinction has increased tremendously in the recent past.

According to IUCN estimates, humans have caused 75% of the extinctions of birds and mammal since 1600. Hunting is responsible for the extinction of 42% of the birds and 33% of the mammals.

Introduction of exotic species (species introduced into a new geographic area) can cause both positive and negative effects. Some inadvertentely introduced species have caused spread of diseases or have become pests or parasites to the existing ones. In nineteenth century, introduction of cats, rats, goats, dogs, insects etc. led to dramatic extinctions. For example, the dodo birds faced extinction because their eggs were easy prey for dogs.

Many species face additional risk of extinction because they are killed for their fur, skin, horns, ivory and other products. The illegal trade in horns of the rhinoceros which are in the list of endangered species, puts them at risk in spite of the legal restrictions in such trade.

Human activities cause alteration and destruction of habitats of species by changes in land use patterns i.e., deforestation, clearing of land for agriculture and urbanization, industrial activities. With increasing pressure of population and its requirements, the plants and animals are being over harvested.

The recent origin of biotechnological techniques has led to many *Genetically Modified Organisms* (GMOs). The impacts of such proliferation are yet to be observed but they would definitely affect the biodiversity since introduction of GMOs would have serious effects on existing species and other varieties with which they may have to compete for their existence.

13.7 Conservation of Biodiversity

Since many of the species are under serious threats of extinction, it is a matter of prime concern that biodiversity should be conserved. Unlike other environmental problems such as pollution or global warming, the loss of biodiversity cannot be reversed. Once a species is lost, it is gone forever and it will not come into existence again.

India is one of the 17-mega diverse countries. It can be divided into 10 biogeographic zones and 26 biotic provinces, which represent all the major ecosystems of the world. Such a rich diversity needs a multifaceted approach for its conservation. Various initiatives have been taken at various levels for the conservation of biodiversity.

The conservation measures include not just marking reserve areas of land but much more than that. Preserving just the endangered species is not enough. Its relation to other species in its habitat, its predators or preys etc. also needs to be studied and a holistic approach is required for conservation. In fact, the ecosystems, natural to the species concerned, should be regenerated and provided for its survival. Thus, one should have a complete knowledge about the types of species present in an ecosystem, their inter-relationships and their type of habitat and other factors affecting their survival. Hence, the study of number and types of different species and documentation of this data over different intervals of time, is an important feature in knowing and conserving the biodiversity.

Since certain areas are richer in biodiversity, their identification for the purpose of conservation is of great interest. Such areas which are extremely rich in species and have high endemism, are facing significant threats to biodiversity. These areas are called **Hotspots**. The term hotspot was coined by Norman Myers (1988). Among the 25 hotspots in the world, 2 are found in India which extend into neighbouring countries, see Fig 13.2.

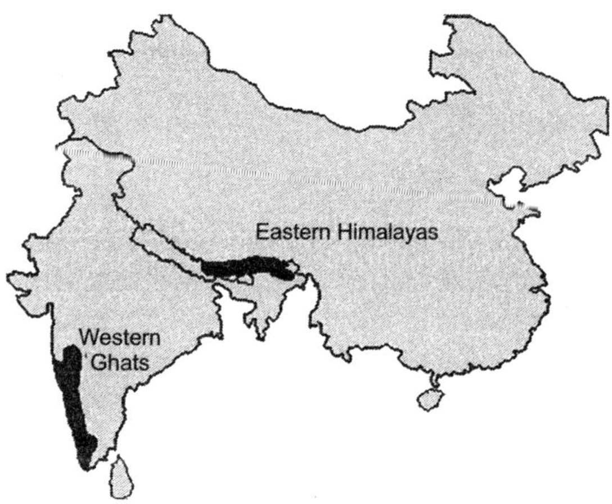

Fig 13.2: Hotspots in India

These hot spots are Western Ghats and the Indo-Burma region covering the Eastern Himalayas. These are rich in floral wealth and are highly endemic with respect to flowering plants, reptiles, amphibians, some mammals and butterflies.

13.7.1 Conventions, Protocols and Acts

Various agreements and conventions have been adopted worldwide to conserve biodiversity and environment. India is also a party to many such international protocols and conventions. Also, at national level, various legislative, participatory and regulatory mechanisms have been initiated to protect biodiversity. These are discussed below:

1. Convention on Biological Diversity (CBD)

In 1992, the biological diversity was put on the international agenda when the Convention on Biological Diversity was signed in the 'Earth Summit'- the United Nations conference on Environment and Development. So far, 180 countries have become a Party to this Convention. The main objectives of the convention are given below:

- the conservation of biological diversity,
- the sustainable use of its components, and
- the fair and equitable sharing of benefits arising out of the utilization of genetic resources by appropriate transfer of relevant technologies, taking into account all rights over those resources and technologies and by appropriate funding.

The convention addresses many issues such as habitat preservation, intellectual property rights, bio-safety and indigenous people's rights. It also covers the ecological, economic and social aspects of biodiversity and related technologies evolved through research and development.

It envisages several actions as given below:

(1) Preparation of a national biodiversity strategy or action plan

(2) Review of biodiversity related institutions and legal frame works existing in the country

(3) Identification of important ecosystems and habitats, their conservation and sustainable use

(4) Preparation of inventories at the species level

(5) Identification of threatened components of biodiversity and implementation of various measures for their conservation and sustainable use

(6) Research and training taking into account special needs of developing countries for conservation and sustainable use of biodiversity

(7) Exchange of relevant information related to technical, scientific and socio-economic research, specialised knowledge, indigenous and traditional knowledge amongst contracting parties

(8) Promotion of technical and scientific cooperation through national and international institutions in the implementation of national policies, development and strengthening of national capabilities through human resource development and institution building

(9) Establishment of joint research programmes and joint ventures for development of technologies

India ratified the Convention on 18th February 1994 and became party to the Convention in May 1994. So far, seven meetings of COP (Conference of Parties) have been held. The Ministerial Meeting of COP-7 was held in Kuala Lumpur where India has taken over Presidency of the Like-minded Mega diverse Group of countries on 19th February 2004.

2. Convention on International Trade in Endangered Species of Wild Fauna and Flora (CITES), 1973

It aims to control or prevent international commercial trade in endangered species or products derived from them. India became party to CITES in 1976. A species may be placed on one of the three lists (Appendices) by vote of members.

Appendix I lists species in danger of extinction that are, or may be, affected by commercial trade.

Appendix II includes species that may become endangered if their trade is not brought under control.

Appendix III lists species that individual treaty nations identify as subject to domestic regulations to restrict or prevent over exploitation.

In India, the international trade in all wild flora and fauna as well as species covered under CITES is regulated jointly through provisions of the Wildlife (Protection) Act, (1972), the Import/ Export Policy and the Customs Act, 1962.

3. The Cartagena Protocol on Bio-safety

The Bio-safety Protocol was signed by 65 countries during the High Level Segment of the 5th meeting of Conference Of Parties (COP) to the CBD. It is the first international regulatory frame work for safe transfer, handling and use of living modified organisms (LMOs).

The Protocol was adopted on 29th January, 2000 and has been signed by 103 countries so far. India signed the Protocol on 23rd January 2001 and acceded to the Protocol on 17th January 2003. The Protocol came into force on 11th September 2003 and 84 countries have already ratified it.

It seeks to protect the biological diversity from the potential risks posed by LMOs. It established an Advance Informed Agreement (AIA) procedure for ensuring that countries are provided with the information necessary to make informed decisions before agreeing to import such organisms into their territory.

The Protocol also established a Biosafety Clearing House for facilitating the exchange of information on LMOs and assisting the countries in implementation of the Protocol.

The first meeting of Conference Of Parties serving as the Meeting of the Parties to the Protocol (COP-MOP) was held at Kuala Lumpur, Malaysia from 23rd to 27th February 2004.

4. Biodiversity Act, 2000

As a follow up action of Party to CBD, the National Policy and Action Strategy on Biodiversity was released by Government of India in May 2000 and Biodiversity Bill 2000 was placed in the Parliament. The bill was finally passed in December 2002.

The Act provides for the establishment of State Level Boards and Local Level Biodiversity Management Committees to deal with matters concerning conservation of biodiversity, its sustainable use and fair and equitable sharing of benefits arising out of the use of biological resources and associated knowledge. The detailed rules are notified in the Gazette on 15th April 2004. A National Biodiversity Authority has been set up at Chennai on 1st October 2003 under this Act.

5. Forest (Conservation) Act, 1980

First Forest Act was enacted in 1927. Since then, many changes have occurred, so relevant amendments have been suggested in the Act. The Act was adopted to protect and conserve forests.

The MoEF has constituted a National Forest Commission on 7th February 2003 to review the working of Forests and Wildlife Sector.

6. The Wildlife (Protection) Act, 1972 and Amendments 1982, 1991

The Wildlife (Protection) Act provides for protection to listed species of flora and fauna and establishment of a network of ecologically important protected areas. It empowers the Central and State Governments to declare any area a wildlife sanctuary, national park or closed area. There is a ban on carrying out any industrial activity in such an area. It provides for regulation of hunting of wild animals, protection of specified plants, sanctuaries, national parks and closed areas, restricting trade or commerce in wild animals or animal articles. The 1982 Amendment introduced a provision permitting the capture and transportation of wild animals for scientific management of animal population. The 1991 Amendment Act made the near total prohibition on hunting more effective.

13.7.2 Other National Initiatives

Besides the above regulatory provisions, other measures to protect and conserve biodiversity are as described below:

(1) An All India Coordinated Project on Capacity Building in Taxonomy (AICOPTAX) had been initiated in the Ninth Plan which will continue up to Tenth Plan. It aims to encourage excellence and research in lower groups of plants and animals, viruses, bacteria, microbes etc.

(2) The Botanical Survey of India (BSI) established in 1890 was revised and reorganized in 1954. Its aims and objectives have been reviewed in 2002. It is entrusted with exploration and documentation of phytodiversity, hot spots, fragile ecosystems etc. and identification and conservation of species with traditional economic value. It is also assigned the monitoring of phytodiversity and compilation of national database of plant species, herbarium and live specimens.

BSI has conserved and maintained germplasm of 25 rare, endangered and wild relatives of ornamental and medicinal plant species. Its collection includes 17 rare and threatened plant species, 362 ornamental plant species and 264 medicinal and economically important plant species.

(3) The Zoological Survey of India was established in 1916. Its headquarters are at Kolkata. It has five major survey programmes.

(4) An ex-*situ* conservation scheme of assistance was initiated in 1992 for the conservation of rare endemic plants by botanical gardens.

(5) The Botanic Garden of Indian Republic (BGIR) has been established at NOIDA in April 2002 as part of Botanical Survey of India for the ex-*situ* conservation of threatened plants and building public awareness for conservation of biodiversity. It is also identified as a project under the Prime Minister's Jai Vigyan Science and Technology Mission Scheme.

(6) The Joint Forest Management Programme has been implemented in all 28 States. The programme covers 17.33 million ha forest area.

(7) A National Board for Wild Life (NBWL) has been constituted on 21st September 2003.

(8) There are 92 National Parks and 500 wildlife sanctuaries having an area of 15.67 million hectares.

(9) There exist many rules for prevention of cruelty to animals. The Animal Welfare Board of India looks after various issues related to animal welfare.

(10) Under the 'Project Tiger' which was launched in 1973, 281 tiger reserves in 17 States have been established. They are spread over 37761 sq km of area and provide a variety of habitats for tigers. The habitats include mountainous terrains of Arunachal Predesh, rainfall areas of Assam and West Bengal, Mangrove of Sunder bans, dry forests of Rajasthan, Central High lands of Madhya Pradesh, Maharashtra and Chattisgarh, tropical and evergreen forests of Orissa, evergreen forests of Kerala and Karnataka and dry and moist deciduous forests of Andhra Pradesh and Tamil Nadu, respectively.

The population of tigers increased from 1827 to 3642 in a period of about last 30 years. Tigers come at the apex of the ecosystem and their numbers indicate the health of the ecosystem.

(11) The 'Project Elephant' was launched in February 1992 for conserving the elephants in their natural habitats. It is under operation in 12 States and covers an area of 60,000 sq km. It aims to save Asiatic elephant besides conservation of habitats, research, public education and veterinary care. There are 24 elephant reserves as on March 2004.

(12) The Wildlife Institute of India (WII) established in 1982 imparts training and carries out research work for conservation and management of wildlife.

(13) A Central Zoo Authority (CZA) has been established in 1992 under the provisions of Wild Life (Protection) Act, 1972. It aims to manage zoos on scientific lines. It also identifies endangered species of wild animals and gives recognition to zoos on the basis of evaluation of their functioning. There are 165 recognized zoos in the country.

In the year 2003, 1830 and 500 star tortoises were rescued and brought back to India from Singapore and Malaysia, respectively and were rehabilitated in different zoos. Similarly lions, tigers, bears, panthers and monkeys have been rescued from circuses and have been rehabilitated; the government bans their performance in the circus.

(14) Forest Survey of India (FSI) was established in June 1981 for generating information and data base on forest cover and forest resources. It also prepares *State of Forest Report*, including National Vegetation Map every two years, as one of its assignments.

Environmental Management

CHAPTER 14

14.1 Introduction

Environmental Management is an important aspect of the environmental conservation. The concern for environment grew all over the world in the second half of the last century. Various acts, treaties, agreements came into existence in several countries. India too has consciously developed several environmental protection related legal provisions and is sincerely implementing them. Besides the legal framework, *Environmental Management planning* is emerging as an important concept which comprises *Environmental Impact Assessment* and *Environmental Auditing*, for ensuring the environment friendly practices. In this chapter, these aspects have been discussed in detail. In addition to these, this chapter also deals with *Environmental Management Systems* including *ISO-14001 Standards*. The final section describes the *agencies* responsible for the implementation of the programmes related to the environmental conservation.

14.2 Important Environmental Laws and Acts

So far various laws, acts and conventions have been dealt at various places in different chapters. Besides them, there are many other legal provisions which exist in India for the protection and conservation of environment. These are as follows.

1. Water (Prevention and Control of Pollution) Act, 1974, Amended in 1988

The Water (Prevention and Control of Pollution) Act, 1974 was passed by the Parliament in 1974 under Article 252 of the Constitution. It came into force on 13th March 1974. The main objectives of the Act are as follows:

(i) To provide for the prevention and control of water pollution and maintaining or restoring of wholesomeness of water (in the stream or well or sewer or land).
(ii) To establish Central and State Boards for the prevention and control of water pollution.
(iii) For conferring on and assigning to such Boards, powers and functions relating thereof and for matters connected there with.

Section 2 of the Act defines terms such as *pollution, stream, sewage effluent, trade effluent* etc. It also describes the functions of the Central and State Boards. The functions of the Central Board include collection, compilation and publishing of technical and statistical data relating to water pollution and take measures for effective control of water pollution. It is also entrusted with the preparation of manuals, codes or guides relating to the treatment and disposal of sewage and trade effluents and dissemination of the related information. It is also responsible for planning and execution of nationwide programme for prevention, control and abatement of water pollution. In addition, there are many more functions assigned to the Board which are carried out in coordination with the State Boards.

Any person empowered by a State Board has right to enter any plant for examining it or its records. The Board also has powers to take samples from the plant and from its effluent and get its analysis done.

Under the *Amendment Act, 1988*, the State Boards are empowered to issue directions, orders of closure, prohibition or regulation of any industry or its operation and to stop its water, electricity or any other services.

There are strict penalties and punishments for not complying with the procedures of the Act.

2. The Water (Prevention and Control of Pollution) Cess Act, 1977

This Act was passed by the Parliament on 7th December 1977 and it came into force on 1st April 1978. It provides for a levy and collection of a cess on water consumed by industries and local authorities. It aims at augmenting the resources of Central and State Boards for prevention and control of water pollution. The Act provides for rebate to any industry/authority, which installs sewage or trade effluents. If any person delays the payment of cess within the stipulated time, then he is liable to pay 12% interest on the total amount per year. There are other penalties for non-payment and furnishing false returns. The Act has been amended in 1991 for augmenting the resources of the Boards by removing the short comings and providing rebate to the industries complying with the consumption and effluent quality standards. The Amendment Act was enforced in May 2003.

3. Air (Prevention and Control of Pollution) Act, 1981, Amended in 1987

The air pollution was raised as a matter of concern at the UN Conference on Human Environment held in Stockholm, Sweden in June 1972 to which India is one of the signatories. The Act was passed for prevention and control of air pollution under Article 253 of the Constitution of India. It provides means for the control and abatement of air pollution. It defines various terms such as *air pollution, air pollutant* etc. The Act also provides for Central and State Boards for prevention and control of pollution which have the powers for planning and executing nationwide programmes as well as providing technical assistance and publishing of technical and statistical data, manuals, codes and guides. They are also assigned the function of specifying standards for air quality.

The Act seeks to combat air pollution by prohibiting the use of polluting fuels and substances, and regulating appliances causing air pollution. The establishment and operation of any industrial plant requires permission from the State Boards. The Boards are also responsible for the inspection of the manufacturing processes and equipments and testing the air quality. These Boards have been re-designated as Central Pollution Control Board and State Pollution Control Boards.

The Central Pollution Control Board has notified the National Ambient Air Quality Standards (NAAQS) for major pollutants in April 1994 for maintaining the necessary air quality. These standards specify the limits for industrial, residential, rural and other sensitive areas. The industrial standards for specific industries such as steel, cement, fertilizer, oil refineries etc. have been notified. The Amendment of 1987 was enacted to empower the Boards for meeting grave emergencies where in the Boards were authorized to take immediate measures and recover the expenses incurred from the offenders. The Amendment also provides for the power to cancel the permission / consent if the industry is not complying with the prescribed standards. The Amendment has, thus, strengthened the enforcement machinery and introduced stiffer penalties. It has also included the noise in the category of air pollution.

The Air (Prevention and Control of Pollution) Rules were formulated in 1982 and these define various procedures.

Various other acts such as Atomic Energy Act of 1982 dealing with the radioactive waste and the Motor Vehicles Act of 1988 for regulating vehicular traffic and transportation of hazardous wastes, compliment the above acts.

The Environmental Protection Act of 1986 also notifies various aspects of pollution due to vehicles. The emission standards, which were notified in 1990, were made more strict in 1996 and were revised in the year 2000. The Supreme Court also notified Euro I and Euro II emission norms for Delhi on 29th April 1999.

4. The Environment Protection Act, 1986

This Act became effective on 19th November 1986. It is an umbrella legislation to provide a framework for coordination of the Central and State Authorities established under the Water (Prevention and Control of Pollution) Act, 1974 and Air (Prevention and Control of Pollution) Act, 1981. This Act empowers the Central Government to take necessary measures for the improvement of the quality of the environment by setting emission and discharge standards, regulating the location of industries, management of hazardous wastes, and protection of public health and welfare.

This Act is superior in approach to the previous Acts. It defines various terms such as *environment, environmental pollutant, environmental pollution, hazardous substance* etc.. It also gives various powers to the Central Government for laying down procedures and safeguards for prevention of accidents and handling of hazardous substances; inspection of any premises, plant, machinery, equipment, materials, manufacturing or other processes; preparation of manuals, codes or guides relating to prevention, control and abatement of environmental pollution and collection and dissemination of related information. The powers also include the closure, prohibition or regulation of any industry, operation or process and stoppage or regulation of supply of electricity or water or any other service. The non- compliance to the provisions to this Act carries the punishment of imprisonment extending to 5 years or a fine upto Rs. 1 Lakh or both. The continued violation envisages an additional fine of Rs. 5000 per day and if the failure is for more than 1 year, then the imprisonment may be extended up to seven years.

Section 17 of the Act provides that any person, in addition to the authorized government officials, may file a complaint with the court alleging an offence under this Act. This provision of "Citizens' Suit" is not contained in any other pollution control legislation.

Several notifications under this Act have been issued which are mentioned below:

1. Doon Valley Notification, 1989

It prohibits the setting up of an industry in which the daily coal or fuel consumption is more than 24 million tonnes per day in the Doon Valley.

2. Coastal Regulation Zone Notification, 1991

It regulates the activities along the coastal zones. It prohibits dumping of ash or any other waste in the coastal regulation zones.

3. Dhanu Taluka Notification, 1991

It declares Dhanu Taluka district as economically fragile region and prohibits the setting up of power plants in the vicinity of this area.

4. Revanda Creek Notification, 1989

It prohibits setting of industries around the Revanda Creek.

5. The Environmental Impact Assesment of Development Projects Notification, 1994 and as Amended in 1997

It requires that all projects listed under schedule I should get an environmental clearance from the Ministry of Environment and Forests.

The clearance is also required by projects under de-licensed categories of the New Industrial Policy and the projects located in fragile regions. Industrial projects with investments above Rs. 500 million need a clearance from the Ministry of Environment and Forests and Letter of Intent (LOI) from the Ministry of Industry and No-Objection Certificate from the State Pollution Control Board. In case the location is in a forestland, then the clearance from the State Forest Department is also required.

6. Ash Content Notification, 1997

It requires the use of beneficiated coal with ash content not exceeding 34% with effect from June 2001, which was later, extended to June 2002. This notification is applicable to all thermal plants located beyond 1000 km from the pithead or thermal plants located in urban area or sensitive area.

7. Taj Trapezium Notification, 1998

It requires that no power plant be setup within the geographical limit of Taj Trapezium assigned by the Taj Trapezium Zone Pollution (Prevention and Control) Authority.

8. Disposal of Fly Ash Notification, 1999

It aims to conserve the topsoil, protect the environment and prevent the dumping and disposal of fly ash discharged from lignite based power plants. It provides mixing of at least 25% of ash by weight, with soil, for the manufacture of clay-bricks or tiles manufactured within a radius of 50 km from a coal or lignite based power plant. The Notification also provides that coal or lignite based thermal power plant shall provide ash for manufacturing products such as cement, bricks, panels or blocks, construction of roads, embankments, dams, dykes or any other construction activity for at least 10 years from the date of above notification.

9. Rules for Manufacture, Use, Import, Export and Storage of Hazardous Microorganisms, Genetically Engineered Organisms or Cell, 1989

Various aspects for the protection of the environment involving the use of gene technology and microorganisms have been covered under these rules. There is a scheme of labeling environment friendly products with 'ECOMARK'. This scheme provides incentives to the manufactures and importers for reducing environmental impacts and rewarding genuine initiatives .for improving the environmental quality.

10. Recycled Plastics Manufacture and Usage Rules, 1999

11. Environment Protection Rules, 1986

These rules prescribe the procedure for setting of standards of emission and discharge of pollutants. They also lay down procedures for sampling, analysis, laboratory reports etc.

12. The National Environment Appellate Authority Act, 1997

It provides for the establishment of a National Environment Appellate Authority for hearing appeals regarding the restriction of location of any industry, operation or process, class of industries and safeguards to be undertaken for getting permission to carry out the above activities.

13. Hazardous Wastes

There are several legislations related to hazardous wastes. The Factories Act, 1948, the Public Liability Insurance Act, 1999 and the National Environment Tribunal Act, 1995, cover various aspects.

The Factories Act, 1948 ensures the safety, health and welfare of workers in their working conditions as well as the employment benefits. There are 29 categories of industries which involve hazardous processes which affect both human health and pollute the environment.

The Public Liability Insurance, Act 1991 deals with the accidents involving hazardous substances and the insurance aspects. In case of an injury or death, the owner is responsible for providing the relief. An amendment of 1992 led to the establishment of Environmental Relief Fund by the Central Government for making relief payments.

The Natural Environment Tribunal Act, 1995 provides for the establishment of a National Environment Tribunal for effective and quick disposal of accident cases occurring by handling of hazardous substances.

There are several notifications related to the management of hazardous waste. These are as follows.

(1) Hazardous Wastes (Management and Handling) Rules, 1989
(2) Bio medical Waste (Management and Handling) Rules, 1998
(3) Municipal Wastes (Management and Handling) Rules, 2004
(4) Hazardous Wastes (Management and Handling) Amendment Rules, 2004

14. The Wild Life (Protection) Act, 1972 and Amendments 1982, 1991

This Act was enacted by the Parliament in 1972. It provides for the protection of listed species of flora and fauna, establishment of sancturies and national parks. The State Wildlife Advisory Boards come under this Act. The Act describes regulations for hunting wild animals and birds, trade in wild animals and animal products. There are several penalties for the violation of the Act. The Amendment of 1982 permits the capture and transportation of wild animals for the scientific management of their population. The Amendment of 1991 makes the near total prohibition on hunting of animals more effective.

15. The Forest (Conservation) Act

It provides that forestland cannot be used for non-forest purposes without the permission of the Central Government. The *non-forest purpose* includes clearing of forestland for cultivation of cash crops, plantation crops, horticulture or any purpose other than reforestation. The 1998 Amendment makes the provisions more stringent for the violators of the Act. The Amendment included the cultivation of tea, coffee, rubber, palm, oil-bearing plants, medicinal plants under the non-forest purposes.

16. The Bio-Diversity Act, 1994

This act was made for the conservation of biodiversity and has already been discussed in Chapter 13.

India is also party to various International Conventions and Protocols which have been dealt at relevant places in different chapters. They are as follows:

(1) The Convention for the Protection of Ozone Layer, 1987

(2) Montreal Protocol

(3) United Nations Framework Convention on Climate Change, 1992

(4) Kyoto Protocol

(5) Convention on Biological Diversity

(6) Cartagina Protocol

(7) Convention on Control of Trans-boundary Movements of Hazardous Wastes and their Disposal – Basal Convention

(8) The United Nations First, Second and Third Conventions on the Law of the Sea (1958, 1960 and 1967)

(9) Convention of the Prevention of Marine Pollution by dumping of waste and other matter - London Convention, 1972

(10) Convention on International Trade in Endangered Species of wild fauna and flora (CITES), 1973

(11) UN Convention on Desertification, 1994

14.3 Environmental Impact Assessment

Environmental Impact Assessment (EIA) is a tool in environmental management. It has been accepted by many countries and is being widely used. Wherever any developmental activity is taken, it definitely affects the environment because the resources are drawn from the environment and during the process of that activity and after its completion, the products as well as wastes become a part of the environment.

EIA studies are useful because they help in carrying out proper planning and hence minimise the damage likely to be caused to the environment. EIA also helps to take care of remedial measures while we plan the project, thereby later saving us from penalties. Also, EIA is a means to test whether the regulatory procedures, guidelines etc. are being met or not, by the way the activity or project is planned to be executed.

Several projects can be better executed if EIA is done in advance. These include development activities such as setting up of an industry, construction of a dam, establishing a power plant, exploration of oil etc. In India, MoEF has made EIA mandatory for 29 specified activities in the categories of mining, industries, transport, power, irrigation, and tourism by issuing a notification on

27th January 1994. An amendment to this was made on 4th May 1994 which required submission of details, technical information and other documents for getting a clearance for the project from MoEF.

The EIA studies include the existing environment, field studies and impact of the proposed activity on land use, water use, hydrology, air and water quality, ecology, atmosphere, socio-economic benefits and losses, financial requirements, conformity to regulatory provision etc.

14.4 Environmental Auditing

Environmental Audit is an effective tool for environmental management and sustainable development. Audit refers to examining and assessment of performance. As far as EIA is concerned, the environmental audit refers to assessment of actual environmental impact, the accuracy of prediction, the effectiveness of remedial or mitigation measures and the functioning of monitoring mechanisms. An audit can be undertaken after the project has been implemented for sometime. Many agencies which provide consultancies and have expertise in this area are coming up.

In India, it has been made mandatory for industries to submit an environmental audit report every year since the Gazette Notification dated 13th March 1992 and the amendments following it.

An Environmental Audit report is to be submitted in form V by every person carrying out an industrial activity, operation or process for the year ending 31st March or before 15th May starting from 1993 to the State Pollution Control Board. The report should clearly mention about the steps proposed and undertaken for waste minimization, recycling and reuse, adoption of clean technologies and investments made for pollution control and protection of the environment. Non-compliance to this may lead to an imprisonment extending upto five years or a fine, which is extendable upto Rs 1 Lakh, or both.

14.5 Environmental Management Systems : ISO-14000 Standards

In the recent years, there has been a total shift in the policy to control industrial pollution. Instead of controlling pollution at end points after the production activities are over, the latest approach is based on 'Environmental Management System' (EMS). It provides a way to minimize the harmful effects of an activity on the environment. It involves the following steps:

- Review of environmental consequences of the operations
- Defining policies and objectives for environmental performance
- Establishing an action plan for achievement of objectives
- Monitoring performance against the objectives
- Appropriately reporting the results
- Reviewing the system and improving it continuously.

Two important EMS have emerged which are ISO 14001 and Eco Management and Audit Scheme (EMAS). ISO stands for International Standards Organisation. It is based in Geneva, Switzerland and was founded in 1947. It has laid down more than 8000 standards.

There are more than 120 countries including India which are full voting members. However, some other countries are only observer members. Only the full member countries participate in the development of standards while other countries are also kept informed about the developments and activities. India is officially represented by the Bureau of Indian Standards (BIS). The concept of EMS emerged after the Rio Earth Summit.

ISO 9000 and ISO 14000 standards are internationally accepted series of standards. These standards are marked on various products and services. ISO 9000 is related to quality management and quality assurance whereas ISO 14000 standards are related to the environmental aspects.

ISO 14000 series contains (a) Organisation and Process Standards, (b) Product-oriented standards and (c) Terms and definitions. Table 14.1 gives some details of ISO 14000 series standards which are applicable at the organizational level.

Table 14.1 Summary of ISO 14000 Series of standards level-wise and area-wise applicable at Organisational Level

Implementing Environmental Management Systems (EMS)	**ISO 14001:1996 ISO 14004:1996** Help an organization to establish a new or improve an existing EMS	**ISO 14001:1996** Specifies the requirements for an EMS that may be objectively audited for self-declaration, second or third-party certification/ registration purposes	**ISO 14004:1996** Provides guidance to help an organization establish and implement an EMS, including guidance that goes beyond the requirements of ISO 14001	**ISO/TR 14061:1998** Contains information that assists in the implementation of ISO 14001 and ISO 14004 by forest management organizations and the forest products industry		
Conducting Environmental Audits and Other related Investigations	**ISO 14010:1996** Provides guidance on the general principles common to the conduct of any environmental audit	**ISO 14011:1996** Provides guidance on the procedures for the conduct of EMS audits, including the criteria for selection and composition of audit teams	**ISO 14012:1996** Provides guidance on the qualifications of internal or external environmental auditors and lead auditors	**ISO 14015:2001** Helps an organization to identify and assess the environmental aspects and associated business consequences of sites and organizations	**ISO 19011:2002** (replaces ISO 14010,14011, and 14012) Provides guidance on the principles of auditing; the management of audit programmes; the conduct of	

Continued.....

				to support the transfer of properties, responsibilities and obligations from one party to another	management system audits as well as on the competence of auditors	
Evaluating Environmental Performance	**ISO 14031:1999** Provides guidance on the selection and use of indicators to evaluate an organization's environmental performance	**ISO 14032:1999** Provides examples from real organizations to illustrate the use of the guidance in ISO 14031	**Communicating results**	**ISO/WD 14063** Will provide guidance on environmental communication related to an organization's environmental aspects and performance	**Understanding terms and definitons**	**ISO 14050:2002** Helps an organization to understand the terms used in the ISO 14000 series standards
Using Environmental Declarations And claims	**ISO 14020:2000** Provides general principles which serve as a basis for the development of ISO guidelines and standards on environmental claims and declarations	**ISO 14021:1999** Provides guidance on the terminology, symbols and testing and verification methodologies an organization should use for self-declaration of the environmental	**ISO 14024:1999** Provides the guiding principles and procedures for third-party environmental labelling certification programs (Type I Environmental Labelling)	**ISO/TR 14025:2000** Identifies and describes elements and issues for consideration when making declarations of quantified product information based on Life Cycle Inventory data		

Continued......

		aspects of its products and services (Type II Environmental Labelling)		(Type III Environmental declarations)		
Conducting Life cycle Assessment (LCA)	**ISO 14040:1999** Provides the general principles, framework and methodological requirements for the LCA of products and services	**ISO 14041:1998** Provides guidance for determining the goal and scope of an LCA study, and for conducting a life cycle inventory	**ISO 14042:2000** Provides guidance for conducting the life cycle impact assessment phase of an LCA study	**ISO/TR 14043:2000** Provides guidance for the interpretation of results from an LCA study	**ISO/TS 14048:2002** Provides information regarding the formatting of data to support life cycle assessment	**ISO/TR 14049:14047** Provide examples that illustrate how to apply the guidance in ISO 14041 and ISO 14042
Addressing Environmental Aspects in Products And product Standards	**ISO Guide 64:1997** Helps the writers of product standards address environmental aspects in standards	**ISO/TR 14062:2002** Provides concepts and current practices relating to integration of environmental aspects into product design and development		**Understanding terms and definitons**	**ISO 14050:2002** Helps an organization to understand the terms used in the ISO 14000 series standards	

As given in the table above, there are many documents in the ISO 14000 series but for EMS certification, an applicant (industry) is audited only against the requirements of ISO 14001. An organization which has obtained ISO 14001 certificate means that it has voluntarily established the procedures and guidelines for EMS and an independent agency called a certifying body or third party, has audited the procedures and guidelines.

Although obtained voluntarily, there are many benefits associated with ISO certificate. These benefits form the basis why industries are keen to obtain it. Some such benefits are as follows:

- Economic benefits resulting from the reduced cost of waste management and / or savings in consumption of energy or materials.
- Consumer recognition leading to strong market position and increased export competitiveness and facilitating trade.
- Improved image of the company amongst regulators, public and the customers.
- Also, the Government of India through Export-Import Bank (EXIM Bank) provides grants upto 50% of the cost incurred in obtaining the EMS certificate to export oriented units.

As per ISO 14001 standards, an organization can self certify or get it certified from a certifying body. A number of certifying agencies, accredited from other countries, are operating in India. They are as follows:

(1) Bureau of Indian Standards (BIS)
(2) Det Norske (DNV)
(3) Korea Management Association Quality Assurance (KMAQA)
(4) Lloyd's Register Quality Assurance (LRQA)
(5) International Certifications Limited (ICL)
(6) Bureau Veritas Quality international (BVQI)
(7) TUV India Pvt. Ltd.
(8) AJAEQS Certification Service Pvt. Ltd.
(9) Standardisation, Testing and Quality Certification (STQC)

Besides ISO 14001, there are two more EMSs. The Eco-Management and Audit Scheme (EMAS) is operating in European Union since 1993 and BS 7750 has been operating in UK since 1992. BS 7750 has been withdrawn after the introduction of ISO 14001 since ISO 14001 has been based on BS 7750. However, EMAS has much more stringent conditions to be met before getting the certification.

Since ISO 14001 does not offer the regulatory compliance, the ISO 14001

Plus is being introduced which merges regulatory compliance and public disclosure along with ISO 14001 certification.

14.6 Agencies and Institutions Involved in the Environmental Management

Though various provisions in the form of regulations, laws etc. exist for the management of environment, their enforcement and implementation requires a network of agencies, their coordination, monitoring, amendments or improvements from time to time. Various such arrangements exist at different levels. There are several organizations set up by the Government, private bodies, NGOs etc. All of these have their unique roles. Some of them are discussed below:

14.6.1 The Central Pollution Control Board

The Central Pollution Control Board was constituted in September 1974 under the provisions of the Water (Prevention and Control of Pollution) Act, 1974. Its main functions spelt out in the Water and Air Act are promotion of cleanliness of streams and wells and improvement of air quality by prevention, control and abatement of pollution. It has various responsibilities which are mentioned below:

(1) To advice the Central Government on any matter concerning prevention and control of water and air pollution and improvement of air quality.

(2) To plan and execute a nationwide programme for the prevention, control or abatement of water and air pollution.

(3) To provide technical assistance and guidance to State Pollution Control Boards, carry out and sponsor investigations and research on water and air pollution.

(4) To collect, compile and publish technical and statistical data relating to water and air pollution and the measures for its effective prevention, control or abatement.

(5) To prepare manuals, codes and guidelines relating to the treatment and disposal of sewage and trade effluents.

(6) To disseminate information related to water and air pollution and their prevention and control.

(7) To lay down and modify the standards for stream or well and air quality in consultation with the concerned State Governments.

(8) To organize training programs.

(9) To perform any other function as and when prescribed by the Government of India.

The Central Pollution Control Board functions as the State Board for various Union Territories.

The Central Pollution Control Board has established 784 stations all over the country in collaboration with State Pollution Control Boards which monitor the water quality of the surface water on monthly or quarterly basis and of ground water on half-yearly basis.

It also monitors the water quality at 784 stations. The trends in the data indicate the effectiveness of the control measures or the need for better methods for pollution control. It has identified ten problem areas and ten polluted river stretches for focusing the pollution control efforts. This has led to the formulation of River Action Plan of the National River Conservation Directorate. The water quality data for more than 70 rivers has been obtained and analysed. Similarly, water quality for 15 lakes and tanks was assessed. The water quality of wetlands has been also studied by the Central Pollution Control Board. The Board also carries out ground water monitoring in various cities such as Lucknow, Ludhiana, Kanpur, Ghaziabad, Agra, Meerut, Jaipur etc. It also monitors the coastal pollution due to various activities. Regarding the air quality assessment, monitoring is done at 183 locations in 64 cities/ towns out of which 83 locations are residential, 98 are industrial and 2 are other sensitive areas. The number of air quality monitoring stations was about 300 in the year 2001.

14.6.2 State Governments

Under the Indian federal system, the environmental management powers are shifting to the States. The States have significant authorities over water supply, solid waste management, sanitation, environmental protection and land use which it can further delegate to the municipalities. The State Pollution Control Boards are India's primary pollution control board enforcement systems. They are authorized to plan, enforce and render advice on comprehensive programmes for prevention, control and abatement of pollution. They may prescribe emission and effluent standards in consultation with CPCB, inspect equipment, industrial plants or manufacturing processes. They may also apply to a Court to restrain emissions which exceed prescribed standards.

There are a large number of small-scale industries whose monitoring is an important task. Also, there is a greater need to build technical capacity for monitoring hazardous wastes streams. Over and above this, there is a slow response from the Courts on enforcement actions which slows down the effectiveness of the regulatory agencies.

14.6.3 Role of Non-Governmental Organisations

There are nearly 1,000 Non-Governmental Organizations (NGO's) actively working in the area of environment and the related issues in India. They play a very important role in the environmental management throughout the nation. Traditionally, they were mainly concerned with agriculture and forestry, but since the Bhopal disaster, increased industrialization and technology change, they are actively participating in raising various environmental issues. Their presence have been felt in recent years quite significantly because of the enthusiasm and activism shown by them. These are real grass root level organizations. Their main concerns are addressing the national needs for self-sufficiency, equitable resource distribution and use of cleaner technologies and environment friendly alternatives for the sustainable development.

14.6.4 Other Bodies

In addition to the institutions and organizations mentioned above, there are several other organizations and institutions attached to the different Ministries, Government of India which work in the field of environment. These are as follows:

- The **Central Ganga Authority** established in 1985 to implement the action plan for the prevention of pollution of the River Ganga.
- The **Bureau of Indian Standards** under the Ministry of Civil Supplies. It was designated as the International organization, primarily entrusted with the responsibility of implementing ISO 9000 and 14000 standards. To carry out this function, it requires certification from the Quality Council of India which is an autonomous body for the creation of National Accrediton Boards in testing and quality management.
- The **Council on Scientific and industrial Research** (CSIR) was founded in 1942. It is a government agency with over 40 national laboratories. It is involved in the research and development activities related to the fields of physical, chemical, biological, engineering and information sciences. The latest on its agenda is the protection of intellectual property and patents.
- The **National Environmental Engineering Laboratory** at Nagpur and other similar laboratories which focus their efforts on industrial sectors such as electronics, leather, food processing, mines, drugs, petroleum, steel etc.

Some Case Studies

CHAPTER 15

15.1 Introduction

So far, many aspects of the environment, its pollution, protection and conservation have been elaborated in detail. There are so many factors affecting the environment that controlling all of them completely is not feasible because many secrets of the nature are yet to be unraveled. However, some of them can be taken care of so that we can save the environment from their harmful effects.

The role of human beings is very important in the conservation of the environment. By wisdom, careful planning and proper management, we can save the environment from various threats and negative consequences. We realize the importance of these aspects only when their neglect leads to disasters. In this chapter, some case studies illustrate the above points. *Bhopal gas tragedy* is one such black spot which can not be erased from the history of India. Besides accidents, the rapid urbanization and the unplanned development can also lead to degradation of the environment. This is clearly evident in case of pollution scenario in the city of Delhi. The two main culprits affecting the environment and the health of residents of Delhi are *vehicular* and *industrial pollution*. Various efforts are being made to control the pollution in Delhi. The results have started showing up and in the due course of time, the pollution is expected to be under control. However, there is a long way to go before the situation is brought back to the state which used to be a few decades earlier.

15.2 Bhopal Gas Tragedy

On the night of 2nd December 1984, the world's worst ever industrial disaster took place in Bhopal, India. It is also known as the Hiroshima of the chemical industry because of the death toll and the suffering it caused, 27 tonnes of methyl isocyanate and other deadly gases were released from the Union Carbide's pesticide manufacturing factory in Bhopal. The methyl isocyanate was used in the manufacture of the insecticide *carbaryl* which was marketed under the commercial name *Sevin*. The sequence of the reactions involved is as follows.

$$\underset{\text{Methyl amine}}{CH_3NH_2} + \underset{\text{Phosgene}}{COCl_2} \longrightarrow \underset{\text{Methyl isocyanate}}{CH_3{-}N{=}C{=}O} + 2\,HCl$$

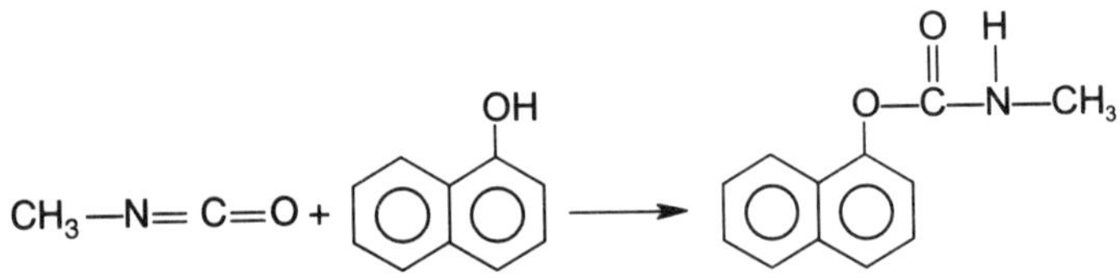

Methyl isocyanate, which was stored in the tanks, leaked and the cloud of the gas gradually spread over a large area of the city killing about 4000 people and affecting thousands of others. Approximately half a million people were exposed to the leaked gases and at least 22,000 have died so far because of the effects of the exposure. It is estimated that about 120,000 people are suffering from serious long-term health effects. The number of deaths as reported at that time is much below the actual number because the deaths continued to occur due to the long-term effects of the exposure to the gas. In fact, one person dies per hour because of these long-term effects of exposure.

The leakage of the gas caused irritation in the eyes, nose and mouth of the people. The people felt the burning sensation and some of them felt as if some one was frying chillies. They started coughing and froth streaking up with blood from their mouths. They started running here and there. Some had convulsions and died while the others lost control over their senses and urine and faeces fell from their legs while they were running. Even, the people ran over the dead bodies of the others.

It seems that water had entered into the largest tank containing methyl isocyanate (MIC). Unfortunately, none of the six safety systems were operational and the plant siren had been turned off, see Fig. 15.1. The vent gas scrubber was also turned off otherwise the leaking gas could have been detoxified. The pipe connecting to the Flare tower had also been removed for maintenance. The refrigeration system for cooling liquid MIC was shut down in June 1984 to save money. Out of the three storage tanks, tank E 619 was empty while the tank E 610 contained 40 tonnes and E 611 contained 15 tonnes of the gas. On that disastrous night, water entered into the tank E 610 and an uncontrollable exothermic reaction began. The tank was blown out of its concrete sarcophagus and a deadly cloud of MIC, hydrogen cyanide, monomethylamine and other chemicals spread out over the city of Bhopal, see Fig. 15.2.

Fig. 15.1: Union Carbide Plant

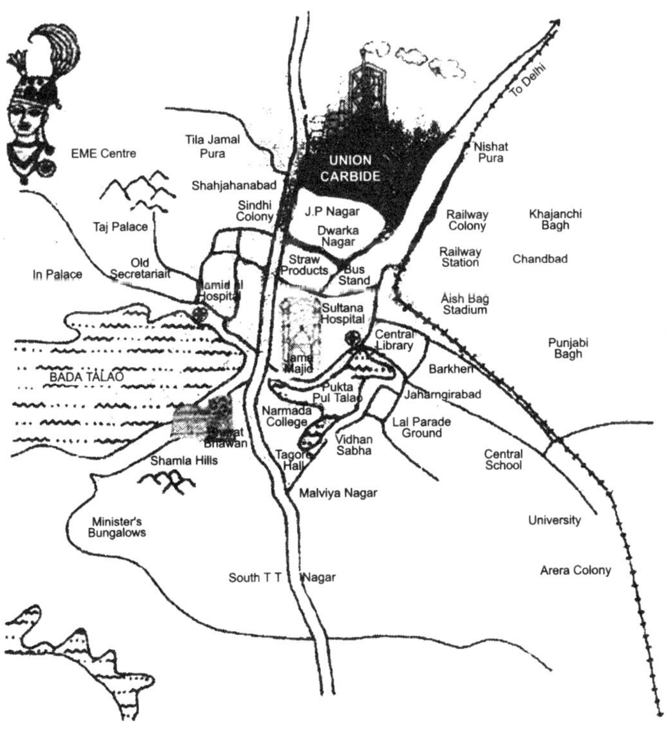

Fig. 15.2: Area affected by Gas Leak

Fig 15.2 shows that the toxic gas affected major part of the city. Out of the 56 wards, 36 wards were affected. Out of the total estimated population of 8,94,539 living in 56 wards in 1984, 5,559,835 people were affected. The livestock were also killed in large numbers. The relief operation were started immediately. Both the State and the Central Governments came into action. The NGOs also helped a lot. It was soon realized that exclusive administrative machinery would be needed to tackle the situation arising out of this disaster. The ad-hoc administrative machinery was immediately created and a Department of Bhopal Gas Tragedy Relief and Rehabilitation was created in early 1985. It is under the charge of a Cabinet Minister of the State Government. A Directorate of Gas Relief and Rehabilitation has also been set up.

The State Government provided an immediate relief of Rs 1 crore. A compensation of Rs. 10,000 per person was paid to the family members of the deceased victims which amounted to Rs. 3.78 crores. Ex-*gratia* payment for the death of animals costed Rs. 0.25 crores. A total of Rs. 36.03 crores was spent by the State Government.

As on March 2003, 10,29,515 cases for claim and compensation were registered. About 1529 crores were spent for the compensation. The claim applications were filed in two phases. Their details are as follows:

Table 15.1 Details of Claims

S.N.	Category	No. of Claim Applications Yearwise		
		1985-89	**1996-97**	**Total**
1.	Claim filed for personal injury (01)	5,97,908	4,03,815	10,01,723
2.	Claim filed for loss of Live stock (02)	612	46	658
3.	Claim filed for losses suffered by commercial/ business establishments (03)	4,745	156	4,901
4.	Claim filed in Death cases (04)	15,310	6,839	22,149
5.	Claim filed for losses incurred by Central/ State/ Governmental/ Non- Governmental Organizations (05)	84	–	84
	Total	**6,18,659**	**4,10,856**	**10,29,515**

The Union Carbide took moral responsibility but never the legal responsibility since the plant was operated by a separate Indian subsidiary, Union Carbide India Limited (UCIL). Union Carbide owned 50.9% of UCIL's shares at the time of the accident. UCIL did some clean up work but the job was not completed. The plant site was later bought by another Indian company, Eveready Industries India Limited. The site is now under the legal responsibility of State Government. In 2001, a US Multinational company Dow Chemical acquired Union Carbide, thereby acquiring its assets and liabilities. However, the Dow Chemical has refused to clean up the site, provide safe drinking water, compensate the victims or disclose information about the health effects of the leaked gases. The Union Carbide agreed to pay $ 470 million as compensation in 1989. However, the victims were not satisfied by this compensation of $300-$500 or about 5 years worth of medical expenses. Since the victims were not consulted in the settlement discussions, they felt cheated by such small compensation.

Even after about 20 years of the accident, thousands of tonnes of toxic waste abandoned by the Union Carbide in and around the plant continue to contaminate the environment. The poisonous and cancer causing chemicals leach into the ground water sources which supply water to about 20,000 people. A study in 2002 by the Fact Finding Mission on Bhopal revealed the presence of dangerous toxins, 1,3,5-trichlorobenzene, dichloromethane, chloroform, lead and mercury in the breast milk of mothers living around the area. The soil samples were also found to contain lead, nickel, copper, chromium, hexachlorocyclohexane and chlorobenzenes. In some samples, the mercury was found to be 6 million times the expected level.

According to some of the workers of the factory, massive amounts of chemicals formulated in the factory such as solvents, catalysts, wastes and pesticides were routinely dumped in and around the factory grounds. These included dangerous substances such as aldicarb, carbaryl, mercury, chlorinated chemicals and poisonous organic substances. The Union Carbide constructed Solar Evaporation Ponds (SEP's) over 14 hectares of area where the toxic effluents and wastes were routinely dumped.

Later, a study in 1990 revealed the presence of di-and tri-chlorobenzenes in the water samples taken from the wells near the factory area. The soil samples from the SEP's also showed the contamination by phthalates, chlorinated benzenes and aromatic hydrocarbons. The contaminated ground water and soil is posing a lot of threat to the people of the Bhopal city. Recently, the task of conducting a survey for the disposal of toxic wastes lying in the factory has been assigned to the Engineers India Limited (EIL). Similarly, the State Government will conduct a fresh survey to test the presence of contaminants in drinking water supply.

The contamination in the soil and water is found to be ten times higher in the area near the factory as compared to rest of the places. Since the contaminated ground water is heading towards the north, in contrast to the movement of MIC towards the south because of the wind direction, the population which was not much affected by the MIC is at a complete risk due to toxic substances present in the ground water. The long-term effects of the accident have also started appearing as the new generations of children with various defects are being born.

Although the civil suit was settled in 1989 for the lump sum of $ 470 million, a criminal trial against the company and its officials is pending for decisions even after 15 years. The process has taken so long that the settlement fund has nearly doubled in the value and amounts to $ 33 million as the interest.

The charges against the Union Carbide's then Chairman Warren Anderson, other top officials and the Corporation were filed in 1991 but they failed to appear in the trial proceedings and have been proclaimed absconders by the Court of Chief Judicial Magistrate (CJM) in Bhopal. After more than 10 years, Anderson was found in August 2002. The Indian Government then submitted an official request for his extradition in May 2003 which was rejected by the US government in July 2004. The request, however, can be resubmitted. Also, on 21 July 2004, the Chief Judicial Magistrate of Bhopal dragged Dow into a criminal case by asking its Indian subsidiary to answer the summons. The Dow Chemical maintains that the liabilities, if any, would still belong to the Union Carbide. However, the settlement with the Union Carbide solely dealt with the disaster related damages caused to the residents of Bhopal and not the environmental liabilities and the pollution caused by the toxic chemicals.

On March 17 2004, the Second Circuit Court of Appeals in the United States sanctioned further property damage, environmental remediation and personal injury claims against the Union Carbide in the lower court. On June 28 2004, the Indian Government in its official letter asked the Court to hold Union Carbide liable for environmental remediation at the site.

This case of hazardous gas leak was not the first in the history of operation of the Union Carbide plant in Bhopal. Right from the beginning, the Company did not manage the affairs properly. It decreased the proposed amount of $28 million to $20.6 million and used unproven and untested technologies. Since the plant did never attain its full capacity, it ran into losses and the Company systematically cut cost by compromising on the safety and maintenance systems.

Several incidents of leaks were reported from time to time. On 25th December 1981, the leakage of phosgene killed one worker while another leakage on 9th January 1982 led to hospitalization of 25 people. Yet another leak on 5th October, 1982 affected hundreds of nearby residents leading to the hospitalization of many people. These issues were raised in the State Assembly and were also brought to the notice of the officials of the company. The State Government at that time decided to shift the plant to a less polluted area. Various reports about the potential dangers from the leakage appeared in the press from time to time but nothing was done seriously and *the accident actually happened* leaving its trail for decades to come.

Such incidents of accidental release of hazardous and toxic substances due to improper handling, lack of safe operations and poor technical maintenance of the plant raise several issues related to the *liability for compensation, management of the long term effects* of such disasters on human health and environment etc. Although the wheel of time cannot be reversed as those who died cannot come back. Various families lost their earning members and face both economic and social crisis. There are yet thousands of people who are not able to earn their livelihood because they cannot move and work properly. They still wait for justice and are there to face the pain throughout their lives.

15.3 Vehicular Pollution in Delhi

The transportation demand in all urban areas continues to rise because of growth in population as well as the changes in the travel patterns.

There is an increasing trend towards ownership and utilization of private and personal motor vehicles for commuting in many cities including Delhi, the reason being the lack of an efficient and convenient public transport system. The more the number of vehicles on the roads, the more would be the congestion and the more would be the air pollution. The increase in number of vehicles results in the consumption of more fuel and is not economic also. The congestion on roads is further increased by a variety of vehicles plying on the roads. These include from bullock carts to cycle rickshaws to cars and buses as well as commercial goods vehicles.

In Delhi, the pollution had increased to an alarming level in the last decade because of the expanding population and increased economic activities. There are many industrial units in and around the city. In addition, there are many office complexes, educational institutes, private companies, busy markets and shopping complexes. Many people land in Delhi in search of better educational, vocational and health facilities and never go back to their home towns.

This trend is visible in population statistics of the city. In 1991, the population of Delhi was 91.21 lakhs which increased to about 132 lakhs in the year 2001. In 1951, the number of industrial units was just about 8,000 which increased to more than 1,25,000 in 1991. Similar growth was registered in the number of vehicles also. In 1975, Delhi and Mumbai had almost equal number of vehicles. But presently, the number of vehicles plying in Delhi is three times more than those in Mumbai. The actual number of vehicles being 2.35 lakhs in 1975 and

26.29 lakhs in 1996. This number is expected to reach 60 lakhs in 2011.

As far as the pollution scenario is concerned, 67% of the total air pollution is due to vehicles and it amounts to roughly 3000 MT per day. The industries and coal based thermal power plants contribute about 25% to the air pollution. The fly ash generated by the three power plants is 6000 MT per day. The industries generate an effluent load of 320 mld. The river Yamuna receives about 1900 mld of municipal sewage and waste water through the 16 drains. The municipal solid waste generated in Delhi amounts to 5000 MT per day.

Table 15.2 shows the contribution of various sources of pollution to the ambient air quality.

Table 15.2 Contribution of various sources of pollution

Source	1970-71	1980-81	1990-91	2000-01
Industrial	56%	40%	29%	20%
Vehicular	23%	42%	64%	72%
Domestic	21%	18%	7%	8%

The above table shows that the contribution of vehicular pollution is major source of air pollution. Table 15.3 shows the vehicular emission load for various pollutants for the last decade.

Table 15.3 Vehicular emission load of different pollutants

Pollutant	Pollution Load (in thousand tonnes)			Percentage
	1990-91	1995-96	1998-99	(%) reduction as compared to 1995-96
Sulphur dioxide	10	15	11	27
Nitrogen dioxide	139	207	182	12
Particulate Matter	19	28	21	25
Lead	0.190	0.362	0.007	97
Carbon Monoxide	243	351	337	4
Hydrocarbons	0.83	113	115	2
Total Pollution Load	493	714	666	–
Emission Load in Tonnes/day	1351	1947	1825	11

The vehicular pollutants have harmful effects on the human health and the environment. They directly affect the respiratory and cardiovascular systems especially high levels of sulphur dioxide and suspended particulate matter lead to increased mortality, morbidity and impaired pulmonary function. The specific effects associated with these pollutants are as follows:

(1) Sulphur dioxide adversely affects the functioning of lungs.

(2) Nitrogen oxides increase the susceptibility to infections, pulmonary diseases and cause impairment of lung function and eyes. They also lead to irritation in nose and throat.

(3) The particulate matter may be toxic itself or can carry toxic substances. They can alter the immune system. The fine particulate matter penetrates deep into the respiratory system and cause irritation of lung tissues which lead to long term disorders.

(4) Lead causes impairment of kidney and liver. In children, it may cause damage of brain which may result in lower IQ, reduced ability for concentrating or hyperactivity.

(5) Benzene is a well-known carcinogen. Its exposure leads to leukemia.

(6) Carbon monoxide affects the cardiovascular system and the nervous system. When the nervous system doesn't work properly, it results in impairment of physical coordination, vision and judgment. Carbon monoxide can also cause nausea and headaches, reduce productivity and increase discomfort.

The type of vehicles plying on Delhi roads is given in Table 15.4.

Table 15.4 Number (in lakhs) and types of vehicles in Delhi

S.No.	Category	31.03.85	31.03.90	31.03.95	31.03.96
1.	**Car/ Jeep/ Station wagon** (Petrol-driven)	1.57	3.45	5.75	6
2.	**Scooter / Motorcycle** (Petrol-driven)	5.79	11.13	16.17	17
3.	**3-Wheeler** (Petrol-driven)	0.30	0.59	0.75	0
4.	**Taxi** (Mostly Diesel)	0.08	0.09	0.12	0
5.	**Buses** (Diesel-driven)	0.14	0.18	0.26	0
6.	**Goods Vehicles** (Diesel-driven)	0.52	0.92	1.20	1
	Total	**8.40**	**16.36**	**24.25**	**26**

The number of two wheelers is about two-thirds of the total number of vehicles. They emit 20-40% unburnt or partially burnt fuel because of drawbacks in the design of their two-stroke engine. The two wheelers are responsible for more than 70% of hydrocarbons and about 50% of carbon monoxide in Delhi.

There are about 80,000 three wheelers. These are also powered by two-stroke engines. They also emit large amounts of carbon monoxide and hydrocarbons, some times beyond the measuring capacity of the instrument. Sometimes back before the option of CNG was available for these vehicles, they used to emit thick black smoke due to adulteration of kerosene in the petrol.

The public transport system in Delhi comprises about 3500 privately operated blue line buses and about the same number of buses under Delhi Transport Corporation. The buses, which are beyond the recommended period of 4-5 years, are being slowly phased out because they emit large amount of smoke and particulate matter. The number of buses definitely needs to be increased keeping in view the increasing population. The overloading of buses, especially in peak hours also leads to higher smoke emissions. There is a need of introducing high capacity buses so that more passengers are carried without causing a burden on the pollution load.

The diesel trucks also emit high levels of smoke and particulate matter. Their permits should be renewed only after satisfactory inspection regarding the maintenance measures for pollution control.

Various norms have been specified and gross emission standards for vehicles have been prescribed from time to time. The Ministry of Road Transport and Highways has notified emission norm *vide* GSR III (E), under the Central Motor Vehicles Rules (CMVR), 1989, amending Rule 115 of CMVR. In this amendment, the emission norms for carbon monoxide and hydrocarbons for new generation vehicles have been proposed which would come into force from 1 October 2004. The mass emission standards for CNG/ converted diesel vehicles have also been notified.

Vehicular Emission Norms

In addition to Delhi, 10 other mega-cities namely Mumbai, Kolkata, Chennai, Bangalore, Hyderabad, Ahmedabad, Pune, Surat, Kanpur and Agra, have already introduced Bharat Stage-II norms which are equivalent to Euro-II norms during April 2000 to 2003. These cities would introduce Euro III and Euro IV equivalent norms by 1st April 2005 and Ist April 2010, respectively. These schedules of implementation would be reviewed in 2006.

For new two and three wheelers, Bharat Stage II and Euro III equivalent norms would be introduced in entire India by April 2008/2010, (Annual Report, MoEF, 2003–2004).

The Supreme Court of India has issued several directions for pollution control in Delhi in the Writ Petition (Civil) No. 13029/1985 in the matter of M.C. Mehta *vs.* Union of India. These are listed below in Table 15.5.

Table 15.5 Various Directions of the Supreme Court of India and their respective Time frames

S.No.	Direction	Timeframe
1.	Strict enforcement of restriction on plying of goods vehicles during the day time	15.08.1998
2.	Elimination of leaded petrol from Delhi	01.09.1998
3.	Phasing out/ ban on plying of old commercial/ transport vehicles	
	(i) Vehicles more man 20 yrs old	02.10.1998
	(ii) Vehicles more than 17 yrs old	15.11.1998
	(iii) Vehicles more than 15 year old	31.12.1998
4.	Ban on supply of loose 2T oils	31.12.1998
5.	Supply of only pre-mix petrol to two-stroke engine vehicles	31.12.1998
6.	Establishment of two independent fuel testing labs	01.06.1998
7.	Only non-commercial vehicles complying with Euro I norms to be registered in NCR	01.06.1999
8.	Only non-commercial vehicles complying with Euro II norms to be registered in NCR	01.04.2000
9.	New ISBT to be built at Delhi's North and South-west borders (to avoid pollution due to entry of inter-state buses)	31.03.2000
10.	Replacement of all pre-1990 autos and taxis with new vehicles using clean fuel	31.03.2000
11.	Expansion of CNG supply outlets from 9 to 80	31.03.2000
12.	Ban on plying of buses more than 8 years old except on clean fuels	01.04.2000
13.	Strengthening of existing air quality monitoring stations and setting up of new stations of critical pollutants	01.04.2000
14.	Replacement (with financial incentives) of post 1990 autos and taxis with new vehicles on clean fuel	31.03.2001
15.	Entire fleet of DTC and private buses to be converted to single fuel mode on CNG	31.03.2001
16.	Augmentation of public transport from existing 6,600 to 10,000 buses	01.04.2001
17.	Permission to ply less than 8 years old autos, taxis and buses on conventional fuels till 30.09.2001 on special permits provided they have placed firm orders for new CNG vehicles or conversion of existing vehicle to CNG mode by 31.03.2001	31.09.2001

Measures taken for the control of vehicular pollution

(1) **Improvement of fuel quality:** The quality of fuel has significantly improved because of the following measures:

(a) Leaded petrol has been completely phased out.

(b) Introduction of low sulphur diesel with effect from August 1997 reducing sulphur content from 0.8% gm/L to 0.25% gm/L. The sulphur content of diesel has been further reduced to 0.05% with effect from March 2001. The quality of petrol has been improved by reduction of sulphur content to 0.05% and benzene content to 1%.

(2) **The pollution standards** prescribed under rule 115 (2) of Central Motor Vehicle Rules, 1989 are as follows:

(a) **For petrol / CNG / LPG Vehicles**

(i) Idling carbon monoxide emission limit for all two and three wheelers should not exceed 4.5%.

(ii) Idling carbon monoxide emission limits for all vehicles other than two and three wheelers should not exceed 3%.

(b) **For Diesel Vehicles:** Smoke density for all diesel vehicles should not exceed 65 hartridge smoke units.

(3) **Use of clean fuels such as CNG:** CNG is emerging as an attractive fuel because of its clean burning characteristics and very low amount of exhaust pollution. It mainly consists of methane (80-90%) which is compressed to a pressure of 200-250 kg cm^{-2}.

The properties of CNG, unleaded gasoline and diesel are given below in Table 15.6.

Table 15.6 Properties of some fuels

Property	Units	Gasoline	Diesel	CNG
Auto ignition temperature	Degree C	320	225	540
Flammability limit	%	1.4-7.6	0.6-5.5	5-15
Flame temperature	Degree C	1977	2054	1970
Octane No.	—	89	—	120
Cetane No.Degree C	—	10	45	–10

In December 1998, Indraprastha Gas Limited (IGL), a joint venture company between Gas Authority of India Limited (GAIL), Bharat Petroleum Corporation Limited (BPCL) and Government of Delhi has been formed. After incorporation of IGL, GAIL has transferred all the activities related to CNG project to IGL for its implementation in Delhi. IGL is implementing the CNG programme as per the directives given by the Hon'ble Court. It is getting all the necessary assistance from the Environment Pollution (Prevention and Control) Authority (EPCA) as EPCA is monitoring the development of the CNG project.

There are 106 CNG filling stations as on March 2003. 1kg of CNG energy is equivalent to 1.39 litres of petrol and 1.18 litres of diesel. The price of CNG is far less than that of petrol which amounts to about 65% of saving on fuel bill.

Table 15.7 gives the emissions of various pollutants from petrol and CNG for different types of pollutants in gram per kilometre of the vehicle run.

Table 15.7 Emission of Various Pollutants

Type of Vehicle	Pollutant	Petrol	CNG	% Reduction
Maruti Omni	CO	19.79	0.55	97
	HC	1.14	1.02	11
Maruti Gypsy	CO	4.94	0.59	88
	HC	1.86	1.42	24
Premier Padmini	CO	18.38	0.94	95
	HC	2.83	2.03	28
Premier 118 NE	CO	15.60	2.04	87
	HC	2.57	1.92	25
Ambassador	CO	52.16	0.78	98
	HC	6.37	4.33	32
Bajaj	HC	3.26	3.99/1.26	63.19
Three Wheeler	CO	5.48	1.57	71.39
	CO_2	47.44	27.60	41.82
	NO_X	0.25	0.20	20.00

Source: Paper by N.K. Verma and A.K. De, Indraprastha Gas Limited

The cost of CNG kit and its installation in a car ranges between Rs. 30000 to 35000. In case, a vehicle runs out of CNG, it can be run on petrol.

(4) Traffic Planning and Management : The existing road capacity can be better utilized by relieving the congestion on the roads. Some such initiatives are as follows:

(i) Traffic management of most polluted intersections/ areas

(ii) Use of synchronized signals

(iii) Display of halting time at major traffic intersections

(iv) Construction of express ways

(v) *Augmentation of Fleet size*

The stage carriage fleet has 7000 routes out of which 2700 are operated by DTC. There are 2100 CNG buses and 4200 mini buses.

Seven corridors- five for high capacity bus system and two for electric trolley bus system have been identified. One corridor of the high capacity bus system involves the operation from Dr. Ambedkar Nagar to ISBT via Moolchand. The first phase involves making this corridor of 5.5 km operational.

(vi) *Start of Metro Rail under Mass Rapid Transport System*

The planning for Metro Rail (Fig. 15.3) in Delhi started in 1950's and various studies were conducted for the feasibility of mass transit system. The first step for construction was the registration of Delhi Metro Rail Corporation (DMRC) in 1995. However, the actual construction project started on 1st October 1998.

Fig. 15.3: Metro Rail

The work of the project is divided into phases. The present **Phase I** is being implemented by the construction of the following three lines:

Line 1	Shahadra- Trinagar- Barwala	28 km	21 stations
Line 2	Vishwa Vidyalaya- Central Sectt.	11km	10 stations
Line 3	Barakhamba Raod- Connaught Palace-Dwarka	22.8 kms	22 stations
	Total	**61.8 kms**	**53 stations**

Line 1 from Shahadra to Tis Hazari has already been commissioned in December 2002. It operates for a distance of 8.5 km and caters to 30,000 passengers per day. A further stretch of 4 km from Tis Hazari to Trinagar was commissioned in October 2003 while the third extension from Trinagar to Rithala covering 9.4 km was commissioned in March 2004. Further from Rithala to Barwala another distance of 6.1 km would be covered under Line 1.

Line 2 (fully underground) -11 kmVishwa Vidhalaya- ISBT (North-South) - 4.0 km by December 2004
ISBT- Central Secretariat (North-South) - 7.0 km by September 2005

Line 3 (elevated/at grade/underground) - 23.5 km
Barakhamba Road- Kirti Nagar (East-west) - 7.2 km by June 2005
Kirti Nagar- Dwarka (East- West) – 16 km by September 2005

Phase II is also divided into two lines and is proposed to be completed by the year 2010. The details of the two lines are as follows:

Line 2 : Vishwa Vidyalaya- New Azadpur- SG Transport Nagar (North extension)- 8.6 km

Central Secretariat – AIIMS- Vasant Kunj (Southern extension) - 18.2 km

Line 3 : Barakhamba Road- Mayur Vihar- NOIDA (Eastern extension) - 15.3 km

The Metro map is shown in Fig.15.4.

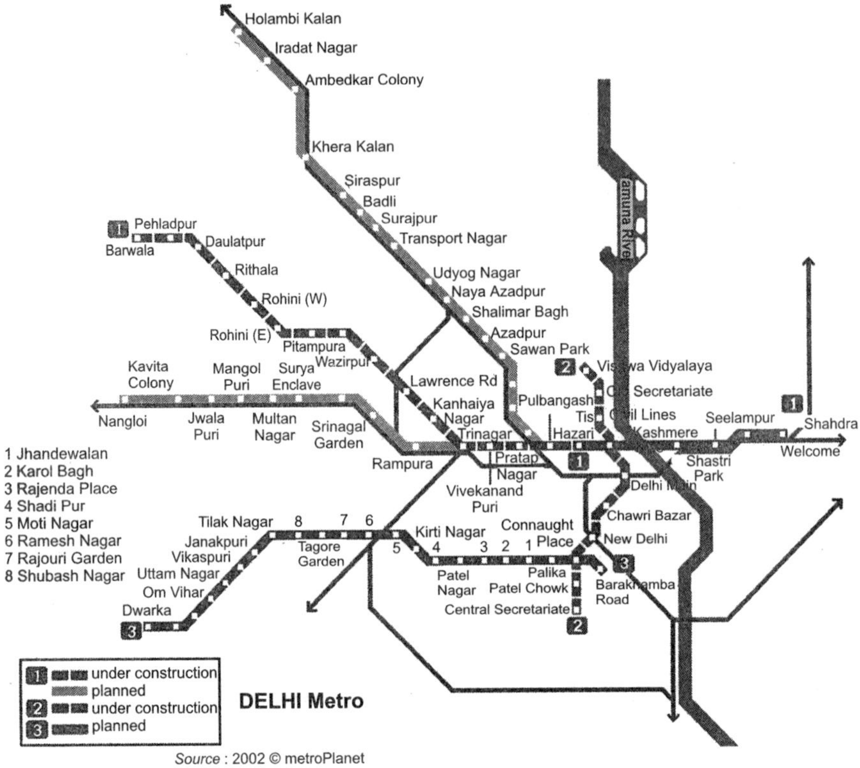

Fig.15.4: Metro Rail Map

The phase I would cover a total of 65.11 km with the estimated cost of Rs. 10,571 crores (Rs 6000 at 1996 price level). However, the cost of the whole project which comprises of 198.5 km was estimated to be Rs. 15,000 crores at 1996 price level. Phase I will have the following benefits:

(1) It will siphon off 21.82 Lakh commuter trips from the road.

(2) The number of buses on roads would decrease by 2,600.

(3) The average speed of buses on roads would increase from 10.5 to 14 km per hour. It will save a fuel cost worth Rs 5 billion per year.

(4) The travel time would be reduced by 50-75%. Thus, there will be a saving of 2 million man-hours per day.

(5) The reduction in atmospheric pollution level would be by 50%.

(6) The net benefit expected for Delhi after phase I would be Rs 437.23 Lakhs per day.

(7) There would be reduction in the number of road accidents. Thus, a safe and more comfortable journey would ensure improved quality of life.

The Delhi Metro Project is an eco-friendly project. It has been awarded OH SAS 18001 (Occupational Health and Safety Assessment Sequence 18001) by Registro Italino Navale India Pvt. Ltd. (RINA), Geneva. It is the only operational metro to receive this certificate in such a short time. It also received ISO 14001 certificate for establishing an environmental management system making it the first metro in the world to receive this certification at the construction stage.

The DMRC Corporate Culture specifies that the construction would not degrade the environment. 30% of the trees in the alignment have been saved and for every free cut during construction, ten trees have been planted in advance as compensatory afforestation.

A detailed Environmental Impact Assessment has been done in 1994 to minimize the negative impacts of the project on the environment. Storm water drains were identified at various sites for drainage of wastewater from construction. Also, locations for temporary and permanent disposal of soil arising out of tunneling and construction works were identified. Efforts were done to keep the noise and air pollution during construction at a minimum. In addition, coordination with other bodies to control traffic, water supplies, electricity and telephone facilities was done to cause minimum inconvenience to the people.

15.4 Industrial Pollution in Delhi

Although industries serve as a backbone to the development of an area in terms of providing employment opportunities, increased income and better living standards, they can also cause poor health of people, congestion of civic amenities, increase in pollution load of the area, if not managed properly.

The **location** of industries is a very important factor contributing to the origin of such problems. Originally, many industries start in the remote non-residential area but the growth of the cities in their peripheries and extension of residential areas to the outskirts makes the industries as part of the town. Since the emission of harmful gases and toxic effluents keep on contaminating (because of improper handling) the water, soil and the surrounding environment, their harmful effects become a matter of serious concern for all.

Similar to the above is the story of the city of Delhi. The industries kept on mushrooming while the city knew no limits for its growth and it kept on expanding to new areas to adjust its ever increasing population. All contributed their share of pollution and the residents faced a crisis in the amenities whether the issue was that of civic facilities, transport, medical facilities, accidents or increasing crimes. Several social, economic and other issues directly or indirectly, visibly or otherwise get associated with such an unplanned and haphazard growth which is becoming cancerous for the city. The situation becomes alarming and several measures to control it need to be initiated by the various sectors of the society. Several steps by the Government, legal actions and participation by the people can bring the situation back to normal.

It was pointed out in the last section that Delhi was among the 12 worst polluted cities with respect to the air pollution according to the 1992 WHO/ UNEP Report on "Urban Air Pollution". The industries account for 20% pollution load to the city. A survey in 1996 by the Delhi Government revealed that 22% of the industrial units were located in the 28 approved industrial areas. The actual number increased by 300% during 1992-2001. The air pollution has increased a lot killing 1 person every hour and 9859 people die prematurely every year. About 30% of the people living in Delhi suffer from severe respiratory diseases; the ratio is 12 times more than the national average. It is also reported that 12% of Delhi's school going children are asthmatic. The poor air quality is reflected by all these evidences.

There are 17 categories of major polluting industries which have come into operation after 1991. They are as follows:

(1) Aluminum Smelter
(2) Caustic Soda
(3) Cement
(4) Copper Smelter
(5) Distilleries
(6) Dyes and Dye Intermediates
(7) Fertilizers
(8) Integrated Iron and Steel
(9) Tanneries
(10) Pesticides
(11) Petrochemicals
(12) Drugs and Pharmaceuticals
(13) Pulp and Paper
(14) Oil Refineries
(15) Sugar
(16) Thermal Power Plants
(17) Zinc

The Supreme Court ordered closure and shifting of various industries located in Delhi. Some such decisions are as follows:

(1) 168 Industries belonging to category 'Ha' and 'Hb' categories under the Master Plan of Delhi-2001 and which were hazardous/ noxious/ heavy and large industries were ordered to stop operating in Delhi with effect from 30.11.1996. These industries had the option of shifting to another industrial estate in the National Capital Region or outside.

(2) 513 Industries belonging to H category were ordered to stop functioning in Delhi w.e.f. 31st January 1997. These industries could relocate or shift to any other industrial estate in the National Capital Region.

(3) 43 Hot mix plants were ordered shifting or relocation on 10th October 1996 and their functioning was stopped in Delhi w.e.f. 28th February 1997.

(4) 246 Brick-kilns falling under H category were ordered to stop functioning w.e.f. 30th June 1997. However, they could relocate or shift in the National Capital Region. And, it was also decided that they could operate on the same site if they change over to modern plants with fly ash-sand lime technology.

(5) 21 Arc/ induction furnaces falling under H category were ordered closure in Delhi w.e.f. 31st March 1997.

(6) 337 Industries falling under H category were ordered closure w.e.f. 30th June 1997.

(7) All non-extensive F category units in non-conforming/residential areas except light service industries were closed down or relocated in 2000-2001. However, the workers interest could not be protected and this lead to a panic in the industrial sector of Delhi. After this, the emphasis shifted to approved industrial areas, since these industries did not have enough space and finances for handling, storage and management of wastes generated by them. There were a total of 21,627 industrial units within the 28 industrial areas. About 1/5th of them were either closed or their plots in the newly established industrial area of Narela and Patparganj were lying vacant. In addition to the industries, there were other establishments such as 29,257 jhuggis and 816 residencies concentrated in the industrial areas of Wazirpur, Anand Parvat, Okhla, GT Karnal Road and Kirti Nagar.

About two thirds of the industries were located in six major areas which are as given below in Table 15.8.

Table 15.8 Location of Industries in Delhi

Place	Percentage of Industries
Anand Parvat	17.23
Okhla	11.34
Wazir Pur	7.70
Kirti Nagar	6.82
Maya Puri	15.10
Narela	9.59

About 57% of these industries used electricity as a single source of energy whereas rest of the units used other fuels like diesel, petrol, coal etc. along with the electricity. These fuels are also responsible for the air pollution in these areas.

The small-scale industries also contributed hazardous wastes like heavy metal sludges, solvents, waste oils, acidic or alkaline wastes etc. Usually, these wastes are dumped into drains or in a common place and such a dumping is difficult to monitor.

About 70% of industrial units are involved in the following 18 industrial activities:

Table15.9 Activities at various industries in Delhi

Product code	Name of the Industrial Activities	Percent of Units
289	Manufacture of other fabricated metal products; metal working service activities	15.30
524	Retail sale of second hand goods in stores	6.11
514	Wholesale of household goods	5.15
252	Manufacture of plastic products	4.97
343	Manufacture of parts and accessories for motor vehicles and their engines	4.77
361	Manufacture of furniture	3.57
291	Manufacture of general purpose machinery	3.56
292	Manufacture of special purpose machinery	3.38
271	Manufacture of Basic Iron and Steel	3.29
181	Manufacture of wearing apparel, except fur apparel	3.22
222	Printing and service activities related to printing	2.98
502	Maintenance and repair of motor vehicles	2.74
272	Manufacture of basic precious and non ferrous metals	2.33
192	Manufacture of footwear	2.20
749	Other business activities	2.17
630	Supporting and auxiliary transport activities; activities of travel agencies	2.11
523	Other retail trade of new goods in specialized stores	2.06
242	Manufacture of other chemical products	2.00

Many of the activities listed in Table 15.9 above yield hazardous wastes.

Table 15.10 shows the percentage of units operating without no objection certificate.

Table 15.10 Percentage of Units without NOC (% of industrial units)

Product Code	Product Name	Units out of Total industrialunits in Delhi	Units operating without valid NOC (Air/water)
181	Manufacture of wearing apparel, except fur apparel	3.20	70.00
222	Printing and service activities related to printing	0.42	70.00
241	Manufacture of basic chemicals	1.37	80.00
242	Manufacture of other chemical products	2.00	80.00
252	Manufacture of plastic products	4.97	77.00
271	Manufacture of basic iron and steel	3.30	81.00
289	Manufacture of other fabricated metal products; metal working service activities	15.29	88.00
291	Manufacture of general purpose machinery	3.60	80.00
292	Manufacture of special purpose machinery	3.40	80.00
313	Manufacture of insulated wire and cable	1.22	76.00
502	Maintenance and repair of motor vehicles	2.74	92.0
749	Other business activities	2.17	87.00

The effluent and emission standards have been notified for various categories of industries. Also, the standards for ambient air and water quality have been notified to keep a check on the industrial pollution. But over and above all these efforts, strict monitoring and implementation of various notifications and guidelines is necessary to control the problem. Even after so many directives, many of the industries do not bother much for the problem of pollution. The real solution lies in the hands of those responsible for causing the pollution. If they honestly and sincerely make efforts and follow the procedure and norms, only then the problem of industrial pollution would be solved.

Towards Cleaner Future

CHAPTER 16

16.1 Introduction to Green Chemistry

It is well known that the quality of life on earth became much better due to the discovery of dyes, plastics, cosmetics and other materials. The life expectancy rose from 47 years in 1900 to about 75 years in 1990's. The world's food supply also increased enormously due to the discovery of hybrid varieties, improved methods of farming, better seeds, use of insecticides, herbicides and fertilizers. The advancement of technology led to new manufacturing processes/industries for the production of newer drugs, medicines and other materials. For all these, chemistry or chemical processes are responsible. It has already been stated (sec. 4.3.1, 6.4.2; 9.3) that virtually all industrial manufacturing processes are responsible for environmental pollution including pollution of air, water and land. This is attributed mainly to the effects of by- products of chemical industry and industrial effluents, which are being discharged into air, rivers/oceans and the land.

The pollution reached such levels that various governments all over the world made laws to minimize it (see chapter 14). This marked the beginning of green chemistry in the middle of 20th century.

Green chemistry is defined as environmentally benign chemical synthesis. The synthetic procedures are designed in such a way that there is virtually no pollution to the environment. We know that as on today, maximum pollution to the environment is caused by numerous chemical industries. The cost involved in the disposal of the waste or by-products is also enormous. In view of this, it is most advantageous to design synthesis for manufacturing processes in such a way that the waste products are minimum and that these should have no effect on the environment. Also, the disposal of the waste products, if any, should be convenient. With this view in mind, the synthetic procedures should be designed in such a way that the starting materials are used up to the maximum extent to give the final product. The reactions should not generate any toxic by-products.

16.2 Designing a Green Synthesis

Any target molecule can be synthesized by a number of alternative routes. The method of choice should not use toxic starting materials and as far as possible, should eliminate by products and wastes. Following considerations should be kept in mind.

16.2.1 Choice of Starting Materials

The starting materials should be selected with extreme care. The synthetic procedure will depend on this. Keep in mind the hazards that may be faced by the workers (chemists carrying out the reactions) and also the transporters who handle the starting materials.

Most of the starting materials make use of petrochemicals (made from petroleum), which are non- renewable. Also, petroleum refining requires lot of energy. It is important to reduce the use of petrochemicals by using alternative starting materials, which may be of biological / agricultural origin. For example, some of the agricultural products such as corn, potato, Soya and molasses can be transformed using a variety of processes into products like textiles, nylon etc. Some examples of materials of biological origin (obtained from biomass) are butadiene, pentane, pentene, benzene, toluene, phenolics, aldehydes, acetic acid, acrylic acid, sorbitol, glucose, furan, tetrahydrofuran etc.

16.2.2 Choice of Reagents

The choice of the reagent should be made on the basis of its availability, efficiency and its effect on environment. Different reagents can be used for a target molecule but the nature of the by- products, percentage yield etc may vary.

16.2.3 Choice of Catalysts

It is well known that certain reactions proceed much faster at a lower temperature giving good yields. Catalysts like heavy metal catalysts should be avoided since they cause environmental problems besides being toxic in nature. The feasibility of a reaction to proceed under photochemical conditions should be explored. If possible, biocatalysts (enzymes) should be preferred for various steps.

16.2.4 Choice of Solvents

Some of the solvents which are used are extremely harmful. For example, benzene –a commonly used solvent is known to cause or promote cancer in humans and other animals. Certain other aromatic hydrocarbons like toluene can damage brain and may have adverse effects on speech, vision or cause liver and kidney problems.

Halogenated solvents commonly used, e.g. methylene chloride, chloroform, polychloroethylene and carbon tetrachloride have been identified as suspected human carcinogens.

The well-known CFCs (chlorofluorocarbons) were used up to 20th Century as cleaning solvents, blowing agents, for moulded plastics and for refrigeration. Though these have very low toxic effects to humans, they are non-inflammable but are responsible for depletion of the ozone layer.

An excellent and versatile solvent is carbon dioxide in liquid state or supercritical CO_2 fluid. The later is most commonly used as a solvent. A gas normally is converted to a liquid state by increasing the pressure exerted on it. However, if a substance is placed at a temperature abovc its critical temperature, T_c (31°C for CO_2) and pressure above critical pressure, P_c (72.8 atm for CO_2), a supercritical fluid is obtained. The T_c of a substance is the temperature above which a distinct liquid phase of the substance cannot exist, irrespective of the pressure applied. P_c is the pressure at which a substance no longer can exist in gaseous state. In supercritical liquid, due to high pressure, the individual molecules are pressed so close together that they are almost in liquid state. Supercritical liquids have density close to that of liquid state and viscosity close to that of gaseous state.

It is best to carry out the reactions in aqueous phase. It is now possible to carry out a large number of reactions, (which were earlier carried out in anhydrous solvents) in aqueous phase. The use of water as a solvent has distinct advantages.

Another way to carry out the reaction is without the use of solvent. In some cases, the starting materials and the reagents serve as solvents. Alternatively, the reactions can be performed in molten state to ensure proper mixing. One well-known example is that of Claisen rearrangement, which is carried out by heating without any solvents. There is still another way in which reactions can be carried out on solid support. All such procedures do not need any solvent.

16.3 Basic Principles of Green Chemistry

Green chemistry as has already been stated (sec. 16.1) is defined as the *environmentally benign chemical synthesis*. Any synthesis performed in small scale or industrial scale should create none or minimum by products which pollute the environment. According to Paul T. Anastas and John C. Warner (Green chemistry, Theory and Practice, Oxford University Press, New York, 1998), the following basic principles of green chemistry have been formulated:

- Prevention of waste/by-products
- Maximum incorporation of the reactants (starting materials and reagents) into the final product
- Prevention or minimization of hazardous products
- Designing safer chemicals
- Energy requirement for any synthesis should be minimum.
- Selecting appropriate solvent
- Selecting appropriate starting materials
- Use of protecting groups should be avoided when ever possible
- Use of catalyst should be preferred when ever possible
- Products obtained should be biodegradable
- The manufacturing process should be so designed so as to eliminate the possibility of accidents during operations
- Reuse of waste byproducts

16.3.1 Prevention of Waste / By-products

The synthesis should be carried out in such a way so that formation of waste (by- products) is minimum. It is very important, since in most cases, the cost involved in the treatment and disposal of waste adds to the overall cost of production. Even the unreacted starting materials form part of waste (see sec. 16.3.2). The waste or by-products, if discharged in the atmosphere, water bodies or land, not only cause pollution but also require expenditure for cleaning up.

16.3.2 Maximum incorporation of the reactants (starting materials and reagents) into the final product

It is believed that if the yield of a reaction is more than 90%, the reaction is considered to be good. The percentage yield is calculated by the expression.

$$\%\ \text{yield} = \frac{\text{Actual yield of the product}}{\text{Theoretical yield of the product}} \times 100$$

A perfectly efficient synthesis in which one mole of a starting material gives one mole of the final product (making the yield of 100%) is considered to be extremely efficient by the above calculations. However, the above procedure may generate significant amount of waste or by-products, which is not visible in the above calculation. Such a synthesis is by no means a Green synthesis. Typical reactions like Wittig reaction and the Grignard reaction illustrate above contention. Both these reactions may proceed with 100% yield but do not take into account the considerable amounts of by-products obtained.

A synthesis or a reaction is considered to be *green* if there is maximum incorporation of the starting materials and the reagents in the final product. One should take into account the percentage atom utilization, which is calculated or determined by the following equation.

$$\% \text{ atom utilization} = \frac{\text{Mol. wt. of desired product}}{\text{Mol. wt. of (desired product + waste products)}} \times 100$$

The concept of atom economy developed by B.M. Trost (Science, 1991, 254, 1471-1477) is a consideration of how much of the reactants end up in the final product. The same concept was determined by R.A. Sheldon (Chem. Ind. (London). 1992, 903-906) and is given as

$$\% \text{ atom economy} = \frac{\text{Formula weight of atoms utilized}}{\text{Formula weight of reactants used in the reaction}} \times 100$$

Let us consider some common reactions, *viz.* rearrangement, addition, substitution and elimination reactions to find out, which ones are more atom economical.

16. 3.2.1 Rearrangement Reactions

In these reactions, there is rearrangement of atoms. The most well known example is Claisen rearrangement.

This is a 100% atom economical reaction, since all the reactants are incorporated into the product.

16.3.2.2 Addition Reactions

The addition of H_2 to an olefin is an addition reaction.

$$H_3C{-}CH{=}CH_2 + H_2 \xrightarrow{Ni} CH_2CH_2CH_3$$

Propene Propane

In this example also, all elements of reactants (propene and hydrogen) are incorporated in the final product (propane). This reaction is a 100% atom economical reaction.

Similarly, addition of bromine to olefins and cycloaddition reactions of olefins are 100% atom economical reactions.

$$H_3C{-}CH{=}CH_2 + Br_2 \xrightarrow{Ni} H_3C{-}\underset{\displaystyle Br}{\underset{|}{C}H}{-}CH_2Br$$

Propene 1,2-Dibromopropane

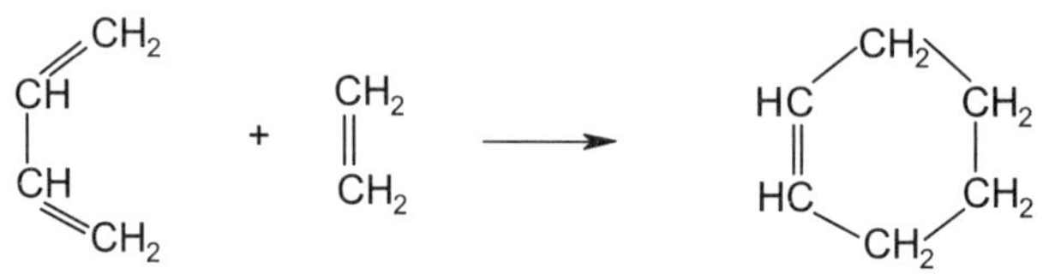

Butadiene Ethene Cyclohexene

16.3.2.3 Substitution Reactions

In these reactions, one atom or group of atoms is replaced by another atom or group of atoms. The atom or group that is replaced is not utilized in the formation of the final product. Thus, substitution reactions are not 100% atom economical; these are less atom economical than rearrangement or addition reactions. Consider the reaction of ethyl propionate and methyl amine.

$$CH_3CH_2\overset{\displaystyle O}{\overset{\|}{C}}{-}OC_2H_5 + H_3C{-}NH_2 \longrightarrow CH_3CH_2{-}\overset{\displaystyle O}{\overset{\|}{C}}{-}NHCH_3 + CH_3CH_2OH$$

Ethyl propionate Methyl amine N-Methylpropionamide Ethyl alcohol

In the above reaction, the leaving group ($-OC_2H_5$) of the ester and one hydrogen atom of methyl amine are not incorporated in the final product. The total of atomic weights of the atoms in the reactants that are utilized is 87.106g mol^{-1} while the total molecular weight including the reagent used is 133.189g mol^{-1} (see Table 16.1). Thus, a molecular weight of 46.069g mol^{-1} remains unutilized in the reaction.

Table 16.1

Reactants		Utilized		Unutilized	
Formula	**FW**	**Formula**	**FW**	**Formula**	**FW**
$C_5H_{10}O_2$	102.132	C_3H_5O	57.057	C_2H_5O	45.061
CH_5N	31.057	CH_4N	30.049	H	1.008
$C_6H_{15}NO_2$	133.189	C_4H_9NO	87.106	C_2H_5OH	46.069

$$\% \text{ atom economy} = \frac{87.106 \times 100}{133.189} = 65.40\%$$

16.3.2.4 Elimination Reactions

In an elimination reaction, two atoms or group of atoms are lost from the reactant. Consider the following Hofmann elimination reaction.

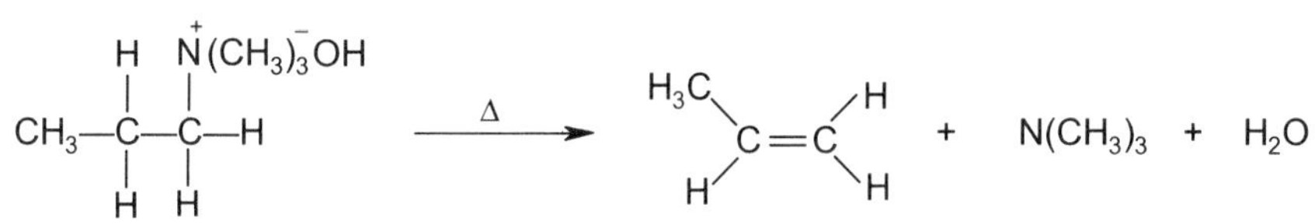

Trimethyl propyl ammonium hydroxide — Propene — Trimethyl amine — Water

The above elimination reaction is not very atom- economical. The percentage atom economy is 35.30%. In fact, this is the least atom economical of all the above reactions.

Let us consider another elimination reaction involving dehydrohalogenation of 2- bromo-2-methylpropane with sodium ethoxide to give 2- methylpropene.

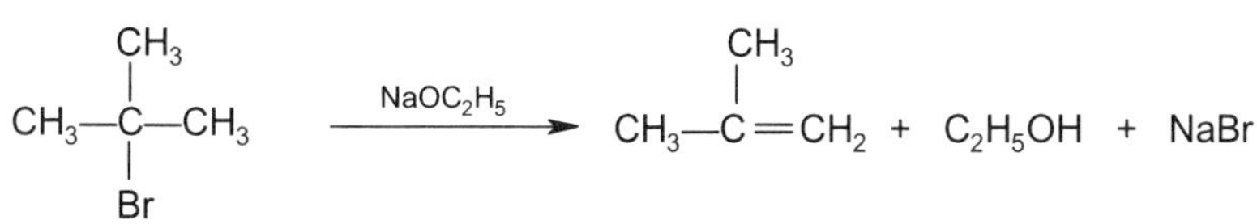

2-Bromo-2-methylpropane — 2-Methylpropene

The above elimination reaction is also not very atom- economical. The percentage atom economy is 27% which is even less than Hofmann elimination reaction.

16.3.3 Prevention or Minimization of Hazardous Products

An important basic principle of Green chemistry is to prevent or at least minimize the formation of hazardous products, which may be toxic and environmentally harmful. These hazardous products may affect the workers. The minimum that should be done is to protect the workers by protective clothing, respirators etc. This, of course, will add to the cost of production. Also, the controls may fail putting the workers at greater risks. Green chemistry offers a scientific option to deal with such situations.

16.3.4 Designing Safer Chemicals

It is extremely important that the chemicals synthesized or developed (e.g. dyes, adhesives, paints, pharmaceuticals etc.) should be safe to use. An example of an unsafe drug is thalidomide (introduced in 1961) for reducing the effects of nausea and vomiting during pregnancy (morning sickness). The off spring born to women taking this drug suffered from the birth defects including missing or deformed limbs. The use of thalidomide was banned and strict laws were formulated for testing new drugs.

With the advancement of better technology, it has now become possible to design and produce safer chemicals. Chemists can manipulate the molecular structure to achieve this objective.

16.3.5 Energy Requirements for Synthesis

For any synthetic procedure, the energy requirements should be kept to a minimum. For example, if the starting materials and the reagents have to be dissolved to bring them into solution and reaction mixture has to be heated to complete the reaction, in such case the time required for completion of the reaction should be minimum so that the energy requirement is least. Use of a catalyst has the added advantage of reducing the energy requirement for a reaction.

In case the final product is impure, it has to be purified by distillation, re-crystallisation etc. for which energy is required. The process should be designed, as far as possible, in such a way that there is no need for separation or purification, so that the energy requirements are minimum.

Energy can also be supplied to a reaction by photochemical means, microwaves or sonication. These procedures take much less time with minimum energy.

16.3.6 Selection of Appropriate Solvent

The solvent selected should not cause any health hazard or any environmental pollution. Use of super- critical CO_2 should be explored. If possible, the reaction should be carried out in aqueous phase or in the solid state without the use of the solvent.

In case of solvents, the volatility is one of the problems which may affect human health and the environment. A novel class of solvents, generally referred to as green solvents, is that of versatile ionic liquids. These are non-volatile and have excellent solvent properties.

16.3.7 Selection of Starting Materials

As far as possible, the starting materials should be obtainable from renewable sources. Thus, petrochemicals are considered to be obtained from non- renewable sources in the sense that it takes million of years for the formation of petroleum oil from vegetable and animal remains. The starting materials which can be obtained from agricultural or biological products, are referred to as *renewable starting materials*. The main problem with these materials is that such products cannot be obtained in a continuous supply due to factors like crop failure etc.

Gases like carbon dioxide (generated from natural sources or synthetic routes) and methane (obtained from natural sources like natural gas or marsh gas) are available in abundance. These are considered as renewable starting materials.

16.3.8 Use of Protecting Groups

In organic synthesis, if there are two reactive groups and we want to use only one of the groups, the other group has to be protected. After the reaction is completed, the protecting group is finally removed. One such example is given below.

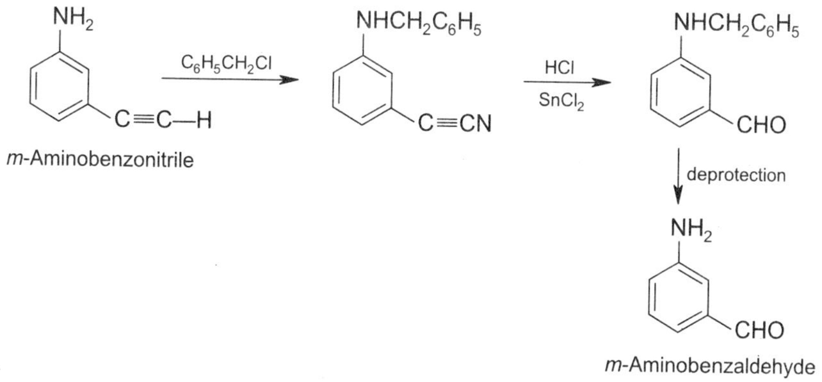

We see that the protecting groups are needed to solve a chemo-selectivity problem. The protecting reagent should be added in stoichiometric amounts only and removed after the reaction is complete. These protecting groups are not incorporated into the final product and hence, their use make a reaction less atom-economical. As far as possible, the use of protecting groups should be avoided.

16.3.9 Use of Catalyst

If possible, a catalyst should be used to facilitate transformations at comparatively low temperatures. The catalyst is not incorporated in the final product and can be reused. Some advantages of the use of catalyst are as follows:

(i) Better Yield

Hydrogenation or reduction of olefins in presence of nickel catalyst

$$\underset{\text{Propene}}{CH_3CH{=}CH_2} + H_2 \xrightarrow{\text{Ni}} \underset{\text{Propane (100\% yield)}}{H_3CCH_2CH_3}$$

Preparation of cyanides

$$\underset{\text{Benzyl chloride}}{C_6H_5CH_2Cl} + \text{aq. KCN} \xrightarrow[\text{Transfer catalyst}]{\text{Phase}} \underset{\text{Benzyl cyanide (90\% yield)}}{C_6H_5CH_2CN}$$

Oxidation of side chain

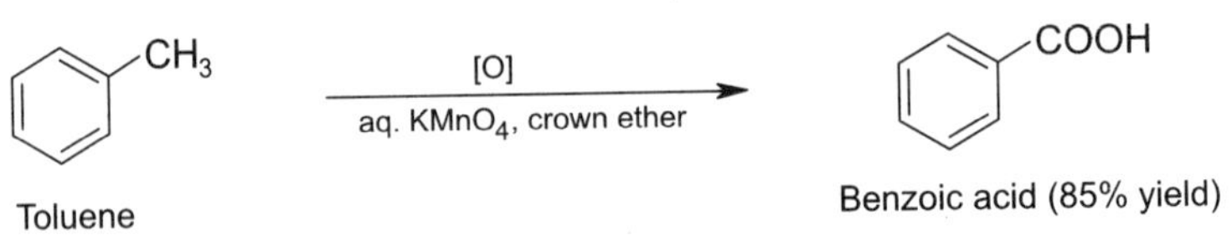

(ii) The reaction proceeds even in those cases where no reaction is normally possible without the use of catalyst.

$$\underset{\text{Acetylene}}{HC{\equiv}CH} + H_2O \xrightarrow[H_2SO_4]{HgSO_4} \underset{\text{Acetaldehyde}}{CH_3CHO}$$

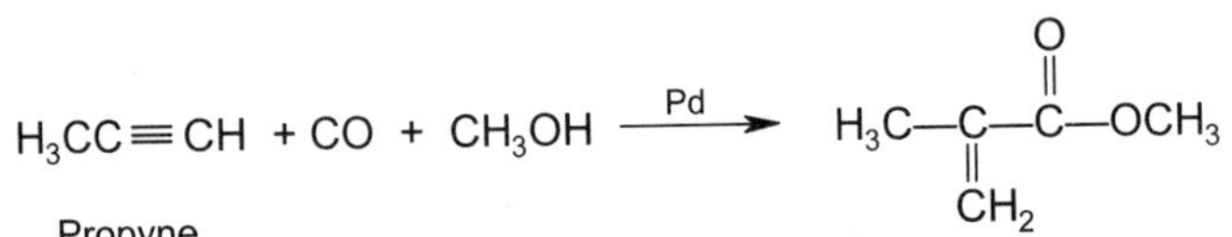

Methyl methacrylate
(shell corporation)

(iii) Enhancement of selectivity

$$\underset{\text{Propyne}}{H_3C{-}H{\equiv}CH + H_2} \xrightarrow{\text{Pd-BaSO}_4} \underset{\text{Propene}}{CH_3CH{=}CH_2}$$

16.3.10 Products Designed Should be Biodegradable

In case, the products obtained are not biodegradable, these keep on accumulating in the land or water bodies and are responsible for the environmental problems besides having other effects. Such a problem was encountered in insecticides and polymers. It is now well known that some of the insecticides (like DDT and organophosphates) are not biodegradable. These accumulate in many plants and get incorporated into food chain and finally into humans and animals causing extremely harmful effects. Also some of the insecticides are responsible for the decline of population of beneficial insects. In view of the above, it is very important that any product (e.g. insecticide) synthesized must be biodegradable. It is equally important that during biodegradation, the products obtained should not have any toxic effects or should not be harmful to human health. It is possible to have a molecule which may possess functional groups that facilitates its biodegradation. The functional groups should be susceptible to hydrolysis, photolysis or other cleavage.

Rohm and Hass Company have developed some diacylhydrazines which are useful as insecticides and are biodegradable.

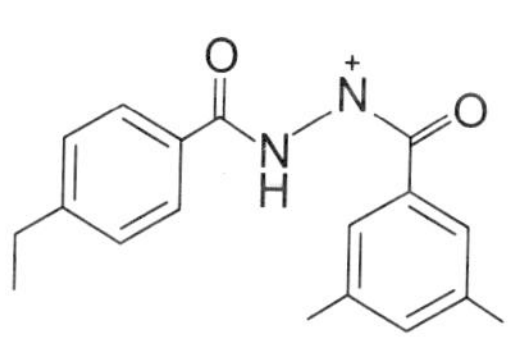

Tebutanozide

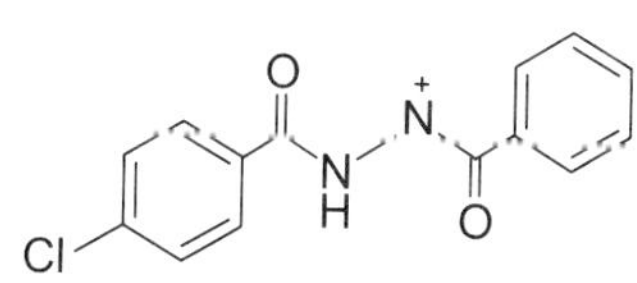

Halofenozide

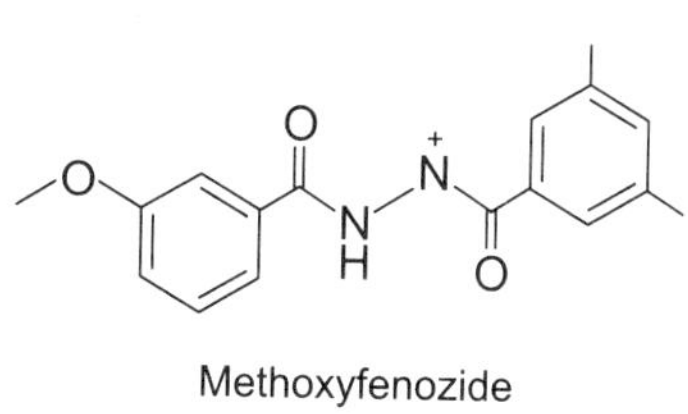

Methoxyfenozide

16.3.11 Designing Manufacturing Plants

It is extremely important to design manufacturing units in such a way that the possibility of accidents is minimum. A large number of accidents have occurred in industrial units. The gas tragedy in Bhopal (Dec. 1984) and several other places has resulted not only in loss of thousands of human lives but have also rendered numerous other people disabled for the rest of their lives. The manufacturing plants should be so designed that it eliminates the possibility of accidents during operation.

16.3.12 Re -use of Waste or By- products

The waste or by- products generated in any industrial unit should be such that they can be recycled or put to some other use.

As on date, numerous examples are reported using basic principles of green chemistry.

16.4 Rain Water Harvesting

Water is one of the most essential components for existence of life. Although it is abundant in plenty on the earth, there is acute water shortage in large areas of the world.

Since ancient times, water had been of immense importance to the human beings. Various civilizations flourished in river valleys. A large population of the world still lives near riverbanks, coastal areas and at places where water is available easily. Water is considered as the gift of God and worshiped even today because of its utmost importance right from the birth to the death.

A large part of the surface of earth is covered with water. But only about 3% of the total amount is fresh water and it is this water, which sustains life on the earth. Again, out of this, 80% is frozen in the form of ice caps, snow, glaciers etc. and rest 20% occurs as liquid in the form of lakes, rivers, ponds and other water bodies. Water also occurs in soil, ground, living beings, atmosphere etc.

Water finds various uses such as irrigation, domestic, industrial, navigation, generation of hydroelectricity etc. The population explosion, rapid urbanization and increasing industrial and developmental activities demand more and more water and have lead to the scarcity of water or *water crisis*. Although some areas in the world are lucky to have sufficient water, some others are deserts, have very scanty rainfall and have very small water resources. Today, more than 1 billion people do not get safe drinking water.

In India, although, we are fortunate in terms of the water received by annual rainfall, the area such as Cherapunji which receives a large amount of rainfall (~1100 mm per year), also suffers from acute shortage of drinking water. India receives an average rainfall of 400 m ham (million hectare metre) out of which 185 m ham is available as surface water, 50 m ham can be stored as ground water and 165 m ham can be stored in soil.

Many factors are responsible for the scarcity of water. The surface water sources have been over exploited leading to a decrease in the water level. The process of urbanization involving construction of houses, markets, roads, pavements and storm water drains, takes a lot of surface area and drains out the rain water as early as possible. Hence, there is no time for the rainwater to percolate to the natural ground to replenish the water sub-surface aquifer. This results in the lowering of the water table.

Since human activities have caused global warming and changes in climatic conditions, rains have also become irregular. The rainfall has reduced and has become erratic. Thus, the need for water conservations is being felt more and more. Also, the scarcity of water affects different sectors of the society to different extents. Rich can still afford to buy mineral water for drinking but they also need water for other activities. The poor, who cannot afford such luxury, have to depend upon whatever type of water is available to them. The poor quality of water leads to the spread of many diseases and epidemics. Also, unsafe water quality leads to poor health and such water is a slow poison.

The clean safe water is the requirement of all. Hence, it is our collective responsibility to make judicious use of available water without wasting even its single drop. Not only this, we have to conserve whatever water is available to us.

Generally, rainwater which reaches us in large quantities, is allowed to drain off and is not utilized. We can save this water and store it for future use. *Rainwater harvesting* is the process of augmenting the natural infiltration of rainwater or surface runoff into the ground by artificial methods. Broadly, the rainwater harvesting involves the following two processes:

(1) Storage of rainwater above or below ground using containers.
(2) Recharging the stored water into the ground for future use.

Any typical rainwater harvesting arrangement has the following four elements:

(1) Catchment area
(2) Conduits
(3) Settlement tank
(4) Recharge facility or storage facility.

Figure 16.1 illustrates this.

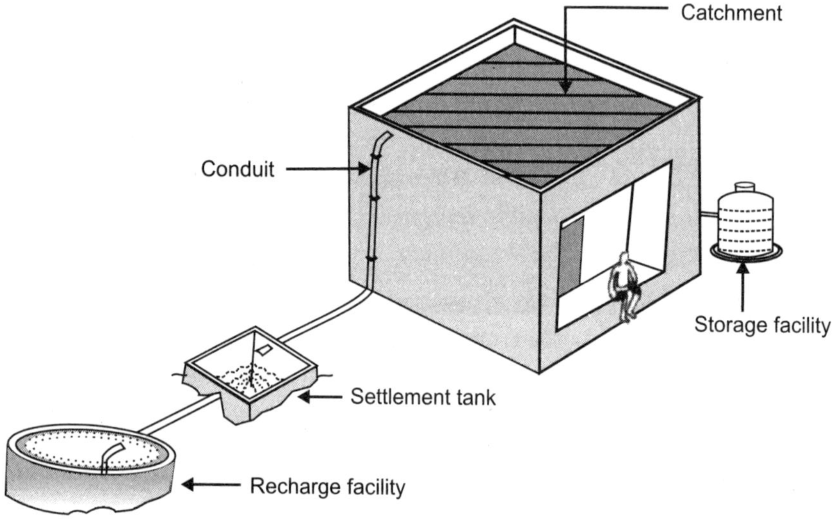

Fig. 16.1: Rainwater Harvesting

The catchment area receives the rainfall directly. It can be any surface such as a terrace or courtyard of a building or any unpaved area like a lawn or an open ground. Conduits are pipes which carry the rainwater from the catchments to the harvesting/storage facility. They can be made of PVC, asbestos, galvanized material or earthen pipes, or bamboos which are used in North-Eastern States.

Traditionally, pits, underground tanks, ponds, check dams, weirs, trenches, dug wells etc. are used for the collection and storage of rainwater.

Several methods are used in different regions of the country for the conservation of water.Some of these are discussed below:

(i) In Thar Desert

In the sandier tracts of the Thar Desert, villagers use an ingenious system of rainwater harvesting known as ***kunds*** and ***kundis***. A ***kund*** is a covered underground tank, see Fig 16.2.

Fig. 16.2: A *kund*

The kunds are popular in the Western arid regions of Rajasthan and in areas having moderate to saline water. They provide convenient, clean and sweet drinking water in these places. Kunds are constructed both by the individuals and by the community and their use is known since the early seventeenth century. There are about 300 kunds in an area called Jalwali in Rajasthan.

Kunds and their catchments areas are kept clean by taking a lot of care. Kunds are circular in shape and have 3 - 4.5 m diameter. Studies have shown that in the 11 desert districts of Rajasthan, kunds provided sufficient water for use.

(ii) Bamboo Drip Irrigation in Meghalaya

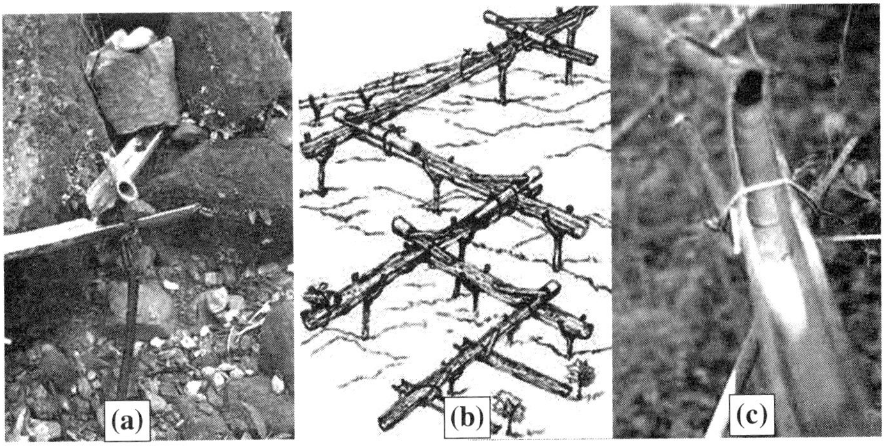

Fig. 16.3: (a), (b) and (c) Bamboo Drip Irrigation

In Meghalaya, bamboo pipes are used for tapping stream and spring water. The system is very perfect and the rate of flow of water can be controlled,see Fig.16.3. It is used by the farmers of Khasi and Jaintia tribes since more than last two centuries. Bamboos of different diameters are used for making channels. It is possible to divert water from the main channel and distribute it according to the requirement. This system is used for the irrigation of betel leaf or black pepper crops planted in areca nut orchards/or in mixed orchards.

(iii) In Trans Himalaya

The Spiti area of Himachal Pradesh is a cold desert area but has managed to get water from distant glaciers by using **kuls**, see Fig. 16.4.

Spiti is a rain shadow area and is snowbound for six months in a year. It receives very less rainfall.

Kuls are basically diversion channels which run along mountain slopes and can be even 10 km long. They have a head at the glacier from which water is to be tapped. The *kul* is lined with stones to prevent clogging and seepage. Kul finally ends up in a circular tank in the village from which water is drawn. People using these kuls have developed socially acceptable system for sharing of water. They have been using this system since very long times without entering into any altercations or enmities.

Fig. 16.4: From Glacier to Kul to Tank
Source: (www.rainwaterharvesting.org).

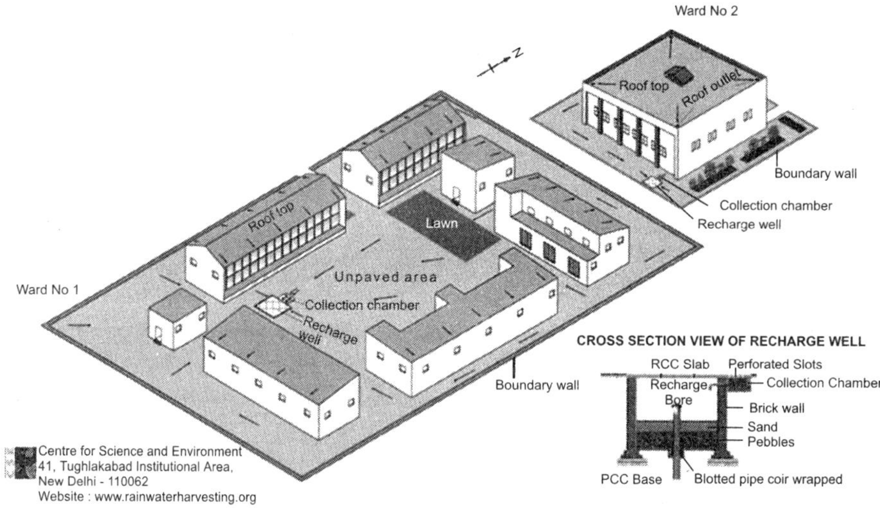

Fig. 16.5: Rainwater Harvesting

Water harvesting is being encouraged in urban areas in these days and a lot of people and institutions have started practicing it. Fig. 16.5 shows layout for rain harvesting in Tihar Jail. Many buildings in cities are proud to use it. Even the Rashtrapati Bhavan has rainwater harvesting facilities Fig. 16.6 (a) and (b). The Presidential Estate has an area of 133 hectare. It has large water requirements both for about 7000 people residing in it and about 3000 people visiting it. The Mughal Gardens also require a lot of water. The total requirement is about 730 million litres per year. About one-third of this requirement is met by groundwater resources which have gone down by 2 to 7 m in last few years. Therefore, rainwater harvesting has been taken up actively in the Rashtrapati Bhavan.

Fig.16.6: (a) Presidental Estate

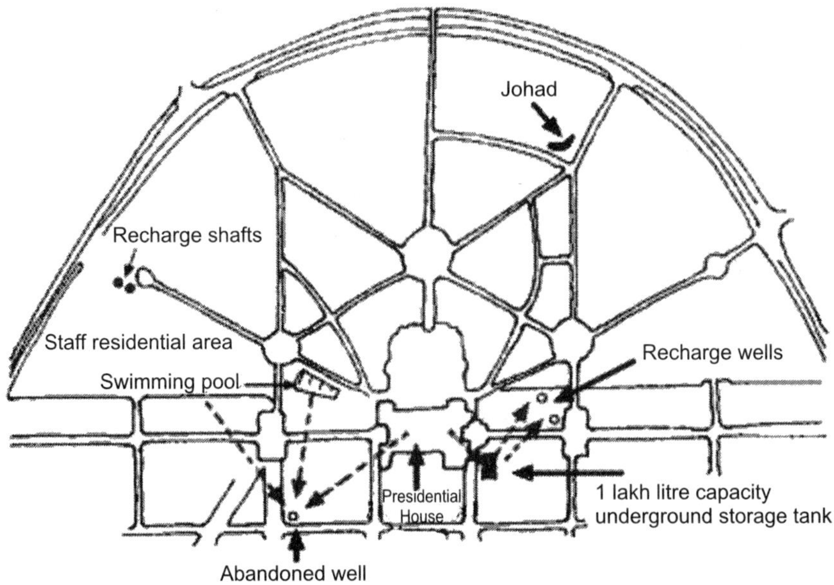

Fig.16.6: (b) Rainwater Harvesting Plan

Rainwater from the northern side of the roof and paved area is diverted to the underground storage tank of 1 lakh litre capacity for low quality use. The overflow from it is diverted to two dug wells for recharging. Rainwater from the southern side of the roof is also diverted to a dry open well for recharging. Runoff from the staff residential area is also diverted to a dry well. The swimming pool having 9 lakh litre capacity is also planned to be connected to the dry dug well. A *Johad* is planned to be constructed near the Mughal Gardens. Water accumulated in the johad percolates in the soil to augment the groundwater.

(*Source*: http://www.rainwaterharvesting.org/methods/modern/rbhavan.html)

16.5 Sustainable Development

Sustainable development means development, which meets the needs of the present without compromising the ability of future generations to meet their own needs. At Rio de Janerio where the Earth Summit was held in 1992, Agenda 21 was adopted by the International Community. It was a landmark achievement which incorporated economic and social concerns. It contained a wide variety of recommendations related to actions on the following issues:

(1) Reducing wasteful use of natural resources

(2) Fighting poverty

(3) Protecting the atmosphere, oceans, plant and animal life
(4) Promoting sustainable agricultural practices for feeding the ever-increasing population of the world.

Thus, Agenda 21 was world's action plan for sustainable development. It offers a better approach for the management of natural resources. There are various adverse effects of globalisation including destruction of natural resources, pollution and loss of biodiversity. These adverse effects have their shadow on both developed and the developing countries. The consequences of mismanagement of the natural resources and the environment threaten each one on the earth.

The following points give a glimpse of this:

(1) About 1.2 billion people worldwide lack sufficient access to fresh water.
(2) Millions of people die every year due to water borne diseases.
(3) Air pollution is major health havoc and causes a large number of diseases and deaths worldwide.
(4) Millions of people die due to diseases such as malaria and AIDS. Had there been better health care systems and proper sanitation, the outbreak of such diseases would have been prevented.
(5) Deforestation and degradation of land leads to poverty and famine and finally results in complex social problems.

Ten years later, the World Summit on Sustainable Development was held in Johannesburg, South Africa from 26th August to 4th September 2002 which brought tens of thousands of participants together. The participants included Heads of State and Government and National Delegates and people from Non-Governmental Organizations, businesses and other major groups. The attention was focused on conserving the natural resources on one hand and providing food, water, shelter, sanitation, energy, health services and economic security to the ever-increasing population of the world, on the other. The Summit was organized by the UN Commission on Sustainable Development whose 10th Session known as CSD 10 acted as the Preparatory Committee for the summit.

The Summit involved various sectors of society such as children and youth, women, workers, trade unions, local authorities, farmers, scientific and technological communities and people from business and industries.

There were around 22,000 participants out of which about 10,000 delegates were accredited by the United Nations. In addition, about 8,000 participants were from the major groups of the society and 4,000 were media people. The Summit was attended by 100 Heads of State and Government.

The three main out comes of the Summit were as follows:

- *Johannesburg Declaration on Sustainable Development* which involved the commitment from Head of State and Government for taking action to make the sustainable development a reality.
- The *Johannesburg Plan of Implementation* which laid down the specific action plans.
- Commitments of various governments and other stakeholders and partnership activities and initiatives for the implementation of sustainable development. Partnerships worth $ 235 million which numbered to 220 were identified before the Summit whereas about 60 more partnerships were announced during the Summit including various major initiatives by the US, UK, Japan, Germany, France and European Union.

Several targets were set:

- To halve the proportion of people lacking access to basic sanitation by 2015.
- To use and produce chemicals in ways not having adverse effects on health and environment by 2020.
- To maintain or restore depleted fish stocks to levels which can produce the maximum sustainable yields on an urgent basis or where possible by 2015.
- To achieve a significant reduction in the rate of loss of biodiversity by 2010.

Agreements were made to help the developing countries through partnership activities for addressing sustainable development issues.

16.6 Some Success Stories

The environment is a complex network of various resources and each one of them has got several governing factors. The environmental crisis of today has alerted everyone. Efforts are going on by various facets of the society for minimizing the environmental damage and restoration of the environmental health. In addition to the role of the State and the Government, various legal and social initiatives have been taken in every part of the world.

India too has conscious citizens and their participation in environmental conservation is unique. Their involvement at individual or community level or in the form of NGOs has brought about visible changes in the state of the environment. Many a times, people have challenged and opposed activities which involved the environmental degradation through legal recourses or by people's movement. On the other side of the action, several people and communities have been tirelessly working for various alternatives and regenerative works for saving the environment. Thus, the approaches may differ but the goal is the same - *conserve the environment.*

People's movements related to construction of dams, Narmada Bachao Aandolan against Sardar Sarovar Project and Chipko movement for controlling deforestation are well known. In addition to these, there have been several grass root level efforts some of which are mentioned below.

(1) **Pani Panchayats** of Maharashtra are now local arrangement where villagers decide the equitable distribution of irrigation water. They have effectively taken care of water scarcity problem by this local decision making approach.

(2) Similar to the above, **Van Panchayats** are working in Himalayan Foot hills for conserving the forest resources.

(3) **Sukhomajri Village –** located near Chandigarh has become a source of inspiration for villagers all over India. This model of development has been replicated in many villages of Haryana, Punjab and Bihar. The Sukhomajri is an example of sustainable development and tells a tale of community forest management. There were severe ecological problems during 1970s. Its barren land, torrential rain washing away the soil, low grass production, poor quality cattle, low economic profile were the characteristic features before the initiation of the project which brought a remarkable change. The soil scientist P.R. Mishra and people's efforts of rejuvenating land using cyclical mode of development brought prosperity to the area. The approach involved enhanced production of biomass for meeting the local needs of the villagers including fuel wood, fodder, food grains etc. and selling the surplus in the market.

The barren land was transformed to a greenbelt. The increased fodder, available after the regeneration of grass and trees in the watershed, encouraged villagers to keep high yielding buffaloes. The tree density increased from 13 per hectare in 1976 to 1272 per hectare in 1992 whereas the grass increased from 40 kg to 3,000 kg per hectare in the same period. The annual house holds income also showed a considerable increase.

(4) **Ralegan Siddhi Village** is located in the drought prone area of Maharashtra where even one regular crop was not assured due to low annual rainfall of 450 to 650 mm. The economic condition of people was bad and irrigated land availability per family was less than one hectare. In early seventies, Krishra Bhaurao Hazare, a retired driver from Indian Army started the construction of storage ponds, reservoirs and gully plugs with the participation of the local people. This increased the

groundwater table which could make possible the plantation of 3,00,000 to 4,00,000 trees in the area. The fallow land was brought under cultivation and total area under farming also increased. Other efforts for equitable access to resources generated were made. Water conservation was given priority.

These steps have made this village a rich and prosperous village and no villager depends upon the drought relief.

(5) **Jhabua** - a poor tribal district of Madhya Pradesh was just a moonscape with barren hills in 1985. With Rajiv Gandhi Mission for Watershed Development in 1994, which focuses on soil and water conservation, the grass, trees and water are being seen again. The implementation of Joint Forest Management and Watershed Treatment have been combined to show a way to success. The water availability and fodder availability have increased and the agricultural production has shown an increase of 7%.

Some more areas showing prominent environment conservation efforts with the involvement of the local people are as follows:

- Dichigam National Park - protection of *hangul* for which Qasim Wani and Abdur Rahman Mir were honoured by 'Sanctuary' Magazine awards in 2002 and 2003, respectively.
- The conservation of snow leopard in Rumbak Village in Ladakh's Hemis National Park where community based efforts helped villagers and the leopard live in peaceful co-existence.

There are several such examples and the list is long. It is clear that once the importance of environmental conservation is understood and people are made aware of it, there is no force on the Earth, which can stop the sincere efforts.

Reference and Suggested Readings

1. New Trends in Green Chemistry. V.K. Ahluwalia and M. Kidwai, Anamaya Publishers, New Delhi, 2004.
2. Green Chemistry, Theory and Practice, Paul T. Anastas and John C. Warner, Oxford University Press, 1998, New York, USA.
3. Real World Cases in Green Chemistry, Michael C. Cann, Mare E. Connelly, American Chemical Society 2000.
4. Green Chemical Synthesis and Processes, Paul T. Anastas, Luren G. Heine and Tracy C. Williamson (Editors), ACS Publication, 2000

5. Pure and Applied Chemistry, Special Topics Issue on Green Chemistry, IUPAC, Vol 72, No. 7, July 2000.
6. Green Chemistry, An Introductory Text, Mike Lancaster, Green Chemistry Network, University of York, RSC, 2002.
7. Organic Synthesis in Water, Paul, A. Grieed (Editor), Blacke Aademic and Professional London, U.K.
8. Organic Reactions in Aqueous Media, Chau-Jun Li and Tak Hang Chan, John Wiley & Sons Inc, New York.
9. Green Chemistry : Environmentally Benign Reactions. V. K. Ahluwalia, Ane Books India, 2005.
10. An Introduction to Environmental Chemistry V.K.Ahluwalia. Longmans, New Delhi, 2005.

Glossary

A

- **Abiotic**: The non-living component such as light, humidity, temperature etc.
- **Acid Rain**: The pouring down of acidic secondary pollutants such as HNO_3 and H_2SO_4 along with rainfall.
- **Aerosols**: Particles having diameter less than 10 μm.
- **Agrochemicals**: The chemicals such as fertilizers, pesticides etc. which are used in agriculture.
- **Air Pollution**: The presence of contaminants such as CO, SO_2, solids and other particles in air.
- **Albedo**: The reflecting capacity of earth.
- **Alkalinity**: The total amount of substances which increase the concentration of hydroxide ions.
- **Atmosphere**: The cover of air that surrounds the earth. It is a mixture of several gases such as nitrogen, oxygen, carbon dioxide etc.

B

- **Bio-Geo Chemical cycle**: The circulation of nutrients between environment and organisms.
- **Biomass**: Organic matter derived from plants materials, animal waste and other wastes generated by timber industry, agricultural crops etc.
- **Biome**: Smaller divisions of the biosphere which are characterized by a typical climate, vegetation, soil and animal life.
- **Biosphere**: Region comprising of lithosphere, hydrosphere and atmosphere where living organisms can exist.
- **Biotic**: The living components such as plant, animals and micro-organisms.
- **BOD (Biological Oxygen Demand)**: Measure of amount of oxygen which is needed by the micro-organism and inorganic pollution load in polluted water.

C

- **Carcinogen**: Cancer causing agent.
- **Chemical Oxygen Demand**: Measure of amount of oxygen required to breakdown the organic matter by chemical oxidizing agents.
- **Climate**: The long term average weather conditions over a period of time for a given place.
- **Consumers**: The heterotrophs which depend on other organisms for food.

D

- **Decomposers**: Micro-organisms which decompose the dead bodies into smaller compounds.

E

- **Ecosystem**: An environmental system comprising a biotic community and its abiotic components.
- **Environment**: The sum total of all living and non-living surroundings.
- **Eutrophication**: The highly productive state of water bodies due to excess of nutrients.

F

- **Fossil fuels**: Fuels formed from the incomplete biological decomposition of dead organic matter such as plants and marine organisms.
- **Fuel cells**: Device to produce electricity by combination of fuel and oxygen by an electrochemical reaction.
- **Fungi**: Plants without chlorophyll. They are parasitic in nature as they cannot make their own food.
- **Fungicides**: Chemicals which inhibit the growth of fungi by fungal diseases.

G

- **Global Warming**: Increase in the average temperature of earth's atmosphere due to Green House effect.
- **Green Chemistry**: The chemistry of environmentally benign products and processes.
- **Green House Effect**: The phenomenon of trapping the heat from the Sun into the Earths atmosphere without letting it escape. Responsible for making the temperature of atmosphere comfortable for living.

- **Ground Water**: The water below the water table where saturation condition exist.

H

- **Hard Water**: Water that forms an insoluble scum on addition of soap or boiling.
- **Hazardous Waste**: The dangerous wastes which is corrosive, toxic, reactive and ignitable.
- **Herbicides**: Chemicals which kill unwanted weeds or interfere with their growth. Examples: pentachlorophenol, 2,4,5-T, 2,4-D etc.
- **Humus**: The organic matter obtained from dead animal and plants.
- **Hydrocarbons**: Compounds containing carbon and hydrogen.
- **Hydrosphere**: All water bodies on the earth such as ocean, seas, rivers.

I

- **Incineration**: Controlled combustion of waste for completely oxidizing it and destroying harmful micro organism present in it.
- **Insecticides**: The chemicals which are used for controlling or eliminating insect pests.
- **Ionosphere**: Region of atmosphere between altitude of 50-100 km.

L

- **Lapse Rate**: The change in temperature with distance in troposphere.
- **Liming**: The treatment of calcium deficient soil with lime.
- **Lithosphere**: The uppermost layer of the surface of earth up to 400 km.

M

- **Macro-nutrients**: The elements required in large amounts such as carbon, nitrogen, oxygen, phosphorus etc.
- **Mesosphere**: The low temperature region above mesosphere which extends up to an altitude of 90 km.
- **Micelle**: The aggregation formed in the presence of surfactant which help in removing oil and grease particle by dissolving them.
- **Micro-nutrients**: The elements required in small or trace amounts such as sodium, cobalt, chlorine, fluorine etc.

- **Municipal Waste**: The total spent water including sewage of the community flowing through a sewer.

N

- **Nitrogen Fixation**: Conversion of atmospheric nitrogen to usable forms such as nitrates or ammonia by various chemical processes.
- **Nuclear Fission**: splitting of atomic nucleus into smaller fragments.
- **Nuclear Fusion**: combining of smaller atomic nuclei to form heavier nucleus.

O

- **Ozone Hole**: The thinning of ozone layer of the atmosphere.

P

- **Particulates**: Finely divided suspended solid particles in the atmosphere.
- **Permanent Hardness**: It is caused by the presence of chlorides and sulphates of Ca, Mg or any other heavy metals.
- **Photochemical Smog**: The combination of smoke, dust and fog with air pollutants such as oxides of nitrogen and hydrocarbons.
- **Pollution**: The presence of harmful substances in the environment which degrades its quality and causes adverse effects.
- **Polymer**: Macromolecule formed by the combination of smaller units called monomers.
- **Preservatives**: The agents used for preventing the undesirable changes caused by bacteria, yeast etc.
- **Primary Particulates**: Particulates having 1-20 μm diameter which are directly injected into the air. Examples: pollens, coal dust, cement dust, fly ash etc.
- **Primary Pollutant**: Pollutant formed as a result of natural phenomenon or other activities which enters directly into the environment. Examples CO, SO_2, hydrocarbons etc.
- **Producers**: The autotrophs which synthesize their own food from simple substances such as CO_2 and water in presence of sunlight. Examples: Green plants, certain bacteria and algae.
- **Putrefaction**: Rottening of organic matter (with bacteria and fungi).

R

- **Radioisotope**: The radioactive isotope of an element which spontaneously undergo radioactive decay by the emission of radiations.
- **Radioactive Fallout**: The radioactive dust falling on the earth after nuclear explosions.
- **Rain Water Harvesting**: The process of augmenting the natural infiltration of rain water or surface runoff into the ground by artificial methods.
- **Reforming**: The process of heating gasoline at high temperature and pressure to increase the octane number. It may involve isomerisation, dehydration, aromatization etc.

S

- **Scrubbing**: Treatment of gases obtained after combustion of coal with $CaCO_3$ to remove SO_2.
- **Secondary Particulates**: Particulates formed from primary particulates. They are formed by combustion processes and vehicular emissions and are responsible for photochemical smog.
- **Secondary Pollutant:** They are formed from primary pollutants by their further reaction. Example SO_3, NO etc.
- **Sewage**: The combined liquid waste discharged from domestic and industrial sources.
- **Smog**: Smoke + Fog
- **Soft Water**: Water which lathers easily with soap.
- **Soil**: The complex mixture of different types of materials derived from the rocks, organic matter and other materials.
- **Stratosphere**: The region above troposphere extending up to an altitude of about 50 km.
- **Surfactants**: The substances which lower the surface tension of water and help in dispersing the soil or dirt particles.
- **Sustainable Development**: Development which meets the needs of the present without compromising the ability of future generations to meet their needs.

T

- **Temporary Hardness**: Hardness due to bicarbonates of Ca and Mg.
- **Total Solids**: Amount of non-volatile substances present in water in a colloidal and molecular dispersed state.

- **Transpiration**: Loss of water from leaves of the plants via evaporation.

W

- **Weathering**: Crumbling and wearing away of rocks.
- **Wet lands**: The lands which are transitional between terrestrial and aquatic systems.

Index

A

% atom utilization 321
Abiotic components 38
Abiotic or non-living components 31
Acid rain 100, 202
Acidulants 168
Active solar systems 11
Addition reactions 322
Aerosols 256
Agrochemical based industries 90
Agrochemical industry 161
Agrochemicals 201, 204
Air (prevention and control of pollution) Act, 1981 279
Air pollutants 63
Algae 148
Alkalinity 132
Alkylation 164
Antidotes 146
Antioxidants 167
Aquatic ecosystems 38
Artificial sweeteners 168
Asbestos 85
Asbestos mining 89
Atmosphere 32, 53
Autoclaving 224

B

Bacteria 148
Basel convention 228
Bhopal gas tragedy 30, 223, 295
Bicarbonate alkalinity 132
Bio-indicators 210
Bioamplification 146
Biochemical oxygen demand (BOD) 130
Biodiversity 259, 260, 262, 267, 270
Bio-diversity Act, 1994 283
Biodiversity Act, 2000 274
Biogas 18
Biomass 17
Biomass energy 18
Biomass gasifiers 17
Biomedical waste 223
Biomes 35
Biosphere 33
Biotic components 38
Biotic or living components 31
Bleaching agents 168

C

Cadmium 144
Carbon cycle 43
Carbon dioxide (CO_2) 68
Carbon monoxide 117
Carbon monoxide (CO) 66
Carbonate alkalinity 132
Carbonates 93
Chemical and photochemical reactions in atmosphere 59
Chemical oxygen demand (COD) 131
Chernobyl 217
Chlorine (Cl_2) 75
Classical smog 108
Clay soils 193
Climate 243, 246, 247
Climate change 243, 251
Climatology 243

Coal mining 88
Composition of atmosphere 57
Composition of soil 194
Conservation of water 123
Consumers 38
Contact herbicides 98
Coral reef 267
Corrosive substance 211

D

Dairy industry 170
Decibel scale 232
Decomposers 39
Depletion of ozone 105
Desalination of sea water 123
Desert 36
Desulfavibrio 50
Desulphurisation of coal 113
Desulphurisation of flue gases 114
Dissolved oxygen (DO) 129

E

Ecosystem 37
Effects of air pollution 100
Elimination reactions 323
Energy component 31
Energy from waste 21
Energy in ecosystem 39
ENSO 256
Environment 29
Environment protection act, 1986 280
Environmental auditing 285
Environmental impact assessment 284
Environmental management 277
Environmental management systems 285
Environmental pollution 37
Eutrophication 51, 182
Exosphere 56

F

Fast breeder reactors 25
Fertilizers 90
Fixation of atmospheric nitrogen 46
Flavours 168
Food processing industries 166
Forest (conservation) act 283
Formation of soil 191
Fossil fuels 6
Fuel cells 23
Fungi 148
Fungicides 95

G

Gaseous cycles 41
Gasification 17
Genetic diversity 260
Geothermal energy 16
Global climate 245
Global warming 8, 104
Grass lands 36
Green chemistry 317
Green house effect 103
Green synthesis 318
Ground water 122
Ground water pollution 153

H

Habitat diversity 260
Hard water 134
Hardness 134
Hazardous waste 175, 211, 221, 229
Hehner's method 136
Herbicides 39, 98
Hormonal insecticides 95
Hydrate alkalinity 132
Hydrocarbons 76
Hydroclaving 224
Hydrogen sulphide (H_2S) 74
Hydrogen– the future fuel 23
Hydrological cycle 41
Hydropower 16
Hydrosphere 32

I

Ignitable material 211
Incineration 223
Industrial effluents 158
Industrial pollution 311
Industrial wastes 200
Insecticides 92
Ion-exchange process 138
Ionosphere 56
Iron cycle 52

L

Latent heat 126
Lead 143, 180
Leather tanning industry 160
Lime soda process 137
Lithosphere 33
Loam soils 193
London smog 30, 109
Los angeles smog 110
Love canal 221

M

Macronutrients 35, 196
Magnesium cycle 52
Major air pollutants 65
Mangroves 266
Man-made environment 31
Marine pollution 183
Mercury 144
Mesosphere 55
Meteorites 77
Methemoglobinamia 180
Mica mining 89
Micelles 174
Microclaving 224
Micronutrients 35, 90, 196
Minamata 222
Minamata disease 144
Mining industries 88
Mining of ores 89
Most probable number (MPN) 132

N

Natural environment 31
Natural pollutants 63
Nino 256
Nitrates 180
Nitrogen cycle 45
Noise 231
Noise pollution 231
Non-conventional sources 10
Non-renewable resources 10
Non-systemic fungicides 96
Nuclear catastrophe 30, 112
Nuclear energy 25
Nuclear fission 25
Nutrients cycle 41

O

Oil shale 9
Organochlorines 92
Organophosphates 93
Oxidation state 133
Oxides of nitrogen 69, 115
Oxygen cycle 48
Ozone (O_3) 75
Ozone hole 105

P

Paper and pulp industry 158
Particulates 76, 117
Passive solar systems 11
Permanent or non-carbonate hardness 135
Permutit process 138
Peroxyacetylnitrate (PAN) 110
Pesticides 92
Petrochemical industry 162
Petrochemicals 165
Petroleum 184
Phosphorus cycle 51
Photochemical smog 77, 109, 111
Photosynthesis 68
Photovoltaics 12

Pollens and spores 77
Pollutants from industry 80
Pollution by petroleum 88
Polymerisation 163
Polymers and plastics 80
Potassium cycle 52
Power plants 89
Preservatives 167
Primary consumers 39
Primary nutrients 90
Primary particulates 77
Primary pollutants 64
Primary treatment 125
Producers 38
Protozoa 148
Pyrethroids 94

R

Radiation balance 58
Radioactive pollutants 187, 201, 204
Radioactive waste 212
Rain water harvesting 328
Reactive substance 211
Rearrangement reactions 321
Reclamation of soil 206
Reducing smog 109
Reformed gasoline 110
Reforming 164
Renewable energy sources 10
Rotterdam convention 229

S

Sandy soils 193
Savannah 36
Sea water pollution 156
Secondary consumers 39
Secondary nutrients 90
Secondary particulates 77
Secondary pollutants 64
Secondary treatment 125
Sedimentary cycle 41
Sequestrants 167
Sex attractants 95
Sludge 8
Smog 108
Soaps and detergents 171
Sodium cycle 51
Soft water 134
Soil 191
Soil erosion 205
Soil pollution 199, 208, 209
Soil profile 193
Solar collectors 12
Solar photovoltaic programme 11
Solar thermal programme 11
Sound 231
Sources of air pollution 80
Sources of energy 2, 5
Sources of noise 236
Sources of water pollution 156
Species diversity 260
Stockholm convention 229
Stratosphere 55
Substitution reactions 322
Sulphur cycle 50
Sulphur dioxide (SO_2) 72
Surface active agents 168
Surface water pollution 154
Surfactants 171, 172
Sustainable development 334
Systemic fungicides 96
Systemic herbicides 99

T

Taiga 36
Temperate deciduous forest 36
Temporary or carbonate hardness 135
Terrestrial ecosystems 38
Tertiary consumers 39
Textile industry 158
Thermal pollution 188
Thermosphere 56
Three mile island 216
Threshold of hearing 233

Tidal energy 20
Tobacco 87
Toxic material 211
Toxic metals in water 139
Transboundary issues 226
Transmission of diseases 177
Transparency 133
Treatment of waste water 181
Tropical rain forest 36
Troposphere 55
Tundra 36

U

Units of hardness of water 135
Universal solvent 126
Use of alternative source of energy 119

V

Vehicular pollution 301
Viruses 148
Volcanoes 77

W

Water (prevention and control of pollution) act, 1974 277
Water cycle 41
Water pollution 151
Water quality parameters 127
Water softening 137
Weathering 191
Wetlands 266
Wild life (protection) act, 1972 283
Wind energy 13